DISCARDED

Alternatives to Confrontation

Alternatives to Confrontation

A National Policy toward Regional Change

Edited by
Victor L. Arnold
The University of Texas

LexingtonBooks
D.C. Heath and Company
Lexington, Massachusetts
Toronto

Library of Congress Cataloging in Publication Data

Main entry under title:
   Alternatives to confrontation.

   Papers presented at a national symposium held in Austin, Tex. in 1977.
   1. Economic zoning—United States—Congresses. 2. Regional economics—Congresses. 3. United States—Economic conditions—Congresses. 4. Regional Planning—United States—Congresses. I. Arnold, Victor L.
HC110.Z6A57   309.2'5'0973   79-2374
ISBN 0-669-03165-8

Copyright © 1980 by D.C. Heath and Company.

All rights reserved. No part of this publication may be reproduced or transmitted in any form or by any means, electronic or mechanical, including photocopy, recording, or any information storage or retrieval system, without permission in writing from the publisher.

Published simultaneously in Canada.

Printed in the United States of America.

International Standard Book Number: 0-669-03165-8

Library of Congress Catalog Card Number: 79-2374

Contents

	Acknowledgments	vii
	Introduction	ix
Part I	General History and Demographics of Regional Change	1
Chapter 1	Unbalanced Growth, Inequality, and Regional Development: Some Lessons from U.S. History *Jeffrey G. Williamson*	3
Chapter 2	Current Demographic Change in Regions of the United States *Peter A. Morrison*	63
Part II	Regional Economic Shifts and Differences	95
Chapter 3	Regional Shifts in Economic Base and Structure in the United States since 1940 *William H. Miernyk*	97
Chapter 4	Regional Differences in Factor Costs: Labor, Land, Capital, and Transportation *David L. Birch*	125
Part III	Financial Impacts of Regional Change	157
Chapter 5	Effects of Regional Shifts in Population and Economic Activity on the Finances of State and Local Governments: Implications for Public Policy *Roy Bahl*	159
Chapter 6	Regional Impact of Federal Tax and Spending Policies *George E. Peterson* and *Thomas Muller*	207
Part IV	Energy and Regional Distribution of Economic Activity	225
Chapter 7	Role of Energy in the Regional Distribution of Economic Activity *Irving Hoch*	227

Part V	*Regional Differences in Metropolitan and Rural Areas*	327
Chapter 8	**Regional Variations in Metropolitan Growth and Development** *Charles L. Leven*	329
Chapter 9	**Rural Conditions and Regional Differences** *Kenneth L. Deavers*	345
Part VI	*Regional Differences in Public Services*	363
Chapter 10	**Public Services and Economic Development** *Robert L. Lineberry*	365
	List of Contributors	383
	About the Editor	384

Acknowledgments

The stimulus for this book was the observation by Walt W. Rostow that the dynamics of contemporary regional change in the United States were not well defined or understood. This point resulted in policy discussion and debate at the national level that was fraught with myth and rhetoric. Dr. Rostow suggested the need for a national symposium that would conduct a dispassionate discussion of factual observations related to this change and would suggest policy options for a national policy toward regional change.

Through the encouragement and generous support of the Lyndon B. Johnson Foundation, the Kerr Foundation, the Economic Development Administration of the U.S. Department of Commerce, the Lyndon B. Johnson Presidential Library, and the University of Texas at Austin, Dr. Rostow's suggestion was implemented. A national symposium, "Alternatives to Confrontation: A National Policy toward Regional Change," was held in Austin in the fall of 1977.

The chapters of this book were prepared, under commission, by prominent scholars and students of regional change to guide the symposium discussion. The topic selection and the work of planning and coordinating the symposium were directed by a University of Texas committee chaired by Dean Elspeth Rostow and composed of Dr. William Hays, Peter Garvie, Dr. Jack Otis, Dr. Alan Campbell, Dr. Jurgen Schmandt, Dr. Melvin Sikes, Professor Jerre Williams, Dean Larrin Kennamer, Dr. William Drummond, and Harry Middleton.

While space does not permit acknowledgment of all who assisted with the symposium or this book, a number deserve recognition. Alan K. Campbell, former dean of the LBJ School of Public Affairs, coordinated the selection and preparation of the technical chapters. Harry Middleton, executive director of the Lyndon B. Johnson Presidential Library, and Mike Naeve, formerly associate director of the Lyndon B. Johnson Foundation, arranged for much of the financial support and helped organize the symposium. Donna Nilsen, Paul Edwards, Ken Leoncyzk, Charles Cockran, and Dorothy Territo managed the logistics of the commissioned chapters, organization of symposium participants, and physical arrangements. Students of the LBJ School of Public Affairs provided assistance in numerous ways during the symposium. Breene Kerr, Garland Hadley, and Larkin Warner, of the Kerr Foundation, and Pat Choate and Victor Hausner, of the Economic Development Administration, U.S. Department of Commerce, provided continued encouragement and assistance.

Special recognition is warranted for Mrs. Lyndon Johnson and Dean Elspeth Rostow. Mrs. Johnson provided warm encouragement throughout the planning process and symposium. Dean Rostow, always gracious and supportive, provided leadership to the process in general and, more importantly, patience and assistance to the editor.

Introduction

Contemporary regional change in the United States, reflected in population movements, changes in the rate of population growth, and economic growth or decline in subnational areas, is of concern to all levels of government. Such change may not, however, be unique to the present. It could be argued that regional change, in these forms, has existed since the nation was founded and is merely an adjustment mechanism, albeit sometimes painful, inherent in our social and economic system.

What may be unique are our current expectations of what ought to be done and who ought to be responsible for doing it. These expectations emphasize not the accommodation of such changes but rather redistribution and equity concerns. Policy discussions are focusing on what ought to be done to redress regional economic shifts and differences, financial impacts of regional change, the distribution of energy resources, the regional differences in metropolitan and rural areas, and associated public services. The dialogue and debate have been laden with rhetoric, emotion, and incorrect perceptions of fact and have promoted divisiveness and sectionalism. The press has coined such phrases as "The Snowbelt-Sunbelt Controversy" and "the new war between the states" to describe the situation.

What is needed is a dispassionate treatment of the elements of regional change. Areas of ignorance with respect to fact and, uncertainty with respect to policy, where further data collection and analysis are required, need to be identified and discussed. Alternatives to accommodation and redistribution policies need to be compared and contrasted. That is the purpose of this book.

Commissioned chapters by leading scholars of regional change address various aspects of regional change and related issues in an attempt to introduce substance into the policy debate. The reader will discover that complete agreement does not exist among the contributors about the appropriate facts to consider or the policy alternatives resulting from that analysis. I trust the reader will be impressed, however, with the objectivity and quality of their work.

Part I addresses the general history and demographics of regional change. Jeffrey Williamson traces historical patterns of regional change from the mid-nineteenth century to the present. While cautioning that one must be careful in drawing lessons from history, Williamson identifies four morals that emerge from analysis of 130 years of regional inequality experience.

First, there have always been pronounced cost-of-living differences in the United States. These differences have been so great that nominal income and nominal wage comparisons have often led to mistaken inferences on the operation of factor markets and regional living standard measures. He argues that increased energy costs and consumer interest in leisure-time outdoor activities and the quality of life will increasingly dominate decisions regarding worker location.

Second, there is little evidence from U.S. history that interregional labor markets were very inefficient even during periods of extraordinary stress.

Third, there is little consistency between regional and national inequality trends. Between-state income variance has never been an important ingredient to total-inequality trends in the United States. Within-state income variance has dominated.

Fourth, unbalanced sectoral rates of technological progress economywide seem to account for much of U.S. regional inequality experience for more than a century. If the remainder of the twentieth century produces similar degrees of technological imbalance, such as national fuel-resource scarcities, we might expect serious future regional implications.

In chapter 2 Peter Morrison notes that population shifts often act as catalysts for larger political issues. He identifies some major contemporary demographic changes under way:

The falloff in the birthrate;

The onset of population decline in many large metropolitan areas;

The revival of nometropolitan population growth;

Changing regional migration patterns.

He observes that, although population growth is slowing, the slowdown is not uniform for every age group or in every place and the number of households is growing at a more rapid rate than the population. He also notes that "migration has become a strong influence in determining which areas of the nation grow and which do not, and consequently has taken on a political weight beyond its demographic and economic impact." Analyzing the twenty-six economic subregions, he concludes that the "geographic locus of the metropolitan population's growth is shifting away from mature industrial subregions," such as the Northeastern Metropolitan Belt and Lower Great Lakes, toward many Southern and Western subregions.

The current net migration toward the South and West and nonmetropolitan areas is altering the status quo along a broad front. Morrison's specific concerns are:

Head counts. Formulas distributing federal funding give weight to the number of people that areas claim as inhabitants.

Introduction

Human capital. Skilled out-migrants in a region may be replaced by less-skilled in-migrants.

Dependency. Some segments of the population are recognized as public burdens and others are thought to be. Their migration imposes real or perceived costs that cannot be ignored.

Undocumented aliens. The majority of aliens locate in a few states and large metropolitan areas.

Part II analyzes regional economic shifts and differences. In chapter 3 William Miernyk points out that the thirty-five years from 1940 to 1975 cover a period of rapid economic growth in the United States. Population grew from 132 million to 213 million, an increase of 61 percent. The gross national product went from $100 million to $1.5 trillion, a fifteen-fold increase. On a per capita basis, real gross national product more than doubled, rising from $2,600 in 1940 to $5,580 in 1975.

Miernyk notes that the center of population continues the westward drift, which has been occurring since the initial settlement of the East Coast. Using per capita personal income as an indicator, he finds that the South and West gained in economic well-being from 1940 to 1975.

Miernyk also observes that since 1970 the national economic growth rate has slowed and the brunt of the slowdown has been borne by the older industrial areas, which have had to pay more for the resources they consume or fabricate and the energy to fabricate them.

In chapter 4 David Birch explores regional differences in labor, land, capital, and transportation. His research indicates that wage rates are lower in the South but that Southern manufacturing wages are approaching those existing elsewhere. He also notes that Southern workers in wholesale trade appear to be gaining on their counterparts elsewhere. "Retail trade, service, and government, which in combination represent almost 70 percent of all employment, are a mixed bag at best, with the South on the lower rather than the high end of the inflation scale in general. The West appears to be moving in just the opposite direction, compensating for its tendency to pay higher-than-average wage rates."

Birch reports that land costs—both agricultural and residential—are "clearly lower" in the South. "The surge in residential land costs in the West has not been paralleled by nearly so rapid a rise in the cost of farm land."

He finds little systematic regional variation in interest rates charged by banks on business loans: "The main differences are attributable to the year and type of loan."

Analyzing transportation, Birch concludes that the Southeastern section of the country holds a distinct advantage in both rail and trucking costs, while the Western states are at a distinct disadvantage, particularly when their remoteness and low density are considered. The remainder of the country appears to be near the average. "For most businesses, the West is a place to avoid if shipping costs are important, and the South is clearly a better choice over most other locations."

The financial impacts of regional change are addressed in part IV. In chapter 5 Roy Bahl offers nine major conclusions on the effects of regional shifts in population and economic activity on the finances of state and local governments.

1. The fiscal structure of Northern-tier states may be characterized generally as local-government-dominated, while that of Southern-tier states may be characterized generally as state-government-dominated. Thus the revenue structure of Southern states tends to be dominated by sales and income taxes, while Northern states are much more reliant on property taxes.

2. In the state and local government sector, the levels of per capita expenditures and average wage per employee are higher in the North than in the South. Both differences would appear to narrow drastically if regional variations in the cost of living could be accounted for. Another substantial part of the per capita expenditure gap between the two regions is due to the higher level of welfare expenditures in the Northern states.

3. Northern states have a greater amount of state and local debt outstanding and a higher level of debt per capita. Because retirement benefits for employees in the North are also higher, a higher level of long-term fixed budgetary commitment is implied.

4. Evidence suggests that central cities are more dominant in the South, although it cannot be argued that there is less metropolitan government fragmentation. Southern central cities account for a larger share of metropolitan population than Northern cities and hold a per capita income advantage over their suburbs, while the reverse is true in the North.

5. Northern states presently maintain a higher level of absolute per capita income, but the gap has been narrowed in the past fifteen years. However, the rate at which the Southern states are narrowing the gap has slowed within recent years. Two other measures of economic strength—population and employment—reveal strong growth in Southern states and almost no growth in the North.

6. Despite the differences in growth in financial capacity, a remarkable similarity exists in the rate of expenditure growth in the North and South.

7. Both demand and supply considerations have contributed to expenditure growth. On the demand side, a greater increase in welfare recipients and public school enrollment in the North may partly explain the high rate of expenditure growth, while in the South higher expenditures may be attributed relatively more to increase population. On the supply side, average public-employee wage levels increased rapidly in both regions but especially rapidly in the North relative to the growth in private-sector income.

Introduction

xiii

8. Northern states were able to maintain a high rate of expenditure increase, primarily because of a greater increase in revenue effort as measured by both an absolute and a per capita basis.

9. The increase in state and local government expenditures over the past decade and a half has been financed primarily by sales and income taxes in the South but much more heavily by property taxes in the North. However, Northern reliance on federal grants increased substantially between 1972 and 1975.

George Peterson and Thomas Muller, analyzing the regional impact of federal tax and spending policies in chapter 6, considered direct job creation through permanent federal employment, direct federal spending on the acquisition of privately produced goods and services, and direct federal investment in capital formation. They conclude that "these are the expenditure categories that have the maximum multiplier effects—in terms of the strength of their inducement to regional migration." He suggests that these federal expenditures "most closely resemble private-sector spending."

Peterson and Muller report that on a per capita basis Pacific states receive two times more federal revenue than Great Lakes states and 80 percent more than Mid-Atlantic states. They assert that 1 to 1.5 million jobs have been redistributed through government spending patterns. Calling defense spending the largest component to total federal spending, they note the effect of the closing of military bases as one agent of regional decline; to cite New England as an example, 100,000 jobs have been lost in ten years in that manner.

Their research demonstrates that during the 1970-1976 period federal grants-in-aid to state and local governments underwent a drastic redesign.

> The per capita grant receipts of a sample of the Sunbelt states have fallen steadily and significantly since 1970 relative to the receipts of a sample of Northern industrial states. In 1970 the average $127 per capita received by the Sunbelt states was 22 percent greater than the average federal grant payment to the industrial states. Although federal grants to all state and local governments climbed drastically over the next six years, by 1976 the Sunbelt states in this sample were receiving 9 percent *less* per capita than their Northern industrial counterparts.

Addressing investment incentives, Peterson and Muller conclude that systematic federal subsidies for new capital investment in preference to replacement and maintenance of existing capital facilities, including housing, have tended to speed up the process of regional and metropolitan adjustment by encouraging the premature scrapping of older facilities.

Irving Hoch addresses the role of energy in the regional distribution of economic activity in part IV. Based on his analysis, he concludes that people in the United States can expect to see considerable population shifts from the Northeast to the Sunbelt in response to increased energy prices. Both consumers and producers will be making the southward and westward migration. Energy-intensive industries and regions will contract their economic activity while fuel-producing states will expand. Hoch also suggests that, because of the increase in energy prices, "urban sprawl will be inhibited, if not stopped in its tracks."

Analyzing federal energy-pricing policies, Hoch contends that in the past they have been counterproductive. Regulating prices of oil and gas below market equilibrium ostensibly will benefit the Northeastern and North Central regions, but the harm to the environment from increased use of coal will balance out the benefits.

Part V addresses the regional differences in metropolitan and rural areas. Charles Leven, analyzing metropolitan areas in chapter 8, argues that the current shift of population from the North to the South is not new but in fact has existed since World War II. Leven concludes that the "problem of the central city" is not a regional problem but a national one, and regional differences as a potential for confrontation are being blunted.

Kenneth Deavers, in his chapter on rural areas, contends that regional-change public-policy discussions tend to ignore rural areas. He calls attention to the fact that, although rural areas lost population during the 1940-1970 period, since 1970 there has been a rapid turnaround. "From 1970 to 1975, net migration has been into nonmetropolitan areas. As a consequence, nonmetropolitan population increased by 6.6 percent in the first half of the decade, compared with a metropolitan growth of 4.1 percent." Deavers identifies the range of nonmetropolitan growth as from 3.4 percent in the North Central states to 13.4 percent in the West.

Observing that nonmetropolitan growth varies from region to region, he finds that in the Northeast it is primarily a movement into picturesque areas and is associated not with economic development but with retirement, acquisition of second homes, and recreational activity. In the South, most of it is associated with employment. In the North Central states, recreation, retirement, and manufacturing growth are associated with much of the population increase. Coal, oil, and mining development produce some growth in the West, but some also stems from retirement and recreation.

In chapter 10 Robert Lineberry analyzes regional differences in public services. He observes that 34 percent of the gross national product is spent for public services by various government units. While some think of those public services as doing more for the rich than for the poor by providing them with better schools and other services, others regard public services as socialism to the poor, especially in such areas as medical care, housing, and food stamps.

Lineberry concludes: "Debate about public services has become intertwined with the pro-growth, no-growth conflicts in cities and states." Turning to public services and the quality of life, Lineberry says, "People in the

Introduction

United States have watched police expenditures increase while crime increases, school expenditures rise while SAT scores decline, and municipal costs escalate while streets get dirtier." In the very cities with the most severe social and economic problems, the cost of public services tends to be high, and their quality is at least debatable. He concludes that a national policy for public services is needed to settle the question of who is to do what.

This question still remains in the analysis of regional change. The contributors and editor look forward to extensions of the analyses presented here, the identification of communities of interest between the North and South, and other efforts to diffuse the rhetoric and engage in serious, factual discussion of national policy alternatives.

Part I
General History and Demographics of Regional Change

1

Unbalanced Growth, Inequality, and Regional Development: Some Lessons from U.S. History

Jeffrey G. Williamson

I. A New Look at an Old Problem

Is the regional growth process equilibrating? If so, what historical forces were responsible for the initial regional disparities from which subsequent convergence emerges? Does regional analysis offer any additonal insight into the determinants of national inequality associated with modern economic growth? There is a long and venerable tradition, both empirical and theoretical, which has confronted these questions, and some very able social scientists have struggled for answers [40, 30, 38, 25, 15, 16, 17, 30, 3]. Except for some vestigial contributions, including those by the present author [47, 48], the topic has lain dormant as conventional wisdom in textbooks for the past decade [41, 36]. Why, then, a new look at this old problem?

There are at least four reasons for exhuming the U.S. regional growth issue. First and foremost, the existing literature is not without flaws, especially in its interpretation of "market failure," in its inattentiveness to cost-of-living issues, and in its failure to embed the regional growth process into a larger theory of national growth and distribution. Second, new evidence has since emerged which may call for a revision in our conventional views of the U.S. regional growth process. Third, the burgeoning literature on why growth rates differ across nations in Europe, North America, and Japan [11, 12, 13, 28, 14, 6] may offer new insights into the U.S. regional experience since the early nineteenth century. Fourth, recent resource-scarcity shocks may introduce a new sensitivity to the politically abrasive forces of regional adjustment in the United States.

The chapter is long, and it begins in section II with some stock taking. I offer a critical review of the existing literature as well as a macromodel of long-run U.S. growth. The emphasis is on the nineteenth century since it is important to understand the forces which produced initial regional disparities to which the U.S. economy subsequently adjusted. Understanding initial conditions is essential to any analytical rationalization of regional growth trends over the past century. In addition, the exercise is not without morals for twentieth-century history. Section III continues this theme, where initial conditions *within* the North and *between* North and South are given due attention. Section IV presents a more detailed accounting of U.S. regional inequality trends than has yet appeared in the literature. Some old pieces of conventional wisdom are reaffirmed, but some are rejected. Section V raises the cost-of-living and price convergence issue in some detail, a topic which has been almost ignored in past contributions. The chapter concludes with a summary of some lessons from history.

II. Unbalanced Growth and Inequality: Searching for a Theory of Regional Development

What Is Wrong with Conventional Wisdom?

The literature on regional per capita income performance during modern economic growth, generated during the 1950s and 1960s, suffers from four weaknesses, each of which helps account for my dissatisfaction with current theories of regonal growth. First, it was built on the foundations of regional analysis with little or no attention given to the national determinants of growth and structurual change. Without a theory of *national* modern economic growth there can be no useful theory of regional divergence or convergence, however elegant the models. This section supplies one paradigm of U.S. growth since the early nineteenth century.

Second, the empirical literature, and theorizing generated to confront that literature, was based primarily on experience since 1880 generally and since 1919 in particular. This is unfortunate since the literature fails to confront the sources of disequilibrium which produced the initial disparities in regional income per capita in the first three-quarters of the nineteenth century, disequilibrium forces which reappear in the first third of the present century. The theoretical explanations are therefore more useful in explaining the convergence of per capita incomes in a national economy previously disturbed by major disequilibrating shocks which presumably started the process in motion. A very special theory emerges, a theory relevant primarily to an economy passing through growth phases which are more characteristic of a system approaching steady state. If we wish to use history to derive morals which are useful to contemporary Third World economies or even to the contemporary U.S. setting of resource-scarcity shocks, then the long-term history between 1880 and 1970 is likely to be far less relevant than the history between 1800 and 1880, or the shorter-term experience between 1900 and 1929.

Third, the theorizing is almost always applied to the United States as a whole, and after the Civil War *two* regional development plays were being acted out, not just one, as the literature implies. In the North, states and regions were playing out a development process initiated by the disequilibrating conditions set in motion by modern economic growth itself. In the South, the spectacular destruction of wealth (through both war destruction and emancipation) created a whole new set of disequilibrium conditions, not unlike conditions in European and Japanese economies created by interwar and World War II events. There is no reason why these two regional dynamics should exhibit the same characteristics at any point in time. After all, the timing of the disequilibrating shocks was quite different, as well as their character. Thus, an effective understanding of U.S. regional growth patterns must begin with separate attention to each of these two processes, and the quantitative decomposition of the "sources of regional growth" in section IV does just that. Furthermore, in what follows I offer a theory of *Northern* regional development in the context of modern economic growth. I then turn to the special case of the North-South problem raised by the disequilibrating conditions produced by the Civil War. The words *special case* do not imply unimportance since it turns out that the North-South dimension of regional development has dominated the United States since the late nineteenth century. In addition, important price, cost-of-living, and "sunshine" effects (let alone slavery) make the North-South problem a bit unique.

Fourth, the literature has almost totally ignored the influence of prices and cost of living. Not only is their inclusion central to welfare and distributional judgments implied by the literature on regional growth, but also they turn out to be an important endogenous ingredient of any theory of regional performance during national development.

National Growth and Distribution:
The Grand Design

What follows is a comprehensive narrative stating my position regarding the forces driving U.S. nineteenth-century experience with growth, accumulation, and distribution. The narrative offers an interdependent view of the process of development from an initial preindustrial agrarian society exhibiting egalitarian attributes across and within regions, to a mature industrialized economy with high and pervasive inequality within regions, but not between. I begin by isolating the exogenous shocks which appear to have set the nineteenth-century economy in motion along a path of modern economic growth. The endogenous response is then traced through ten subsequent decades. The formal general equilibrium attributes of the paradigm can be found elsewhere [53, 49, 50], so I shall rely here on a verbal statement of the model.

Disequilibrium Shocks and the Early Ninteenth Century

The primary initial disturbances to the early national economy were two, and they were distinctive by their magnitude as well as by their persistence over the half century following 1820: rising rates of labor force growth, especially among the unskilled, and sharply unbalanced rates of total factor productivity growth across sectors.

The U.S. economy had been characterized by high rates of population growth throughout the seventeenth and eighteenth centuries, and thus the gentle upward drift of the rate after the first decade of the nineteenth century looks modest by comparison. Labor force growth is another matter entirely since it rises from a per annum rate of roughly 2 percent in 1800-1810 to approximately 3.8 percent in 1840-1850. This acceleration appears to have taken place in two discrete jumps—the first between the 1800s and 1810s and the second between the 1830s and 1840s. The acceleration is persistent, however, and the antebellum rates never retreat to the lower levels obtained early in the century. An important part of this labor force acceleration can, of course, be explained by immigration. While the surge in foreign-born numbers is well known, it is important to recall its character. There is a long empirical tradition in the historical literature which debates the assignment of "push" and "pull" factors to transatlantic migrations, and the most recent econometric research [48, chapter 11] documents that the vast majority of observed nineteenth-century immigration to the United States was of the "push" variety, suggesting that changing U.S. economic conditions were never very influential in accounting for long-term trends. The finding of low "wage elasticity" on this side of the Atlantic has tremendous relevance at this point in our argument. It strongly supports the view that immigration, and thus labor force growth, was exogenous to the early nineteenth-century economy. There are abundant qualitative accounts which confirm the econometrics, among them those which emphasize European blight, famine, and demographic "transition," as well as the sharp decline in direct and indirect steerage costs during the antebellum period.

Not only did the exogenous surge in antebellum immigration foster an acceleration in labor force growth, but limited evidence suggests that it also had an unskilled labor bias since the share of high-status and high-skill occupations among immigrants reporting their occupations declines steadily from almost 70 percent in the 1820s to about 25 percent in the 1850s [34, chapter 8]. In short, after 1810 labor force growth accelerates and the unskilled labor force appears to have undergone a more dramatic acceleration than did the skilled labor force.

The second exogenous shock lay with technological change—if not its overall rate, at least its character. The conjectures offered by modern economic historians appear to document an unambiguous acceleration in total factor productivity growth in the manufacturing sector in particular and "modern" machine-cum-skill intensive sectors in general. During the 1800-1835 period, these conjectures imply a modest rate of total factor productivity growth somewhere in the neighborhood of 0.75 percent per annum. The rate accelerates thereafter, reaching something like 3 percent per annum during the last two decades of the antebellum period. Coincidentally, "traditional" unskilled labor-cum-resource intensive sectors lag. Indeed, the evidence suggests that agricultural total factor productivity growth underwent retardation over the period, from about 0.5 percent per annum prior to 1835 to no improvement at all after the 1830s. In short, the antebellum period can be characterized by high and sharply rising *imbalance* in sectoral rates of total factor productivity growth. The unbalanced character of technological progress is emphasized since there is little evidence of either high rates of total factor productivity growth economywide or a marked acceleration in that rate.

These two main exogenous shocks, in my view, were central in setting the U.S. economy in motion along a path of impressive modern economic growth. What was the nature of the endogenous, economywide response?

Laws of Motion: The Antebellum Response

First and foremost, unbalanced productivity growth induced a striking decline in the relative price of modern goods [53, chapters 5, 9, and 10]. Nowhere in nineteenth- or twentieth-century experience is the decline more rapid than during the antebellum decades. This supply-induced relative price decline had an inevitable impact: it produced a dramatic shift in output mix out of traditional goods production. Nowhere in the U.S. experience is the rate of industrialization more dramatic or agriculture's relative demise more impressive. Some of the output-mix response can be attributed to domestic demand and price-elastic shifts into modern goods, and some of it to import substitution and the displacement of foreign manufactured imports. The net effect was an extraordinary rate of unbalanced output growth favoring modern industrial activities [53, chapter 7]. While this shift in output mix was further reinforced by Engel effects associated with per capita income improvement, its unusual rate compared with the late nineteenth century must be assigned instead to the unbalanced character of technical progress.

Unbalanced output growth such as this had an inevitable consequence for the growth in factor demands. It raised the relative demand for factors used intensively in the favored modern sectors, namely machines and skills. At the same time, the acceleration in labor force growth tended to generate food and other resource-intensive demands which outstripped even an accelerating land stock expansion. The net impact of all this was unskilled "labor savings" in the aggregate. Inequality indicators rose markedly and along a broad front. Pay ratios drifted upward, urban skill premiums rose, profit shares increased, and the share of national income accruing to the impecunious unskilled dropped sharply [35]. What had been an economy characterized by "dear labor" in the 1820s had become an economy of abundant labor and scarce skills by the Civil War. What had been an agrarian society of relative egalitarianism in the eighteenth century had become an economy dominated by high urban wealth concentration by the late antebellum period. While the exogenous forces of accelerating unskilled labor supply and unbalanced sectoral total factor productivity growth were producing these endogenous nominal distribution trends, they were also tending to produce distributional influences on the expenditure side. Both increasing land scarcity and technological imbalance, but especially the latter, were tending to drive the relative price of traditional consumption goods upward. Since the low-wage unskilled worker was a heavy consumer of traditional and resource-intensive wage goods—fuel, light, food, and rents—he failed to share fully in the benefits of declining modern consumer goods prices, a benefit the high-income family was able to capture. Thus, while the low-wage worker found his nominal earnings lagging, he also found his relative income position deteriorating even further in real terms [50, 52]. While the urban worker found his relative standard of living being eroded on the expenditure side, the producer of those traditional goods, located primarily in backward regions, received a benefit in the form of rising output prices. Thus, the "price" forces were more deleterious on distribution *within* regions than between them.

Now this impressive change in the structure of the U.S. economy obviously could not have taken place without a high and rising rate of capital formation from the low rates typical of both the Colonial and early national periods. And rise it did. The gross and net investment shares drifted upward across the antebellum period. Furthermore, Abramovitz and David [1] have estimated that the net reproducible capital stock rose from a 4.5 percent per annum growth rate prior to the mid- to late 1830s to 5.5 percent afterward. Both figures exceed by a factor of 5 or 10 the rates typical of the two centuries before 1805. What were the sources of this notable secular acceleration in accumulation rates [51]?

Part of the explanation for the surge in accumulation rates after 1805 lies with the impact of labor force and population growth on population-sensitive investment in the form of housing and the like. Part of it appears to be explained by induced inequality trends. The major part, however, seems to lie with the unbalanced character of technological change itself. The shift in output mix toward more machine-cum-skill intensive activities tended to raise rates of return to conventional capital in the modern sector. Saving and accumulation rates rose as a consequence.

Now, then, what was the impact of high and accelerating rates of conventional capital accumulation on the inequality drift already set in motion by economywide labor saving? This elastic factor supply response served to suppress the rise in profit rates and, presumably, profit shares as well. But such a view is too simplistic since there were additional and supportive forces at work. First, the high rate of capital formation fostered the expansion of the modern sectors where raw labor was used least intensively. Unbalanced output growth was exacerbated, and labor saving in the aggregate was reinforced. Second, since skills and machines appear to be complementary in production, the relatively dramatic growth in conventional capital tended to raise the relative demands for skills. This must have induced a further scarcity of skills and an upward drift in pay ratios. Thus, while the rapid rise in conventonal reproducible capital stocks served to hold down the rise in rates of return and profits, it inflated the relative demand for skills and thus augmented earnings inequality. Third, the rise in the investment share in gross national product implied another source of unbalanced output growth which saved on unskilled labor. Since producer durables are capital-cum-skill intensive, the relative expansion of capital goods production tended to augment the returns to machines and skills.

All these forces were reinforced by an additional influence which becomes much more apparent late in the antebellum period. Whether because of scale economies or pure technological forces, total factor productivity growth in capital goods sectors in general, and in the producer durable sector in particular, accelerated at an extraordinary rate [51].

The resulting cheapening of machines fostered capital deepening, mechanization, and labor saving at the firm level throughout the economy, even on the farm. It also gave additional impetus to the rise in accumulation rates and investment shares. The net result was to reinforce the inequality and accumulation trends listed above.

The U.S. economy was thus set in motion across the antebellum period. Unbalanced total factor productivity growth and labor force growth acceleration set in motion the classic correlates of modern economic growth. The relative price of traditional, resource-intensive consumption goods tended to rise, fostering a deterioration in the relative income position of the low-wage unskilled worker on the expenditure side. Nominal inequality trends were even more apparent. Distribution changed across a broad front which included lagging unskilled wages, declining unskilled labor shares, rising pay ratios, and earnings inequality. Conventional wealth concentration followed apace. Physical capital accumulation surged to unprecedented rates, and the industrial capital goods sector increased in relative importance. And behind it all was the underlying drama of rapid industrialization and unbalanced output growth.

*"Equilibrating" Shocks and the Late
Nineteenth Century*

The motion of the U.S. economy was altered around the Civil War decade as the exogenous variables driving the economy changed in character and rate.

First, overall labor force growth retreated from the peak rates achieved in the 1840s. If we ignore the Civil War decade itself, the rate of labor force growth declined systematically and dramatically from 3.8 percent per annum in the 1840s to 2.1 percent per annum in the late 1890s. There is also evidence of an even more spectacular decline in the rate of growth in total private man-hours employed, from 3.2 percent per annum between the early 1870s and the early 1880s to 2.2 percent per annum around the turn of the century.

Second, sectoral disparities in rates of total factor productivity growth contracted sharply over the late nineteenth century, and technological imbalance fled the U.S. scene. This can be documented most graphically by the behavior of agricultural total factor productivity growth compared with industry [21]. The rate in industry declines while that of agriculture rises, so that the two sectors exhibit approximate total factor productivity growth balance at 1 percent per annum over the late nineteenth century. This characterization also seems to apply to the capital-consumption goods dichotomy. The downward drift in the relative price of both producer durables and investment goods slows down and almost ceases over the late nineteenth century [51]. These "price dual" calculations supply the inference that the measured rate of total factor productivity growth imbalance must have sharply diminished between these two sectors as well.

The contrasting ante- and postbellum behavior of labor force and productivity growth behavior had an inevitable impact on the motion of the late nineteenth-century U.S. economy. The rate of change in the commodity price structure slowed down markedly [53, chapters 5, 9, and 10]. This induced a diminution in the rate of industrialization, a rate more dependent on endogenous Engel effects than on a mix response to exogenous technological imbalance. On these grounds alone, the aggregate rate of labor saving diminished and the drift toward inequality was retarded. In addition, the rise in the saving rate slowed down as the favorable forces generating its increase dissipated. Since the shift into capital goods production in general, and producer's durables in particular, halted, another source of economywide labor saving was sharply curtailed. Furthermore, the trend acceleration in the rate of accumulation slowed down markedly. The explanations lay with the decline in labor force growth, a retardation in the rightward shift in investment demand, a downward drift in rates of return to conventional assets, a cessation in the upward drift in inequality, and a deceleration in the rate of price decline of investment goods.

The structural transformation of the U.S. economy certainly did not halt after the 1850s. Nor did the downward drift in the relative price of modern goods cease. Nor did the rate of accumulation fall back to the levels of the early nineteenth century. Nor did the economywide bias in the rate of labor saving disappear. Rather, as the *intensity* of these forces dissipated, so too did their influence on relative factor demands which had served to raise pay ratios, the nonlabor share, earnings inequality, and wealth concentration for some six decades of modern development. What resulted, therefore, was a high plateau of inequality quiescence which persisted to the turn of the century.

Lessons for the Twentieth Century?
1900 and beyond

Around the turn of the century the United States witnessed another change in the motion of the key exogenous forces which had been driving inequality trends for the previous century. Disequilibrating forces reappear, although perhaps not with quite the same intensity as in the antebellum period. From the late 1890s to shortly before World War I, the downward trend in labor force growth halted. In fact, there was a modest rise in growth rates up to 1910 or 1913. Furthermore, from the late 1880s to the 1920s there was an abrupt fall in the "average quality" of the immigrant population, at least to the extent that the new immigrants came from countries with lower per capita income and thus, presumably, lower average levels of human capital. Certainly the extensive historical literature would concur that the rate of expansion in unskilled labor accelerated up to World War I. Most importantly, however, there is persuasive evidence that technological imbalance reappears between 1900 and 1929, centered with special drama on the 1920s [27, 49].

Since similar exogenous forces seem to reemerge around the turn of the century, it seems plausible to expect the "nineteenth-century model" to account equally well for the rise in U.S. inequality up to 1929. Whether the nineteenth century supplies lessons for the twentieth, however, depends on two factors: the magnitude of the exogenous shocks compared with early nineteenth-century experience and the sensitivity of the economy to those shocks after a century of transformation which so markedly reduced the relative importance of traditional sector activities. Since both influences were more modest than in the antebellum period, it turns out that the impact of imbalance and unskilled labor force growth was less potent as well.

Finally, we note that following 1929, "technological balancedness" returns to the U.S. scene, especially in the form of impressive total factor productivity growth in the farm sector. Furthermore, the rate of labor force growth also declines following the 1929 "turning point."

Why Regional Growth Rates Differ:
Some Hypotheses

National versus Regional Inequality Trends. Our growth-distribution narrative stresses sectoral Total Factor Productivity Growth (TFPG) imbalance as a key underlying force driving long-term trends in U.S. inequality. At no time was I very explicit regarding the likely regional implications of such technological perturbations. This section offers a simple general equilibrium model which may be helpful in sorting out the potential impact of technological imbalance on *within*-region inequality trends, on the one hand, and *between*-region inequality trends, on the other. The model relies heavily on Ronald Jones' exposition of more than a decade ago [26].

Imagine a world consisting of only two fully mobile means of production, unskilled or raw labor L and reproducible assets K, a composite of productive wealth, including skills, machines, and land improvements. For the moment we exclude factor market imperfections which might otherwise introduce wage gaps and rate-of-return differentials. The economy is broken up into two sectors. One sector uses unskilled labor much more intensively and exhibits very slow rates of TFPG. This sector is by far the largest, and it contains such traditional activities as food production, raw material processing, and trade. Label this traditional sector A. The second sector, M, uses unskilled labor much less intensively, has far higher capital-cum-skill requirements, and exhibits very rapid rates of total factor productivity growth. The modern sector is initially much smaller, and it includes manufacturing, modern transportation, and communications. Furthermore, suppose we allow commodity prices to be determined endogenously in the model, and specify the simplest demand conditions possible. Let the elasticity of substitution between commodities on the demand side be written as

$$\sigma_D = \frac{\overset{*}{A} - \overset{*}{M}}{\overset{*}{P}_M - \overset{*}{P}_A} > 0$$

where the asterisk denotes rates of change and P_j are commodity prices. It is a simple matter to analyze the impact of unbalanced total factor productivity growth on prices, factor rents, and output mix with such a model. First, does technical change favoring the modern sector lower the relative price of modern goods? Equation 1.1 supplies an unambiguous answer:

$$\overset{*}{P}_A - \overset{*}{P}_M = \frac{\theta_{LA} - \theta_{LM}}{\sigma} \left(\lambda_{KM} - \lambda_{LM} \right. \\ \left. + \frac{\delta_L + \delta_K}{\theta_{LA} - \theta_{LM}} \right) (\overset{*}{T}_M - \overset{*}{T}_A) > 0 \qquad (1.1)$$

Some Lessons from U.S. History

That is, the relative price of traditional products (call them agricultural goods hereafter) should drift upward over time in response to technological imbalance. Why unambiguous? Well, the economywide elasticity of substitution σ is always positive. In addition, unbalanced rates of (neutral, disembodied) total factor productivity growth favor the modern (hereafter manufacturing) sector, so $\overset{*}{T}_M - \overset{*}{T}_A > 0$. Furthermore, by assumption, unskilled labor's share in total costs is higher in agriculture, so $\theta_{LA} - \theta_{LM} > 0$, where θ_{ij} is the share of factor payments to input i in sector j. Also, since manufacturing has the higher capital, and lower unskilled labor, requirements, it follows that $\lambda_{KM} - \lambda_{LM} > 0$, where λ_{ij} is the share of input i employed in the jth sector. Finally,

$$\delta_L = \lambda_{LM} \theta_{KM} \sigma_M + \lambda_{LA} \theta_{KA} \sigma_A > 0$$
$$\delta_K = \lambda_{KM} \theta_{LM} \sigma_M + \lambda_{KA} \theta_{LA} \sigma_A > 0$$

where σ_j are sectoral elasticities of substitution. To repeat, unbalanced rates of neutral, disembodied technical progress which favor industry should tend to foster an upward drift in the relative price of agricultural goods.

How about the rate of structural change and industrialization? Equation 1.2 supplies the answer:

$$\overset{*}{M} - \overset{*}{A} = \frac{\sigma_D (\theta_{LA} - \theta_{LM})}{\sigma} \left(\lambda_{KM} - \lambda_{LM} \right.$$
$$\left. + \frac{\delta_L + \delta_K}{\theta_{LA} - \theta_{LM}} \right) (\overset{*}{T}_M - \overset{*}{T}_A) > 0 \qquad (1.2)$$

Since $\sigma_D > 0$, manufacturing will always expand relative to agriculture, in constant prices, if productivity growth is higher in industry. The *rate* of industrialization, however, is another matter, and we note its dependence on two influences in particular: the size of the TFPG gap and the elasticity of substitution on the demand side of the commodity market. The latter influence implies that the more responsive is demand to relative price changes, the greater is the induced industrialization rate. I have more to say about these demand conditions in a moment, but note the following: In an economy open to trade and which has as a result very elastic demand conditions (supply can be easily "vented" abroad without depressing price), industrialization will be all the more dramatic. In a closed economy, where in addition σ_D is very low, the rate of industrialization will be lower.

The impact of unbalanced TFPG on factor income distribution economywide is *not* unambiguous, even in this simple model. Since we are exploring the impact of changes in TFPG imbalance, the initial endowments of K and L are held fixed. Thus, the behavior of factor rents dictates trends in distribution. Will profit rates or rents *r* rise more rapidly than wage rates *w*, thus inducing inequality? Equation 1.3 supplies the relevant statement:

$$\overset{*}{r} - \overset{*}{w} = \frac{\sigma_D - 1}{\sigma}(\lambda_{KM} - \lambda_{LM})(\overset{*}{\hat{T}}_M - \overset{*}{\hat{T}}_A) \gtreqless 0 \qquad (1.3)$$

Alternatively, we can write the relative growth in sectoral per capita incomes $y_j = Y_j/L_j$ as

$$\overset{*}{y}_m - \overset{*}{y}_a = \frac{\sigma_D - 1}{\sigma}(\lambda_{KM} - \lambda_{LM})(\theta_{KM} - \theta_{KA})$$

$$(\overset{*}{\hat{T}}_M - \overset{*}{\hat{T}}_A) \gtreqless 0 \qquad (1.4)$$

In short, the relative price of unskilled labor and/or relative per capita farm incomes will decline in response to unbalanced TFPG favoring industry only if $\sigma_D > 1$. Thus, demand conditions are crucial.

If $\sigma_D < 1$, we would have the classic "immizerizing growth" case which at one time was popular in the trade development and agricultural economics literature, but with a twist. Technological progress concentrated in manufacturing bids down the relative price of industrial goods so extensively that nominal unskilled wages are raised relative to rents on capital and premiums on skills. This "perverse" case would result from low price elasticity on agricultural goods demand, a case which we view with skepticism. It is *not* low price elasticities of domestic commodity demands that we challenge, but rather low price elasticities of total demands for the nineteenth-century U.S. economy, an economy actively engaged in European trade and surely a "price taker," if not for cotton, at least for most traded commodities. Nonetheless, even though $\sigma_D > 1$, the lower are those price elasticities on the demand side, the less potent is the impact of TFPG imbalance on national inequality trends. The smaller σ_D is, the larger is the induced decline in manufacturing's terms of trade, and thus the more modest the rise in the *value* of marginal products of *K*, implying less inequality nationwide in response to TFPG imbalance. The same is true of sectors: A portion of the benefits associated with manufacturing's TFPG spills over to the farm sector as relative farm prices rise in response to the imbalance. The same is true of regions: Given regional specialization, farm states lag in response to TFPG imbalance, but the extent of the lag will be less as the induced improvement in farm terms of trade is greater. The moral, of course, is that we must examine U.S. regional inequality experience with careful attention to such induced price responses since they should have tended to moderate the distribution impact of TFPG imbalance.

How about the *relative* magnitudes of regional (between) inequality compared to national (within) inequality? We note that the only difference between equations 1.3 and 1.4 is the presence of $0 < \theta_{KM} - \theta_{KA} < 1$. Since that term is less than unity, it follows that national inequality trends should tend to be *more* pronounced than regional inequality trends, even if regions were fully specialized. But regions were never fully specialized, so that even farm states should have gained some benefit from unbalanced TFPG, the size of the benefit determined by the relative magnitude of their manufacturing sectors. This issue is central to the regional inequality literature, and thus I return to it at length below. On the other hand, we have ignored the historical realities of factor immobility between sectors and regions. Since unbalanced TFPG favoring manufacturing should produce "short-run" wage and rate-of-return gaps (for example, $w_m > w_a$ and $r_m > r_a$), inequality is given further impetus. Whether factor immobility tended to have a more pronounced effect on between-region or within-region incomes variance is an empirical issue reserved for section III, and I shall dwell on the theoretical implications in a moment.

Regional Divergence and the Conventional Literature. The development and the regional growth literature both have found it useful to decompose aggregate economic activity along dualistic lines, the "traditional" sector being agriculture. It might prove helpful to do the same here. Regional output per worker can be decomposed into its agricultural and nonagricultural parts:

$$y_j = \alpha_j^M y_j^M + \alpha_j^A y_j^A$$

where $\alpha_j^M = 1 - \alpha_j^A$ = share of jth region's nonagricultural labor force in the jth region's total labor force

y_j^M = nonagricultural income per nonagricultural worker

y_j^A = agricultural income per agricultural worker.

Suppose we select as our statistic of regional inequality the weighted coefficient of variation V, or its square V^2. Then the regional variance in y_j can be decomposed into the following "sources of regional growth" expression:

$$V^2 = \frac{1}{\bar{y}^2} \sum_j w_j (\alpha_j^M)^2 (y_j^M - \bar{y}^M)^2 \tag{A}$$

$$+\frac{1}{y^2}\Sigma_j w_j(\alpha_j^A)^2(y_j^A - y^A) \qquad (B)$$

$$+\frac{2}{y^2}\Sigma_j w_j(\alpha_j^A \alpha_j^M)(y_j^M - y^M)(y_j^A - y^A) \qquad (C)$$

$$+\frac{1}{y^2}(y^M - y^A)^2 \Sigma_j w_j(\alpha_j^M - \alpha^M)^2 \qquad (D)$$

$$+\frac{2}{y^2}(y^M - y^A)\Sigma_j w_j(\alpha_j - \alpha^M)$$

$$[\alpha_j^M(y_j^M - y^M) + \alpha_j^A(y_j^A - y^A)] \qquad (E) \qquad (1.5)$$

where w_j is region j's share in the national labor force. This expression can also be written, perhaps more transparently, in terms of relative income "gaps":

$$V^2 = \Sigma_j w_j(\alpha_j^M)^2 \left(\frac{y_j^M - y^M}{y}\right)^2 \qquad (A)$$

$$+ \Sigma_j w_j(\alpha_j^A)^2 \left(\frac{y_j^A - y^A}{y}\right)^2 \qquad (B)$$

$$+ 2\Sigma_j w_j \alpha_j^A \alpha_j^M \frac{y_j^M - y^M}{y} \frac{y_j^A - y^A}{y} \qquad (C)$$

$$+ \left(\frac{y^M - y^A}{y}\right)^2 \left[\Sigma_j w_j(\alpha_j^M - \alpha^M)^2\right] \qquad (D = F \cdot G)$$

$$+ 2\frac{y^M - y^A}{y}\Sigma_j w_j(\alpha_j^M - \alpha^M)\left(\alpha_j^M \frac{y_j^M - y^M}{y}\right.$$

$$\left.+ \alpha_j^A \frac{y_j^A - y^A}{y}\right) \qquad (E = 2 \cdot H \cdot J) \qquad (1.6)$$

Each of these five terms, $V^2 = A + B + C + D + E$, has a specific interpretation as a statistical "source of regional inequality." It will be useful to decompose it still further into

$$V^2 = A + B + C + F \cdot G + 2 \cdot H \cdot J$$

Furthermore, we may wish to focus on *changes* in regional inequality between two points in time such that

$$dV^2 = dA + dB + dC + G\,dF + F\,dG + 2(J\,dH + H\,dJ)$$

or, finally,

$$dV^2 = dA + dB + dC \qquad (1)$$
$$+ G\,dF + 2J\,dH \qquad (2)$$
$$+ F\,dG \qquad (3)$$
$$+ 2H\,dJ \qquad (4) \quad (1.7)$$

These four components of dV^2 can be given the following labels and in order of appearance: (1) the contribution of changing variance in sectoral incomes per worker, (2) the contribution of changes in economywide sectoral productivity "gaps," (3) the contribution of changing variance in industrial structures, and (4) a residual, changing covariance term.

The first term in our V^2 equation, A, denotes the contribution of nonagricultural income per worker variance. If sectoral total factor productivity growth were truly region-specific, at least in early growth stages, then we would expect that some portion of the rise in aggregate regional inequality over time would be explained by increased variance in y^m_j, if indeed the regional specificity of modern sector total factor productivity growth persisted over long periods. The question is an empirical one, but it certainly seems reasonable to expect increasing Northern variance in y^m_j during periods of especially dramatic rates of TFPG in the nonfarm sector. This would include the antebellum period as well as the period between 1900 and 1929, especially the 1920s. It may also include the South over the Civil War decade. After all, the war may have created conditions during which the interregional transfer of new "modern" technologies was interrupted, creating increasing gaps between best and average practice in Southern nonagricultural sectors.

The second term in equation 1.5, B, denotes the contribution of agricultural income per worker variance. Agricultural labor productivities are likely to exhibit far more extensive regional variance, for two reasons: first, the quality of immobile resources, land "quality," matters far more in this case; second, the mobility of agricultural capital and farm labor is likely to be far lower between regions than is true of urban-specific factors of production. In addition, to the extent that farm capital formation per worker is associated more generally with regional scarcity of reproducible capital, low agricultural productivities are likely to be associated with low regional per worker productivity in general. On these grounds, the variance in y_j^a is likely to be positively correlated with variance in y_j^m, and furthermore the variance in y_j^a would certainly be far greater than that of y_j^m at any point.

Yet, it is the *changing* variance in y_j^a which is our focus. Since agriculture never registered dramatic rates of total factor productivity growth in the nineteenth century, it might be inferred that change in the variance of y_j^a across regions is unlikely to have contributed much to trends in aggregate regional inequality over time. This inference fails on two grounds. First, the sheer size of the agricultural sector prior to 1860 ensures that any change in the interregional variance of y_j^a will have a magnified impact on overall regional inequality trends. Second, interregional factor immobility within agriculture may well cause increased disparities in farm capital availabilities, the richer regions enjoying the higher capital intensities for long periods in the presence of continued disequilibrating growth conditions associated with early modern economic growth. Outside of the North, the Civil War most assuredly had the same disequilibrating impact. Since the war was especially destructive of agricultural capital stocks, and given the reduction in voluntary labor supplies associated with emancipation, we expect even greater variance in y_j^a shortly after the Civil War when the South is included.

The fourth term in the "sources of regional growth" expression has attracted the most attention in the literature. The contribution of the variance in "industrialization" levels to aggregated productivity variance across states is measured by D. Note, however, that the term has two components, both of which may have a profound impact on regional inequality trends: the average productivity gap between the two sectors and the variance in industrialization levels itself. To be even more precise, equation 1.7 for dV^2 denotes the impact of changing variance in industrialization levels by $F\, dG$, while $G\, dF + 2J\, dH$ denotes the *total* impact of changing productivity gaps ecnomywide. These two effects are called *differential shift* and *proportionality shift* in the regional literature [41, pp. 344-5; 3, chapter 5], but they became a staple in the literature long ago when Kuznets introduced them in 1955 to account for inequality patterns over time. If national demand or technology forces tend to raise $y^m - y^a$, those regions already committed to industry will find their incomes favored. The previous section stressed the impact of unbalanced TFPG in generating these "gaps," but subsequent Engel effects and the rising capital formation shares would tend to reinforce the impact of technological imbalance on the demand side.

Similarly, tariff policy would also favor the expansion of the gap, although recent analysis suggests that the Metzler effect may have served to benefit Southern cotton producers through terms-of-trade changes. There are other potentially offsetting forces to consider as well. If unbalanced total factor productivity growth favoring nonagriculture was indeed typical of the antebellum period, then the relative decline in nonfarm commodity prices would produce an offsetting improvement in the agricultural terms of trade and thus a benefit to incomes in regions specializing in farm products. The behavior of transport costs may also tend to offset the impact of technological imbalance on the sectoral income gap. Transatlantic and internal transport cost reductions tended to have a far greater impact on commodity price differentials for bulky agricultural products, serving therefore to raise the relative price of farm products at the "farm gate."

While these induced and endogenous price effects may have played an offsetting role, systematic demand and technological forces associated with early modern economic growth must, on net, serve to drive the sectoral productivity gaps upward. In our model, therefore, rising gaps are inevitable as long as technological progress continues to favor the nonfarm sector. During periods of balanced sectoral TFPG, on the other hand, the gap may stabilize or even decline as increasing land scarcity raises labor productivity on the farm, excess labor migrates off the farm, and rising capital intensity follows in the wake of the outmigration. Since technological imbalance is characteristic of both the antebellum and the 1900-1929 periods, rising gaps should be typical of those historical episodes and regional inequality forces should be most apparent during those periods on these grounds alone. Obviously, there may have been other exogenous forces at work also driving the gap, two of which will be emphasized in sections III and IV: the destruction of agricultural capital associated with the Civil War and the exogenous behavior of cotton prices.

What about "differential shift"? Given the sectoral income gaps in D, any increase in the variance of industrialization levels across regions would tend to raise aggregate regional inequality. This source of regional divergence has been appreciated for some time. Since early industrialization is region-specific, and both scale economies and specialization serve to reinforce that specificity, more rapid rates of industrialization are most likely to prevail in the initially advanced states where, after all, the rapid modern sector TFPG is centered.

Extending the Argument to Convergence. We have yet to confront the forces fostering convergence following the unbalanced TFPG and demand shocks. It is a simple matter to do so by reference to our decomposition equations for V^2 and dV^2.

The first and second terms in equation 1.5, A and B, should certainly decline during periods of subsequent "technological quiescence." The analysis which follows should apply equally well to the variance in either nonagricultural or agricultural incomes per worker. Somewhat arbitrarily I focus on the nonagricultural sector isolation. What, then, do we expect to find during periods of "technical quiescence" following previous disequilibrating unbalanced national TFPG and demand shocks?

A simple aggregative model of two regional economies N and S, each producing a composite of nonagricultural commodities and services Y, should suffice. As before, let Y be produced by only two inputs, unskilled labor L and a composite of human and nonhuman wealth K. While the two regions may differ in size either in their respective labor forces or in regional GNP, my focus here is solely with per capita performance indicators. Ignore the size dimension and further consider an initial state of regional "egalitarianism" where per capita nonagricultural incomes are equal. If production functions in the two regions are identical, (nonagricultural) GNP per worker y will also be equated across regions. Of course, regional per capita incomes may diverge even in this simple case if wealth holders tend to concentrate their residences in one of the two regions. Productivity is my focus, however, so I restrict my attention to the behavior of y. This initial "egalitarian" case can be seen by reference to the marginal productivity functions f_K and f_L in figure 1-1. The equality of $k_S = k_N$ ensures equal per capita incomes, but also equality of wage rates $w_S = w_N$ as well as rates of return on wealth $r_S = r_N$.

There are two ways this regional "idyllic equality" might be disturbed. The simplest case can be seen diagramatically in figure 1-1, the case of war destruction. Suppose Civil War (and/or emancipation) changes the relative endowments in the regional economies. While emancipation would clearly serve to lower voluntary postbellum labor supplies in the defeated South, presumably the more relevant U.S. case is one where the capital stock suffers sufficient destruction so as to lower capital-labor ratios drastically below their antebellum levels, to, say, k_S^*. The net result is to lower per worker nonagricultural incomes in the South, to raise rates of return there, and to create wage gaps between the regions. (The war also has a profound equalizing impact on the distribution of incomes *within* the defeated region, of course, but that is not the focus here.) We would observe regional inequality in the postbellum period, and subsequent developments need not rely solely on the speed of interregional factor market adjustments. Even in the absence of interregional capital flows, the higher Southern rates of return would generate an accumulation response, more rapid rates of capital stock growth there, and thus equalization of wages, rates of return, and per capita GNP. In addition, classic equalizing forces would produce migration from low-wage to high-wage regions and capital flows to high regional rates of return. To repeat, equalization forces would almost surely

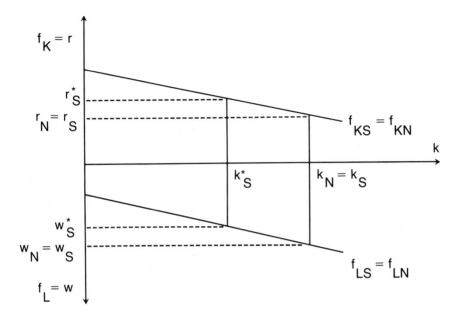

Figure 1-1. War Destruction

be set in motion, and the only interesting historical issue would be the role of institutional arrangements influencing the speed with which the adjustment took place. Of course, other macroeconomic forces may be at work which might influence the process, but we would expect the North-South regional growth pattern to be dominated by the war-induced disequilibrating forces over subsequent decades. This "catching up" view of late nineteenth-century regional development obviously has its counterpart in explanations of post-World War II "miraculous" growth rates in Europe and Japan.

A more interesting regional story emerges when a second potential disequilibrating force is considered, unrelated to Civil War. Let the rate of nonfarm total factor productivity growth undergo a dramatic surge in the North; furthermore, let these technological forces have a distinct bias which tends to save on raw labor. As long as these technological forces are centered in the North, the initial impact is to drive a regional wedge between rates of return and real wages. But note in figure 1-2 that r'_N has risen by far more than w'_N because of the postulated bias. The initial disequilibrating impact, therefore, is to produce a discrepancy in regional nonfarm per worker incomes, but the discrepancy in rates of return to capital is far more extensive than in real wages. The difference, of course, implies that income inequality trends have appeared in the North, at least in the form of declining

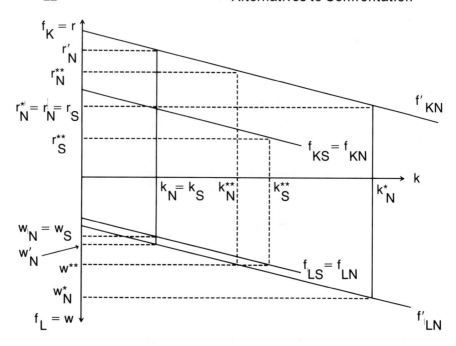

Figure 1-2. Unbalanced Growth

unskilled labor shares. Rates of return to skills would also be higher in the North, and earnings inequality trends would also be characteristic of the disequilibrium induced by these "labor-saving" technological forces.

The disequilibrating forces raise per capita GNP in the North, but, in contrast with the war destruction story in figure 1-1, the disequilibrating shocks do not cease there. Even if the two regions were isolated, or if the magnitude of the disequilibrium were sufficiently large that migration response was initially too weak to erode these gaps, forces would be set in motion to widen regional inequalities still further. Clearly, the higher rates of return in the North would induce a capital formation response, and eventually the capital-labor ratio would rise there, perhaps even to k_N^* in figure 1-2.

It might be useful to point out exactly why capital stock growth is likely to be higher in the North as these disequilibrating forces have their play. Some very simple algebra will do. Let the rate of capital stock growth in region j be G_j^k. Thus, the differential between the two regions can be written as

$$G_N^k - G_S^k = (I/Y \cdot Y/K)_N - (I/K \cdot Y/K)_S$$

where I/Y is the share of (net) investment in income and Y/K is the average capital-output ratio. Even in the absence of interregional capital flows, I/Y will be higher in the North, for three reasons: the rate of return is higher, at least initially, thus fostering a savings response according to neoclassical theory [51]; per capita incomes have risen, fostering a rise in savings rates according to Keynes [42, pp. 103-105]; and the distribution of income has shifted to favor property income recipients, thus fostering rising saving rates according to classical theory [51; 42, pp. 107-111]. Furthermore, the average productivity of capital has also risen in the North relative to the South, at least initially, thus fostering an additional differential in capital stock growth rates. Obviously, location-specific improvements in the quality of capital goods (or a decline in their relative price) would have a reinforcing effect on the interregional capital stock growth differentials. In short, interregional divergence in nonfarm per worker incomes would be fostered during the disequilibrating phase on two grounds: the acceleration in technical change is initially centered in the North where it is embodied in modern sector inputs, and the rate of accumulation, human and non-human, rises as well.

To summarize, in the absence of classic interregional migration responses, we would observe higher k in the North, per capita income disparities between the regions, real wage differentials for unskilled labor, but, at the end of the disequilibrium phase, similar rates of return on capital. To the extent that skills are inelastic in supply, and given their complementarity with conventional capital, rates of return on human capital may remain high in the North and thus skill premiums may be higher there as well.

Just how long these disequilibrium forces may persist depends on two factors. First, the size of the initial shock and its persistence over time will surely matter. Second, the speed of interregional adjustment through factor markets will also matter. In our example, the rate of growth of the unskilled labor force in the North should tend to exceed the national average either through interregional migration to the North or by foreign immigrants selecting Northern regions for location. Either historical force would tend to have the same influence: the capital-labor ratio in the South would rise toward the Northern level, while the capital-labor ratio in the North might even fall back from k_N^*. Note, however, that if wage equalization takes place in figure 1-2, we would observe the odd historical result that capital-labor ratios would be higher in the technologically lagging South, but rates of return would be lower there after the technologically dynamic region had undergone rapid change. The "unbalanced regional accumulation" case seems the more likely one, in which event we would have the $k_N^* - k_S'$ differentials discussed above.

There is an additional force for equalization to consider: the diffusion of new technologies. The diffusion can take two forms. The new technologies which were initially centered in nonfarm sectors in the North may spread to other nonfarm "footloose" sectors. The process is bound to be slow, of course, but it would manifest itself by a rise in the rate of total factor productivity growth in those sectors initially location-specific to the South, and a retardation in these rates in those sectors initially location-specific to the North. More importantly, the new technologies must have spread *within* sectors to other regions in much the same fashion that the gap between best and average practice has been eroded among post-World War II developed economies. The rate of diffusion, however, would be influenced by the ability of the South to shift its capacity into the dynamic sectors, and underlying the process is accumulation in the dynamic sectors located in the "backward" region. The net result would be the *simultaneous* convergence of per worker incomes, capital-labor ratios, rates of return, and wages.

Since the underlying process *is* accumulation, we had best pause for a moment and examine it in more detail. What conditions would ensure that capital stock growth would be higher in the South after the initial disequilibrium phase and in response to the subsequent spread in new technologies? Again, a little algebra will be helpful. Is it necessary, as Romans implies [42, chapters 4 and 5], that investment per worker be higher in the backward region to yield higher capital stock growth rates there? Certainly not. Regional capital stock growth differentials can be written as

$$G_S^k - G_N^k = (I/L)_S k_S^{-1} - (I/L)_N k_N^{-1}$$

Based on Soltow's [43] estimates of wealth per active male, for example, k_N exceeded k_S by a factor of about 2 in 1870. It follows that investment per worker in the South need only have exceeded 50 percent of the Northern figure to have enjoyed a higher capital stock growth rate and thus convergence. Anything higher than that but still below $(I/L)_N$ would have fostered a greater rate of convergence.

To summarize, following episodes of technology and demand shocks, classical equalization forces predict the decline in the variance of both y_j^q and y_j^m over time.

What about "differential shift"? Given the sectoral income gaps in D, any reduction in the variance of industrialization levels across regions would tend to diminish aggregate regional inequality. The literature does not always clearly state the mechanism by which "differential shift" operates. To the extent that the more rapid rates of total factor productivity growth are embodied in modern sector machines and skills, spatial diffusion of productivity gains will in large measure be constrained by the speed of industrialization in backward regions. Thus, convergence of industrialization levels and output mixes will tend to equalize regional per capita incomes. The source of the convergence of industrialization indicators, however, would be the diffusion of industrial technologies, the convergence of capital-labor ratios, and their joint impact on comparative advantage. This supply-oriented approach would therefore view convergence in industrialization levels as simply a proxy for convergence in factor endowments *and* technologies. It supplies the inference that classic equalizing factor endowment forces were at work, namely higher rates of accumulation in poor regions either because of an internal response to higher rates of return or because of immigration of capital seeking out the higher rates of return, or because of both. This interpretation thus stresses the forces of disequilibrium in interregional factor markets creating regional disparities in relative endowments, and the subsequent classic equalizing forces are viewed as the prime determinant of positive "differential shift" effects in the form of higher industrialization rates following the disequilibrating episode.

The regional and postwar sources of growth literature view the "differential shift" process in a very different light. Figure 1-3 offers an *intra*regional "labor market failure" view of the "differential shift" effect. It

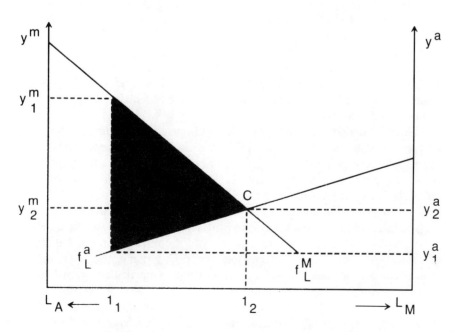

Figure 1-3. Intraregional Labor Market Failure

is central, for example, to Borts and Stein's analysis [3, chapter 2]. It is also central to Denison's [12] and Kindleberger's [28] decomposition of the sources of international differences in the rates of postwar growth. The argument is familiar. The two sectors diagrammed in figure 1-3 illustrate the actual distribution of employment l_1 and the equilibrium distribution l_2 for a given region. We observe large sectoral labor productivity gaps, or service income gaps, between agriculture and nonagriculture early in U.S. development. Borts and Stein and Denison infer that there are too many laborers in agriculture. Subsequent success in transferring workers to nonagricultural employment produces regional income gains as the *intra*regional "labor market failure" is removed. The more rapid the rate of industrialization given the size of the initial market failure (say, given the sectoral income gap and the $l_2 - l_1$ gap), the larger will be the regional growth. Regions already industrialized will gain little from the reallocation of labor resources since the scope for reallocation already has been largely exhausted, and even though intrasectoral gaps are as large. The key to this analysis is the doubtful assumption that sectoral income gaps can be equated with gaps in marginal productivities of homogeneous labor and their wages. Intraregional labor market imperfections are the initial villain of the piece, and the success of a region in eliminating those imperfections will determine the speed with which the poor agricultural region approaches the national income per capita average. As we see in section III, there is little in the history of U.S. labor markets to support such a view. *Intra*sectoral differences in wage rates for homogeneous labor were never large, *intra*sectoral labor markets were typically quite efficient, and the alleged gains in regional GNP from the reduction in the size of the triangle *ABC* in figure 1-4 implied by the regional literature are illusory. The evidence suggests instead that service income gaps are to be explained by differences in human capital inputs into the two sectors.

There is an alternative view of "market failure"—McKinnon's "capital market fragmentation" view [37]—which would instead stress intraregional *capital* immobility. The alternative view is also consistent with the *inter*regional analysis of figure 1-2, where the labor-saving bias of new industrial technologies serves to raise interregional rates-of-return differentials by far more than interregional wage differentials. The alternative view of intraregional factor market failure is offered in figure 1-4, where rates of return replace service income indicators, and the intrasectoral distribution of capital Ψ is at issue. If it can be established that rates of return in nonagriculture in general, and manufacturing in particular, were higher with the advent of modern economic growth and that they persisted over long periods, then we would have an alternative vehicle for convergence even in the absence of significant intrasectoral wage differentials. As we see in section III, the historical evidence is indeed consistent with this alter-

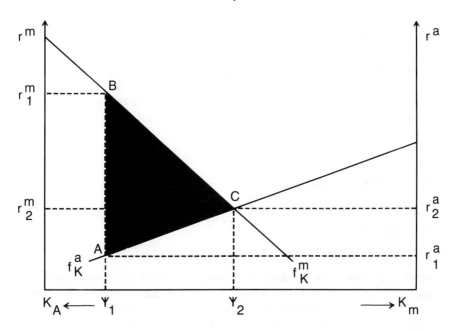

Figure 1-4. Intraregional Factor Market Failure

native position. Since physical capital stocks are, after all, immobile once in place, the intraregional rate-of-return differential can be exploited only by new capital formation activities centered on those sectors with the higher rates of return.

In summary, the "differential shift" effect associated with the interregional convergence in industrial output mixes is to be explained by human and nonhuman capital formation forces rather than by the extent of initial "labor market failure" in backward regions and subsequent success in removing such alleged market misallocations. If I am correct, then convergence of "industrialization" levels across regions simply serves as a proxy for capital-labor ratio equalization, and as a result, we must fall back on those analytical devices previously discussed which explain the behavior of interregional capital-labor ratios over time. This view has another attractive attribute: it appears to be consistent with Easterlin's findings [17] that convergence in property income per capita after 1880 was a main statistical component of the convergence in personal incomes.

The final force producing regional convergence lies with the economywide behavior of the sectoral "productivity gap." After disequilibrating forces have driven the gap upward, what influences would reverse that trend, thus contributing to regional income convergence? First, the disappearance of unbalanced TFPG would tend to halt those forces which drove the gap upward in the first place. Second, the outmigration of labor from the farm sector would tend to raise land and capital endowments among those farm workers left behind. The combined effect, of course, is to induce a downward drift in the sectoral productivity gap over time. Clearly, the poorer regions specializing in agricultural activities derive the greater benefit, and thus regional convergence is fostered on this score as well.

III. Initial Conditions

Disequilibrium and Regional Inequality in a Market without Conflict—Antebellum Initial Conditions in the North

Sectoral Income Gaps. Easterlin's [16, 17] state per capita and per laborer income estimates have been available for some time, and there is no need to reproduce those figures here. It might be helpful to remind the reader, however, that Easterlin's antebellum estimates document unambiguously that regional inequality was already a part of the U.S. scene as early as 1840. An earlier work of mine [47, table 4, p. 25] showed that the weighted coefficient of variation across states in 1840 was higher than that of 1948, a century later. Furthermore, the 1840 figure for the United States was about the same as for France in either 1864 or 1951, a country where *"pôle de croissance"* had its intellectual beginnings. The figure exceeds by far that of Canada in 1935, or 1960, a country now beset with regional conflict bordering on secession. The figure is, in fact, almost comparable to that of interwar Italy, a nation with perhaps the most studied North-South problem. In short, the United States was already beset by regional inequalities in 1840, and they were marked either by present U.S. standards or by the international standards of the late nineteenth or early twentieth century. Furthermore, what is true for the United States as a whole is also true for the Northern states by themselves [53, table 3.4, pp. 3-28].

An explanation of regional divergence in early development was offered in part II, and a central prediction of that thesis was the rise in sectoral productivity gaps. Table 1-1 shows that these gaps were on the rise early in the nineteenth century and reach a peak in the 1850s. Agricultural labor productivity growth lagged during this period of dramatic imbalance in sectoral rates of total factor productivity growth. As a consequence, nonfarm income per gainful worker (1879 prices) was from 2.5 to 2.8 times that of farm labor productivities in the 1850s. The size of the gap is very large com-

Table 1-1
Economywide Sectoral Income Gaps, 1839-1879

	1879 Prices		Current Prices		
Year	Gallman: $(y_m - y_A)/y$ (1)	Gallman: $(y_{CA} - y_A)/y$ (2)	Gallman: $(y_m - y_A)/\hat{y}$ (3)	Gallman: $(y_{\hat{CA}} - y_A)/\hat{y}$ (4)	Easterlin: $(y_m - y_A)/y$ (5)
1839	0.766	0.807	1.16	0.33	1.158
1849	1.269	0.944	1.30	0.54	—
1859	1.106	1.279	1.00	0.77	—
1869	0.824	1.027	0.88	0.85	—
1869	0.782	0.944	0.84	0.78	—
1879	0.760	0.947	0.78	0.97	0.886

Source: Columns 1 through 4 are from R.E. Gallman, "Commodity Output, 1839-1899," in *Trends in the American Economy in the Nineteenth Century* (Princeton, N.J.: Princeton University Press, 1960), pp. 16, 31. Column 5 is calculated from E.A. Easterlin, "Interregional Differences in per Capita Income, Population, and Total Income, 1840-1950," in *Trends in the American Economy*, appendix tables A-1 and A-2, pp. 97-100.

Notation is the following:

$y_{\hat{m}}$ = income per worker in mining and manufacturing
$y_{\hat{A}}$ = income per worker in agriculture
$y_{\hat{CA}}$ = income per worker in construction (variant A)
y = average income, weighting sectors M, A, and CA
y = average income, weighting sectors M and A

pared with that of 1839. It is also large when judged by the experience of other countries said to have suffered severe "dualism." The U.S. gap in 1859 was larger, for example, than the Italian one in either the 1870s or the 1950s, than the Canadian gap at the turn of the century, than the French in the 1870s, than the Japanese in early Meiji, and is approximately the same as that of Soviet Russia on the eve of the First Plan [30, table 3.4, p. 117]. Clearly, sectoral productivity gaps were already very much part of the U.S. scene by the 1850s, and they reflected the disequilibrating influence associated with three previous decades of unbalanced total factor productivity growth. Once again, what was true for the United States as a whole was also true of the Northern states separately.

Intraregional Wage Gaps and Labor Market Failure? Sectoral income gaps and wage gaps are hardly the same thing, although the regional and development literature would seem to suggest otherwise. An example is offered by New England in 1890. Coelho and Shepherd [9, table 2] estimate that nominal wages of urban common labor were only 9 percent higher than farm labor during that year in New England. For the same region Lee et al. [33, tables Y-3 and Y-4, pp. 755-756] estimate the ratio of service income per worker in the nonagricultural compared to the agricultural sector to have been 3.68 in 1880 and 3.45 in 1900. In the late nineteenth century at least, approximate nominal wage equalization intraregionally in New England was consistent with enormous differentials in average labor productivities (or service income per worker). The latter is an irrelevant index for the former. Is the same true of the early and mid-nineteenth century?

Table 1-2 suggests that intraregional wage differentials were trivial in the late antebellum United States, both North and South. With the excep-

Table 1-2
Intrasectoral Wage Differentials, Nominal Daily Earnings, Urban Common Relative to Farm Labor

a Intrasectoral Wage Differentials and Nominal Daily Earnings

Region	1850 Farm Labor (with Board) (1)	1850 Common Labor (with Board) (2)	1850 Ratio (1) ÷ (2)
New England	$0.72	$0.77	1.07
Maine	0.76	0.76	1.00
New Hampshire	0.72	0.63	0.88
Rhode Island	0.70	0.72	1.03
Connecticut	0.69	0.76	1.10
Mid-Atlantic	0.62	0.60	0.97
East North Central	0.58	0.58	1.00
Indiana	0.56	0.55	0.98
West North Central	0.61	0.56	0.92
Missouri	0.61	0.55	0.90
South Atlantic	0.47	0.48	1.02
Georgia	0.52	0.50	0.96
East South Central	0.55	0.49	0.89
West South Central	0.65	0.70	1.08

b Ratio Urban Common to Farm Nominal Daily Wage

Period	Massachusetts	Vermont
1840-1844	—	1.244
1845-1849	1.050	1.340
1850-1854	—	1.260
1855-1859	—	1.118
1860-1864	1.070	1.084

Source: Table 1-2a is calculated from S.L. Lebergott *Manpower in Economic Growth* (New York: McGraw-Hill, 1964). His farm labor average monthly earnings, with board (table A-23, p. 539), is converted to a daily basis using the conversion factors implied by his table A-30, p. 546. The common labor series is taken from his table A-25, p. 541.

Table 1-2b is taken from J.G. Williamson and P.H. Lindert, "A Macroeconomic History of American Inequality," mimeo., 1976, appendix table AS.3, pp. 3-39).

tion of Vermont during the 1840s and 1850s, no region exhibited pronounced farm-nonfarm wage gaps *for labor of comparable skill*. The average of these ratios for the North and for the United States as a whole is 0.99, evidence which unambiguously rejects the thesis of intreregional market failure. Furthermore, since both of these low-wage occupations are quoted including board, cost-of-living differentials are unlikely to matter much, a supposition consistent with the negligible nominal wage differentials for unskilled labor. In contrast, Coelho and Shepherd [9, table 2] offer 1890 intraregional wage data *without* board and for the same pair of occupations: for New England and the Mid-Atlantic, the ratios of urban common to farm daily wage rates were 1.09 and 1.22, respectively. While this late nineteenth-century evidence may seem to offer modest confirmation of the "intraregional labor market failure" thesis, section V shows that the cost-of-living differential between farm and nonfarm in the early 1890s would have produced an *equilibrium* nominal wage gap of 23 percent based on national averages.

In summary, we can find little evidence to support the view of intraregional labor market failure, either in the antebellum United States or in the postbellum North.

Interregional Wage Gaps and Labor Market Failure? Evidence of efficient intraregional labor markets does not necessarily imply that *inter*regional labor markets also worked efficiently. Table 1-3 supplies some evidence on regional wage equalization, for labor of occupationally homogeneous attributes. There *is* evidence of interregional labor market failure in 1850,

Table 1-3
Interregional Wage Differentials, Nominal Daily Earnings for Common Labor, 1850-1890
(U.S. = 100)

	Lebergott				1890	
	1850	1860	1869	1880	Census	Aldrich
New England	116	97	101	104	102	107
Mid-Atlantic	103	100	102	103	99	104
East North Central	100	95	102	106	103	106
West North Central	89	92	100	116	101	106
South Atlantic	78	79	68	78	82	73
East South Central	78	87	83	89	90	79
West South Central	109	120	104	—	99	102

Source: S.L. Lebergott, *Manpower in Economic Growth* (New York: McGraw-Hill, 1964), table A-25, p. 541; P. Coelho and J. Shepherd, "Regional Differences in Urban Prices and Wages," table 2, Paper presented to the Western Economic Association, San Diego, Calif., June 1975.

both within the North and between the North and the South,. These wage differentials persist even after regional cost-of-living adjustments are performed [10, table 4, p. 213]. It appears, however, that interregional wage differentials in the North had disappeared by the Civil War decade (table 1-3), so that real wage differentials between Northern regions were pretty much nonexistent by 1890. While real wage equalization was typical of the North after the 1850s, the North-South wage gap persists up to 1890 [10, Table 4, p. 213] and does not disappear until well into the mid-twentieth century [7].

Capital Markets and Rate-of-Return Differentials. It appears that conventional and human capital markets worked far less effectively than did unskilled labor markets, both intraregionally and interregionally.

Table 1-4 supplies some regional indicators on skill premiums and thus the potential rate of return to skill acquisition within a given region. The data would appear to confirm the model predictions offered in part II.

In 1851 and 1855, the earliest years for which such data are available, the skill premiums are far higher in the Northeast, an observation consistent with the relatively rapid growth in demands for skills in the region where modern economic growth was initiated. It might also be noted [53, pp. 3-23] that the skill premiums in Massachusetts had been on the rise since 1825, and most of the upward surge took place by 1837, long before the great influx of Irish immigrants. Note, too, that these conditions had reversed by 1860: skill premiums were then highest in the relatively

Table 1-4
Skill Premiums by Regions: Nominal Daily Wage, Engineers to Common Labor, 1851-1890

	North				South		
Year	New England	Middle Atlantic	East North Central	West North Central	South Atlantic	East South Central	West South Central
1851	2.08	1.50	1.26	—	—	—	—
1855	1.84	1.54	1.63	1.49	—	—	—
1860	1.66	1.57	1.97	2.11	—	2.41	—
1865	1.56	1.50	1.59	1.70	—	1.52	—
1870	1.65	1.55	1.69	1.91	—	1.54	—
1875	1.81	1.75	1.74	2.00	—	1.86	—
1880	1.87	1.76	1.82	1.99	—	1.83	—
1890	1.59	1.79	1.63	1.72	2.12	2.26	1.92

Source: Calculated from: P. Coelho and J. Shepherd, "Regional Differences in Urban Prices and Wages," table 3, Paper presented to the Western Economic Association, San Diego, Calif., June 1975; and P. Coelho and J. Shepherd, "Regional Differences in Real Wages," *Explorations in Economic History* 13, 2 (April 1976), table 7.

"backward," skill-scarce, and presumably rapidly industrializing North Central and South Central regions. This predictable consequence of technological diffusion remains an attribute of the Northern economy throughout the late nineteenth century. Although the regional skill premium differentials are almost erased by 1890, Barry Chiswick [4, chapter 4] shows that rates of return to schooling are still higher in poorer Northern states as late as 1960, and that the variance in these rates is central to variance in incomes. In contrast, note that Southern skill premiums declined sharply during the Civil War decade, presumably in response to the destruction of complementary physical capital inputs. Southern skill premiums regain Northern rates by 1875, but they do not begin to creep above Northern levels until about 1890. For the next seven decades, however, skill premiums remain relatively high in the backward South, and even in 1960 rates of return to schooling are far higher in the South [4, chapter 4].

Antebellum rates of return to nonhuman manufacturing capital are presented in table 1-5. These estimates are the result of recent efforts by Bateman and Weiss [2], and they suggest that rates of return were

Table 1-5
Antebellum Rates of Return, Intraregional and Interregional Estimates

a Manufacturing

Region	Mean Rates of Return, Census, All Firms		Mean Rates of Return, Census, Large Firms	
	1850	1860	1850	1860
South	0.25	0.28	0.19	0.22
East	0.18	0.22	0.13	0.20
West	0.26	0.25	0.25	0.21
United States	0.22	0.25	0.17	0.21

b Southern Commercial Agriculture

Investment Type	Mean Rate of Return
Cotton, "antebellum average," 1830-1860	5.7
Cotton, "representative year," 1850s	10.0
Rice, "representative year," 1850s	−3.8
Cotton, land, and capital, 1850s	10.6

Source: Tables 1-5a and b are both taken from F. Batement and T. Weiss, *Industrialization and the Slave Economy* (Greenwich, Conn.: Johnson Associates, 1977), tables 6-4 and 6-5, pp. 6-23 and 6-26. The manufacturing estimates are based on Bateman and Weiss' recent working of manuscript Census data. The Southern commercial agriculture estimates are a collection of estimates which have appeared in the literature, including those offered by R. Fogel and S. Engerman, *Time of the Cross* (Boston: Little, Brown, 1974), p. 70.

somewhat lower in the East during the 1850s. While these interregional differentials are certainly consistent with the forces of convergence through capital migration, note that approximate equalization was already the case in 1860 among large firms (where imputation problems are less troublesome to the observer). On the eve of the Civil War, therefore, approximate rate-of-return equalization seemed to characterize U.S. manufacturing both within the North and between North and South.

*Intra*regional capital markets are another matter entirely. Table 1-5 documents antebellum rates of return on commercial Southern farms, farms where agricultural rates of return must surely have been highest, at least judging by the slave plantation "efficiency" arguments made by Fogel and Engerman [19]. It seems apparent that *intra*regional rate-of-return differentials were everywhere large in the antebellum United States, but they are likely to have been largest in the *North*. Judging by the research of Lau Christensen [5, p. 581], these intraregional rate-of-return differentials were still pronounced as late as 1929, while they finally disappear in the post-World War II decades.

Summary. During the 1850s and on the eve of the Civil War, the U.S. economy was already characterized by wide disparities in regional per capita incomes, so much so that the extent of inequality was comparable to that of many other economies which historically have had pronounced North-South problems. Furthermore, the sectoral income gap was on the rise economywide, and even more so in the North, as unbalanced total factor productivity growth favored the nonagricultural sector over the antebellum decades. United States "dualism" in the 1850s was at least as great as the classic examples of dualism offered by the literature, for example, Italy, Japan, or Soviet Russia during comparable stages of early modern economic growth. However, there is no evidence of *intra*regional labor market disequilibrium in the North on the eve of the Civil War. Wage equalization seems to be the more effective description of Northern *intra*regional labor markets. In addition, any evidence of *inter*regional wage differentials had more or less disappeared by the Civil War decade. Capital markets are another matter entirely. Intrarregional capital market failure is confirmed by abundant evidence of higher rates of return on nonfarm investments, especially in the North. In contrast, interregional rates of return in manufacturing seem to have almost equalized by 1860.

Meanwhile, Down South: The Revisionist
Position and Antebellum Performance

There are two reasons for the exclusion of the antebellum South from the quantitative analysis in section IV. First, criticism of the Southern data base has yet to be answered in the literature [18, 23]. Second, Civil War events suggest the wisdom of restricting the analysis of longer-run nineteenth-century regional experience to those areas which emerged relatively unscathed by the conflict. Nonetheless, we would be remiss were we to ignore the recent debate on the "myth of Southern backwardness."

It has always been obvious that Southern development lagged between 1840 and 1880, so much so that the North-South "problem" becomes a durable part of the U.S. scene for a century after. What is not so obvious, however, is the extent to which those trends would have prevailed in the absence of the conflict. While no explicit counterfactual is attempted here, table 1-6 supplies some relevant estimates of antebellum North-South performance. Based on these data, Fogel and Engerman remind us that per capita incomes (*including* slave populations) in the antebellum South exceeded all Europe except Great Britain. The South can hardly be characterized as backward if these average productivity indices are taken at face value. Nor was her growth any lower than that of either the Northeast or the North Central area in the two decades prior to 1860, though it might, of course, be argued that it is the (unobservable) growth performance from 1820 to 1850 which matters. It is true that growth rates *within* the South Atlantic, *within* the East South Central, and *within* the West South Central regions were all far below the rates within both Northern regions. Nonetheless, it is the weighted performance that matters, and the westward population shift within the South was an important source of productivity gains.

Table 1-6
Regional per Capita Income Estimates, 1840-1860

Region	1840 (1860 Dollars)	1860 (1860 Dollars)	Per Annum Growth, 1840-1860 (Percent)
United States	$ 96	$128	1.4
North	109	141	1.3
Northeast	129	181	1.7
North Central	65	89	1.6
South	74	103	1.7
South Atlantic	66	84	1.2
East South Central	69	89	1.3
West South Central	151	184	1.0
Great Britain	—	130	1.6
France	—	84	1.1
Germany	—	69	1.2
Italy	—	50	0.4
Sweden	—	42	0.5

Source: From *Time on The Cross* by R.W. Fogel and S.L. Engerman, (Boston: Little, Brown and Co., 1974), pp. 249-250. Copyright © 1974 Little, Brown and Co. Reprinted by permission.

Postbellum Initial Conditions

We need not delay long here since the impact of the Civil War is obvious and well documented. First consider nonhuman wealth endowments per worker reported by Lee Soltow [43, table 3.2, p. 67, adult male, free and slave]. The ratio of per worker wealth, North to South, rose sharply from 1.34 in 1860 to 2.02 in 1870. Second, personal income per capita in the South fell from 72 percent of the U.S. average in 1860 to 51 percent in 1800 [17, p. 528]. Finally, we have already noted the emergence of intraregional and interregional wage gaps in the postbellum South.

IV. A Century of Regional Development

Trends in Regional Inequality

Tables 1-7 and 1-8 confirm conventional wisdom. Since 1880, the U.S. economy has undergone regional convergence in per capita and per worker incomes. More to the point of this chapter, however, note the episodic deviations from that trend and the contrasting behavior over the 1840-1880 period.

First, pronounced regional convergence prevailed within the North between 1840 and 1880, a result consistent with the predictions of our model following the technology-induced unbalanced growth of early modern economic growth. Second, and in contrast, regional divergence was true of

Table 1-7
Regional Inequality Performance, 1840-1970: Coefficient of Variation in State Income per Capita

Year	Region		
	North	North and South	United States
1970	.102763	.155367	.148668
1950	.115894	.234063	.225804
1929	.238490	.385408	.373722
1920	.203628	.330943	.326310
1900	.127632	.300265	.322179
1880	.160444	.310221	.358517
1880*	.151168	.306694	—
1840*	.273146	.283737	—

Note: See appendix 1A for definition of regions and data sources used. Weighted coefficient of variation used throughout. 1840* and 1880* refer to comparable regional groups with state observations in both years (for example, excluding Minnesota, the Dakotas, Nebraska, Kansas, Texas, and Oklahoma).

the North and South combined over the same period, the well-known Civil War impact. Third, convergence took place everywhere in the United States between 1880 and 1900, a result consistent with the relatively balanced character of sectoral technological progress over the late nineteenth century. We note, however, that the convergence was far more pronounced in the North, suggesting that additional forces were at work in the South which inhibited its convergence toward national averages. Fourth, this long-term convergence was interrupted between 1900 and 1929. Indeed, these three decades were ones of regional divergence in per capita incomes, especially so during the 1920s. Once again, this result is consistent with the evidence of technological imbalance across sectors during the first third of the twentieth century. Yet, the divergence was far more pronounced in the North, suggesting that there were additional forces at work specific to the South which tended to offset the impact of unbalanced growth. Fifth, the 1929-1970 period records convergence everywhere. Once again, this pronounced convergence appears to be consistent with the balanced growth typical of the mid-twentieth century.

What were the statistical components of this regional inequality experience?

Labor Supply Response

To what extent might the regional income per capita trends be explained by purely demographic forces?

Table 1-8
Regional Inequality Performance, 1840-1970: Coefficient of Variation in State Income per Worker

	Region		
Year	North	North and South	United States
1970	.086807	.118959	.116590
1950	.086735	.183323	.176559
1929	.178174	.325026	.311995
1920	.148501	.275126	.268160
1900	.139739	.380030	.383271
1880	.172866	.381191	.395519
1880*	.155364	.380491	—
1840*	.254981	.312895	—

Note: See appendix 1A for definition of regions and data sources used. Weighted coefficient of variation used throughout. 1840* and 1880* refer to comparable regional groups with state observations in both years (for example, excluding Minnesota, the Dakotas, Nebraska, Kansas, Texas, and Oklahoma).

What might be called the "conventional labor supply model" would predict that high-wage regions tend to have higher labor participation rates (LPRs). If so, then the variance in per capita incomes should exceed the variance per worker incomes. The evidence in tables 1-7 and 1-8 is mixed. While there does appear to be a positive correlation between per worker incomes and LPRs during most of the present century, that was not the case prior to 1920. Per capita income variance was *smaller* than per worker income variance everywhere in the nineteenth century except in the North in 1840. During the nineteenth century, low per capita incomes in backward regions were not the result of low LPRs, and regional differences in labor productivities were more pronounced than in average incomes.

The conventional labor supply model would also predict that those forces which produce convergence in per worker incomes should produce even larger convergence in per capita incomes. Given the discussion above, it is unlikely that these predictions will survive a nineteenth-century test, but the following questions are surely relevant for twentieth-century experience: To what extent did changes in regional LPRs contribute to trends in regional income inequality? Did the behavior of these demographic forces ever play a systematic and dominant role, or can the analysis of U.S. regional inequality focus solely on the causes of labor productivity differentials across regions? Table 1-9 supplies the answers. There we have decomposed changes in the variance in regional incomes per capita into (1) changes in the variance in incomes per worker, (2) changes in the variance in LPRs, and (3) changes in the covariance in these two variables.

The conventional labor supply model appears to have been at work in two phases of U.S. history, everywhere between 1929 and 1970 and in the North between 1840 and 1880. Indeed, it was the 1919-1954 period which led Frank Hanna [24, chapter 7] to conclude that demographic forces were a very important component of the convergence of per capita incomes in the United States. Table 1-9 confirms Hanna's finding for the 1929-1970 period since we also find the convergence of per capita incomes to have been far more extensive than for per worker incomes. For example, the convergence of LPRs themselves accounted for 7 percent of the Northern per capita income convergence between 1929 and 1950. (It accounts for *all* the convergence between 1950 and 1970.) Since per capita incomes and LPRs have been positively correlated in recent decades, the change in the covariance term also contributed to convergence, an additional 37 percent. If we, like Hanna, restricted our attention to the 1929-1950 period, we would conclude that "only" 56 percent of the Northern per capita convergence was attributable solely to convergence in labor productivities during those two decades. Similar results are forthcoming for the North and South combined as well as for the United States as a whole. Furthermore, post-1840 Northern experience would appear to offer further (but mild)

Table 1-9
Sources of Changing Regional Inequality, 1840-1970: Decomposing Changing Regional Income per Capita Variance

Decomposition	1950-1970	1929-1950	1920-1929	1900-1929	1880-1900	1840-*1880*	1880-1970
				North			
Total ΔV^2, of which	−.002871	−.043446	+.015413	+.025174	−.009452	−.051757	−.015182
Due to:							
Δ Var Inc per Wkr	+.000049	−.024548	+.009583	+.003105	−.011250	−.041837	−.023061
Δ Var LPR	−.000582	−.002998	−.000045	+.000084	−.002152	−.002605	−.005693
Δ Cov	−.002337	−.015900	+.005875	+.021985	+.003949	−.007316	+.013572
				North and South			
Total ΔV^2, of which	−.030646	−.093755	+.039017	+.019364	−.006078	+.013554	−.072098
Due to:							
Δ Var Inc per Wkr	−.018746	−.075690	+.030968	−.078111	+.002364	+.056691	−.139215
Δ Var LPR	−.001099	−.001512	−.000100	−.010248	−.001252	+.000939	−.014211
Δ Cov	−.010801	−.016553	+.008148	+.107724	−.007190	−.044076	+.081328
				United States			
Total ΔV^2, of which	−.028885	−.088681	+.033190	+.002679	−.024735	n.a.	−.106432
Due to:							
Δ Var Inc per Wkr	−.016836	−.069024	+.025946	−.086604	−.009179	n.a.	−.155697
Δ Var LPR	−.001193	−.002183	+.000115	−.009921	−.002461	n.a.	−.015643
Δ Cov	−.010855	−.017474	+.007128	+.099204	−.013094	n.a.	+.064909

Note: The calculation is derived from the formula

$$V^2 = \left(\frac{1}{y^*}\right)^2 \frac{\Sigma}{j} w_j l_j^2 (y_j - y)^2 + \left(\frac{1}{l}\right)^2 \frac{\Sigma}{j} w_j (l_j - l)^2 + 2\left(\frac{1}{y^*}\right)\left(\frac{1}{l}\right) \frac{\Sigma}{j} w_j l_j (y_j - y)(l_j - l)$$

where we denote y^* income per capita, y income per worker, l the labor participation rate, and w_j the jth state's share of population in the "national" total; V^2 is the square of the weighted coefficient of variation across states reported as V in table 4-1. Thus, V^2 refers to the change in V^2 over the periods indicated, and LPR refers to the labor participation rate.

support for the conventional labor supply model. In the North, 5 percent of the convergence in per capita incomes following 1840 can be explained by the convergence in LPRs, the poorer regions having enjoyed a relative rise in their LPRs. Evidence supporting the conventional labor supply model stops there, however. Between 1840 and 1880, regional divergence in per worker incomes for the North and South combined was far more dramatic than in per capita incomes. Demographic forces were serving to mute the forces of regional labor productivity divergence. The same is true of the long-term convergence over the century following 1880. For all three regions, the convergence of state per worker incomes was far more pronounced than was per capita incomes between 1880 and 1970.

It would take us far afield to offer a full accounting for these conflicting demographic influences, and demographic historians have yet to suggest a convincing explanation for these trends. We are content to stress the following key finding: Long-term U.S. experience with regional inequality is not simply a demographic phenomenon, although it has become increasingly so in recent decades. Thus, explanations for differing regional income per capita growth must center on the behavior of differing regional labor productivity growth, experience for which our theorizing is best equipped.

Why Regional Productivity Growth Rates Differ: Looking for "Sources"

Part II offered a decomposition formula by which changes in the variance in regional per worker incomes can be broken up into meaningful and additive component "sources." Since the changing variance itself is the result of differing regional growth rates, the decomposition offers an empirical accounting of the sources of these disparate regional growth rates. How much of the U.S. long-term regional experience is explained by differing regional productivity growth within sectors? How much by the convergence and divergence in regional industrial structures? How much by the behavior of economywide experience with productivity gaps between sectors? The answers are central to any effort in modeling long-term regional inequality experience, and tables 1-10, 1-11, and 1-12 present a convenient summary of the U.S. experience since 1840.

Consider first Northern experience during the four decades following 1840. Table 1-10 documents the sources of the convergence in state per worker incomes following the disequilibrating shock of early industrialization. Changing industrial structures contributed nothing to that convergence, since it appears that the richer and more industrialized states departed even further from the national average. Differential shift was serving to widen regional per worker differentials, although it must be empha-

Table 1-10
Sources of Changing Regional Inequality in the North, 1840-1970: Decomposing Changing Regional Income per Worker Variance

Decomposition	1950-1970	1929-1950	1900-1929	1880-1900	1840*-1880*	1880-1970
Total ΔV^2, of which Due to:	+.000012	−.024217	+.012210	−.010347	−.040891	−.022342
1. ΔVar Sectoral Incomes per Wkr Combined	+.000959	−.010503	+.002553	+.001662	−.016384	−.005329
ΔVar NFI per Wkr	+.001275	−.010518	+.012023	+.001757	−.002611	+.004537
ΔVar FI per Wkr	−.000391	+.000122	−.005499	−.001903	−.008286	−.007671
ΔCov	+.000075	−.000107	−.003971	+.001808	−.005487	−.002195
2. ΔVar Industrial Structure	−.000891	−.000603	−.011880	+.000372	+.006147	−.010257
3. ΔSectoral Productivity Gap	+.000863	−.009225	+.003726	−.000446	−.021909	−.004043
4. ΔResidual Cov Forces	−.000919	−.003886	+.017811	−.011935	−.008745	−.002713

Note: See part II for decomposition formula. Rows 2 and 3 use Paasche weights, except for 1880-1970 which uses 1929 weights.

Table 1-11
Source of Changing Regional Inequality in the North and South Combined, 1880-1970: Decomposing Changing Regional Income per Worker Variance

Decomposition	1950-1970	1929-1950	1900-1929	1880-1900	1880-1970
Total ΔV^2, of which Due to:	−.019449	−.072053	−.038747	−.000900	−.131149
1. ΔVar Sectoral Incomes per Wkr Combined	−.010065	−.020496	−.009936	+.005461	−.035036
ΔVar NFI per Wkr	−.003876	−.019199	+.030192	+.003614	+.010731
ΔVar FI per Wkr	−.001398	−.000165	−.028926	−.005991	−.036480
ΔCov	−.004791	−.001132	−.011202	+.007838	−.009287
2. ΔVar Industrial Structure	−.000934	−.003711	−.021746	+.007093	−.018699
3. ΔSectoral Productivity Gap	−.002788	−.031558	−.013892	−.001870	−.055852
4. ΔResidual Cov Forces	−.005662	−.016288	+.006827	−.011584	−.021562

Note: See part II for decomposition formula. Rows 2 and 3 use Paasche weights, except for 1880-1970 which uses 1929 weights.

Table 1-12
Sources of Changing Regional Inequality in the United States, 1880-1970: Decomposing Changing Regional Income per Worker Variance

Decomposition	1950-1970	1929-1950	1900-1929	1880-1900	1880-1970
Total ΔV^2, of which Due to:	−.017594	−.066157	−.049537	−.009559	−.142847
1. Δ Var Sectoral Incomes per Wkr	−.009853	−.019068	−.017806	+.004776	−.041951
Δ Var NFI per Wkr	−.002937	−.018043	+.024654	+.003360	+.007034
Δ Var FI per Wkr	−.001538	−.000130	−.028930	−.006749	−.037347
Δ Cov	−.005378	−.000895	−.013530	+.008165	−.011638
2. Δ Var Industrial Structure	−.000834	−.002787	−.020371	+.004237	−.025762
3. Δ Sectoral Productivity Gap	−.001558	−.030642	−.015179	−.002957	−.046794
4. Δ Residual Cov Forces	−.005349	−.013660	+.003819	−.015615	−.028340

Note: See part II for decomposition formula. Rows 2 and 3 use Paasche weights, except for 1880-1970 which uses 1929 weights.

sized that the influence was minor. In short, while the poorer regions were indeed industrializing over the period, and thus shifting labor into high-productivity sectors, they did so at a slower rate than the richer regions. Sectoral labor productivities, however, did converge. On average, the poorer regions enjoyed a higher rate of growth in sectoral labor productivities. These forces combined to account for 40 percent of the convergence in state per worker incomes (−.016384 versus −.040891). Which sectors made the greatest contribution to the convergence? Agriculture accounted for the lion's share: the changing variance in nonfarm incomes (NFI) per worker accounted for only 6 percent of the total (−.002611 versus −.040891). Since the farm sector was, after all, a very large share of total economic activity, the result may not appear surprising. Yet, it is also consistent with the view that rate-of-return differentials between regions were far more pronounced in the farm sector in 1840, and thus the result offers support for the operation of classic equalizing accumulation forces along the lines of part II. Furthermore, since farm and nonfarm labor productivities were positively correlated, the changing covariance term is also negative. If differential shift fails to account for any of the Northern convergence and if converging sectoral per worker incomes account for only 40 percent, we are left with a very large share of the convergence explained by the regionwide behavior of the sectoral productivity gap. Indeed, 54 percent of the convergence in Northern state per worker incomes following 1840 is explained by the decline in the gap between agricultural and nonagricultural labor productivities (−.021909 versus −.040891). Those states heavily committed to farm activities (the backward states) derived a potent benefit from "proportionality shift."

Northern regional growth experience after 1840 can be summarized in the following terms: there was no convergence in industrial structure, converging nonfarm labor productivities played a minor role, converging farm labor productivities played a far larger role, and the decline in the farm-nonfarm labor productivity gap played the major role.

Is this pattern repeated for the North and South combined following 1880? Table 1-11 supplies confirmation. The regionwide decline in the sectoral productivity gap accounts for 43 percent of the convergence in state per worker incomes between 1880 and 1970, while the convergence of sectoral incomes per worker accounts for only 27 percent. Furthermore, it is the convergence of farm income per worker which explains all the sectoral income convergence effect since the variance of NFI per worker *rose* over the century. Finally, we note that the differential shift effect made a positive but small contribution, 14 percent (−.018699 versus −.131149). Once again, we conclude that the main source of long-term convergence lies with behavior over time in farm-nonfarm productivity gaps and interstate differentials in farm labor productivities. Roughly similar findings emerge when the U.S. as a whole is analyzed (table 1-12).

*Demand Shocks, Technology Shocks, and
Unbalanced Growth*

We turn now to a more challenging task—accounting for departures from long-term regional inequality trends within the past century. For this purpose, we have found it useful to divide the century into four subperiods of roughly equal length: 1880-1900, 1900-1929, 1929-1950, and 1950-1970.

Consider first the Northern regional inequality trends in table 1-10. The 1880-1900 period records a continuation of the previous four decades of convergence, but at slower rate as the states approach parity in per worker incomes. The next three decades record a reversal in those long-term trends. Obviously it is important to understand the sources of this dramatic reversal, and we shall dwell on them shortly. Between 1929 and 1950, the regional convergence resumes, to be sure, but what is more astounding is the enormous rate of convergence. Indeed, the total convergence in state per worker incomes between 1929 and 1950 ($-.024217$) exceeds the convergence for 1880-1970 as a whole ($-.022342$). Once again, it is important to understand the special conditions which produced this spectacular convergence. Between 1950 and 1970, Northern regional inequality indices remain relatively stable, but since there is little significant difference in state per worker incomes in the North by 1950, especially in *real* terms, we shall make no effort to analyze their trends over the past two decades.

Why the reversal in trend during 1900-1929 and the spectacular resumption in 1929-1950? The source of the reversal lies with two influences: a rise in the sectoral productivity gap and an increase in the regional variance in NFI per worker. Both of these forces can in combination account for all Northern regional divergence over the first third of the twentieth century. Converging farm productivities and converging industrial structures tend to produce income per worker convergence, but the other two divergence influences dominate. It seems to me evident that both influences are precisely those associated with technological imbalance favoring the nonfarm sector. It served to raise relative nonfarm productivities, increasing the gap, and it also increased the variance in NFI per worker since the rapid rates of TFPG were centered in those sectors location-specific to high-income states, that is, in the skill and machine-intensive sectors generally and the durable goods producing sectors in particular. These forces reverse sign following 1929, a result fully consistent with the evidence of technological and demand balance which characterizes the two decades up to 1950. Changes in the variance in state industrial structures play *no* role in accounting for the reversal after 1900 or for the resumption of convergence after 1929. The same holds for changes in the variance of farm incomes per worker. Indeed, both variables were operating in the opposite direction.

When one turns from Northern regional inequality experience over the past century to economywide regional inequality experience, some interesting contrasts emerge. First, however, let us review the similarities. The overall convergence in state per worker incomes between 1880 and 1970 was even more pronounced in the United States as a whole than in the North, −.142847 (table 1-12). Furthermore, the "sources" of that long-term convergence appear to be similar to those already discussed for the North. The nationwide decline in the productivity gap accounts for 33 percent; the convergence of sectoral incomes per worker accounts for 29 percent, all of which explained by the convergence of farm incomes, and none by nonfarm productivity behavior; convergence in state industrial structures accounts for a much smaller share, 18 percent. As in the case of the North, therefore, explanations of differences in long-term regional rates of productivity growth must lie primarily with the behavior of the economywide productivity gap and with interstate differentials in farm productivity growth. Similarities with the Northern regional experience appear within shorter-term episodes as well. The dramatic convergence between 1929 and 1950 reappears economywide, and the sources of that convergence are roughly the same. That is, 46 percent of the per worker incomes' convergence between 1929 and 1950 is attributable to the sharp decline in productivity gaps, a notable departure from the much lower contribution from this source in the 1900-1929 period. The convergence of state industrial structures following 1929 accounts for only 4 percent of the total convergence in per worker incomes, an even smaller contribution than in the preceding 1900-1929 period. Finally, it appears that almost all the contribution of converging sectoral incomes per worker after 1929 is attributabe to the declining variance in NFI per worker. In contrast, the 1900-1929 period records a divergence in state NFI per worker, a result fully consistent with trends in the North.

What, then, are the differences? There are two, and they both center on the 1900-1929 episode. First, the convergence in state farm productivities was much more dramatic economywide than in the North. This offset the divergence in state NFI per worker, thus producing a net contribution toward convergence in sectoral incomes, variance combined. Since this result appears for the North and South combined (table 1-11), we conclude that the 1900-1929 period contained unusually favorable farm productivity forces specific to the South. Second, while farm-nonfarm productivity gaps rose in the North after 1900, they declined economywide. True, the decline was far less extensive than in the 1929-1950 episode, but the contrast with Northern experience during the first third of the twentieth century is still striking.

What is it about Southern agricultural productivity performance which makes U.S. regional inequality experience between 1900 and 1929 different from that taking place within the North?

I believe the answer may lie in large part with cotton prices. Table 1-13 documents an extraordinary rise in the relative price of cotton from 1900 to

Table 1-13
Relative Price of Cotton, 1900-1950
(1900 = 100)

Year	Farm Products (1)	Raw Cotton (2)
1900 (1898-1902)	100.0	100.0
1920 (1918-1922)	110.7	148.6
1929 (1927-1931)	111.3	120.5
1950 (1949-1951)	114.9	150.6

Note: Column 1 refers to the Bureau of Labor Statistics (BLS) farm products wholesale price index deflated by the BLS all-commodity wholesale price index. Column 2 refers to the raw cotton price index deflated by the BLS all-commodity wholesale price index. All prices are from U.S. Department of Commerce, *Historical Statistics of the United States*, pt. 1 (Washington: GPO, 1975), pp. 200, 203.

1920, far in excess of farm products generally. To some extent the same is true of the 1929-1950 period. In contrast, cotton prices collapsed during the 1920s, but the total rise from 1900 to 1929 still exceeds that of farm products generally. This relative price behavior may well explain the contrasting regional inequality behavior between the North, on the one hand, and the North and South combined, on the other: farm incomes per worker converge at a far less dramatic rate within the North in 1900-1929; sectoral productivity gaps rise in the North but fall economywide over the same period; thus, regional divergence occurs in the North but not economywide. Furthermore, these relative price trends suggest that contrasting regional inequality behavior should have been centered on the 1900-1920 period, while comparable behavior should have characterized the 1920s. Table 1-14 confirms this prediction. It seems plausible, therefore, that the exogenous trends in cotton prices may account for the U.S. departure from the "Northern model of regional divergence" operating during the first third of the twentieth century.

V. Cost-of-Living Impact and Converging Price Structure

Incomes, Prices, and Denison-Gilbert-Gerschenkron Effect

One of the "stylized facts" emerging from cross-national and time-series studies seems to be that expenditure or output shares are inversely correlated with relative prices. That is to say, large expediture or output shares of a given commodity coincide with low relative prices for that commodity.

Table 1-14
Sources of Changing Regional Inequality, 1900-1950: Decomposing Changing Regional Income per Worker Variance, North versus North and South Combined

Decomposition	1900-1920		1920-1929		1929-1950	
	North	North and South	North	North and South	North	North and South
Total ΔV^2, of which	+.002523	−.068706	+.009687	+.029959	−.024217	−.072053
Due to:						
1. ΔVar Sectoral Incomes per Wkr	−.000819	−.024718	+.003372	+.014782	−.010503	−.020496
$\quad \Delta$Var NFI per Wkr	+.007035	+.012064	+.004988	+.018128	−.010518	−.019199
$\quad \Delta$Var FI per Wkr	−.005278	−.027395	−.000221	−.001531	+.000122	−.000165
$\quad \Delta$Cov	−.002576	−.009387	−.001395	−.001815	−.000107	−.001132
2. ΔVar Industrial Structure	−.006741	−.012635	−.000755	−.004673	−.000603	−.003711
3. ΔSector Productivity Gap	−.001410	−.029816	+.004041	+.010720	−.009225	−.031558
4. ΔResidual Cov Forces	+.011493	−.001537	+.003029	+.009130	−.003886	−.016288

Note: See part II for decomposition formula. Rows 2 and 3 use Paasche weights, except for 1880-1970 which uses 1929 weights.

Thus, Milton Gilbert and his associates [21, 22] found that when U.S. relative prices were used to evaluate Italian postwar GNP, Italian GNP rose relative to the United States GNP, compared with the case when Italian relative prices were used in the calculation. Symmetrically, Soviet prewar growth rates were found to be far lower when relative prices of 1940 were used rather than those of 1926-1927, a differential which came to be called the "Gerschenkron effect." Edward Denison [12] illustrated this effect once again by showing us that the "miraculous" growth in postwar Western Europe and Japan was far less impressive when U.S. relative prices were used to calculate their growth performance. The Denison-Gilbert-Gerschenkron (DGG) effect seems to matter a great deal in time-series and cross-national accounting comparisons, and the explanation is simply that goods which decline rapidly in price across time and space are accompanied by relatively rapid output and expenditure growth. Furthermore, the analysis in part II suggests the unbalanced total factor productivity growth had a similar impact on output mix and price structures in nineteenth-century United States. There is no a priori reason for believing that the DGG effect is less important across regions.

Since comparisons of labor productivities and income per capita differentials across U.S. regions are used to make both welfare and labor market efficiency judgments, the relative prices that matter most are cost-of-living differentials. That being the case, we begin with the assertion that "conventional wage goods"—food, fuel, light, and rent—should have lower relative prices in backward regions. If this assertion holds, then it follows that the deflation of regional nominal performance measures by local cost-of-living indices should tend to raise the measured real income of the backward region relative to the more affluent. What explanations can we offer for this version of the DGG hypothesis? Two underlying explanations would tend to give the hypothesis analytical support. First, backward regions tend to be heavy producers and consumers of consumption goods, while the more affluent specialize in investment and intermediate goods. Second, backward regions tend to be heavy producers and consumers not only of consumption goods, but also of "necessities" in particular. If consumption goods in general and necessities in particular also tend to incur high transport costs, then their relative prices can indeed differ across space, with low prices prevailing in backward, low-wage regions. This characterization would appear to hold given that backward regions tend to be abundant in natural resources and unskilled labor while poorly endowed with skills and conventional capital. Thus, resource- and unskilled labor-intensive products should tend to have low relative prices in the backward region. These would include food and fuel, but also nontradeables like housing which, ignoring site rents, tend to be both labor- and materials-intensive. Given these factor intensity characterizations and the presence of high transport costs, it follows that low-wage, backward regions should also be regions of low cost of living. It also follows, of course, that comparisons of nominal wages, labor productivities, and per capita incomes exaggerate the extent of true regional disparities.

What does the DGG hypothesis suggest regarding trends in regional inequality over time? We would expect regional cost-of-living differentials to converge over time, for two reasons. First, the impact of economic growth economywide is to diminish the relative expenditure shares devoted to necessities, consumption goods with the highest embodied transport costs and thus with the highest regional price variance. Economywide, therefore, we should observe a shift into those commodities which exhibit less regional price variance, and thus regional cost-of-living difference should collapse on these grounds alone. (The emphasis here is on *commodities*. Services are another matter entirely, as we shall see below.) Second, the relative price structure itself should tend to converge as transportation costs across regions tend to decline as nominal wage costs converge (thus precipitating a convergence in prices of labor-intensive nontradeables), and as individuals in the backward region migrate from the farm to urban areas (thus moving to nonfarm urban employment where necessities have higher distribution costs). This prediction, after all, is nothing more than a straightforward corollary of the DGG hypothesis: patterns of both prices and expenditures converge as the incomes of poorer regions rise relatively.

If the DGG corollary holds, then as regions undergo "true" convergence, the price-expenditure effect diminishes. The implication seems to be that wide regional income variance is in part spurious, but that the magnitude of the bias declines over time as regional relative prices converge. (The convergence should appear both *among* consumption goods and *between* consumption goods and all other investment and intermediate goods.) In short, if the DGG effect holds for the nineteenth- and twentieth-century United States, then real incomes should exhibit less regional variance than nominal incomes. Similarly, the DGG thesis predicts that some of the observed regional convergence over time since the early to mid-nineteenth century must be due to the convergence of prices themselves.

Does the DGG thesis hold for U.S. regions over the past century? While published cost-of-living data for the twentieth century have always been available for the regional economist, such a test has, to my knowledge, never been performed. In addition, Phillip Coelho and James Shepherd [8, 9] have now supplied the necessary data to perform the test even for the nineteenth century, years for which the DGG thesis is likely to have the most empirical relevance.

Table 1-15 offers some initial evidence. The table reports what we call "Koffsky-adjusted" cost-of-living indices across nine Census divisions and for the 1840-1970 period. Each regional cost-of-living index is expressed in terms of New England = 100. As appendix 1A shows, these indices convert urban regional price relatives to regionwide cost-of-living indices by apply-

Table 1-15
Koffsky-Adjusted Regional Cost-of-Living Indices, 1840-1970
(New England = 100)

Region	1840	1880	1900	1920	1929	1950	1970
New England	100.0	100.0	100.0	100.0	100.0	100.0	100.0
Mid-Atlantic	98.5	95.7	99.1	101.3	101.3	100.0	98.4
East North Central	80.2	82.5	88.0	96.1	99.0	100.0	95.9
West North Central	100.8	93.4	93.0	98.8	92.7	95.9	94.3
South Atlantic	123.3	100.4	95.7	99.5	91.6	95.9	91.0
East South Central	83.5	80.6	90.6	97.2	84.0	94.5	88.5
West South Central	92.1	81.9	87.3	94.6	86.1	91.8	87.7
Mountain	—	—	—	117.6	94.0	98.6	95.1
Pacific	—	—	—	103.1	97.7	100.0	97.5

Source: Appendix 1A.

ing estimates of farm-nonfarm cost-of-living differentials to the urban prices. Employment shares are used as weights in constructing the aggregate regional indices. It does seem apparent that much of the wide variance in regional cost of living observed in 1840 had dissipated a century later, but considerable variance persists even today. The convergence never took place smoothly and without a hitch, but some impressive examples stand out in the historical record: the abrupt decline in the South Atlantic relative between 1840 and 1880; similar declines in the Mountain and Pacific regions after 1920; and the steady upward drift in the relatives for the East North Central. Table 1-16 documents the price convergence for some key commodity groups, quoted in urban centers, and for three points in time. While space precludes presentation of the complete data here, table 1-16 supplies some confirmation of the position that all prices tended toward convergence after 1840, but that the behavior of rents, fuel and light, and food mattered most since these were the items whose prices varied the most at every point (and in the order given).

Price Convergence and the Cost-of-Living Impact

The regional price data in tables 1-15 and 1-16 do not, of course, supply an explicit test of the DGG hypothesis, at least where it applies to *weighted* measures of income dispersion. Have backward states always undergone the relative rise in living costs, approaching the high living costs in rich states, and has it always been the poorest states undergoing the most dramatic cost-of-living increase? The evidence in table 1-15 is mixed. For example, between 1880 and 1970 while most Southern regions experienced a rise in relative living costs, it was not true of the South Atlantic. Between 1920 and 1970, the South Central and South Atlanic states all underwent a further

Table 1-16
Urban Cost-of-Living Indices, by Region and Commodity Group, 1851-1935
(New England = 100)

Region	Food			Clothing			Fuel and Light			Rent			Other		
	1851	1890	1935	1851	1890	1935	1851	1890	1935	1851	1890	1935	1851	1890	1935
New England	100	100	100	100	100	100	100	100	100	100	100	100	100	100	100
Mid-Atlantic	98	97	98	127	105	97	123	106	88	88	112	112	93	110
East North Central	83	95	95	127	103	104	32	79	85	91	107	87	114
West North Central	96	85	94	127	110	100	—	92	94	66	104	88	104
South Atlantic	—	106	100	—	110	95	—	85	85	—	105	94	102
East South Central	85	98	94	111	104	92	35	75	78	134	87	87	100
West South Central	—	95	95	—	105	94	—	104	78	—	90	91	101
Mountain	—	107	98	—	118	105	—	160	98	—	96	112	100
Pacific	—	104	95	—	113	110	—	163	98	—	87	103	112

Source: Appendix 1A. Observations for rents in 1890 and "other" in 1851 are unavailable.

decline in relative living costs. Thus the issue can be resolved only by the explicit introduction of relative regional weights; that is, the question can be answered only by applying the cost-of-living data directly to the state (nominal) income per worker estimates themselves.

If the DGG hypothesis holds, we should find nominal income convergence steeper than real income convergence. Table 1-17 suggests the contrary. From 1880 to 1970, convergence is steeper in *real* terms. Thus, nominal incomes understate the "true" convergence in the United States over the past century. Furthermore, the result is not simply a North-South phenomenon since the same finding appears within the North itself. Consistent with this result, we note that trending regional divergence for the North and South combined between 1840 and 1880 is greater in nominal than in real terms. there are two significant exceptions to our general result, however: 1929-1950 economywide and 1840-1880 in the North. It seems apparent that much more remains to be done to sort out the causes and impact of these regional cost-of-living trends, as well as to improve their quality.

Finally, consider the regional inequality indices presented in table 1-18. For *every* year documented in table 1-18, the variance in real regional incomes per worker is smaller than the variance in nominal incomes, and in some years the variance is much smaller. Furthermore, it is *not* a North-South "sunshine" effect we are observing since the same holds within the North. The essential point is that cost-of-living advantages in low-income areas have always been quite pronounced, and these advantages persist to the present.

VI. Balance versus Confrontation: Looking backward from the 1970s

One must take care in drawing lessons from history, but four morals emerge from our analysis of the past 130 years of regional inequality experience which bear repeating.

Table 1-17
Cost-of-Living Impact: Changing Coefficient of Variation in State Income per Worker, Nominal and Real, 1840-1970

	Region					
	North		North and South		United States	
Period	Nominal	Real	Nominal	Real	Nominal	Real
1950-1970	0	−.007	−.064	−.075	−.060	−.070
1929-1950	−.091	−.080	−.142	−.117	−.135	−.106
1900-1929	+.038	+.034	−.055	−.091	−.071	n.a.
1880-1900	−.033	−.037	−.001	−.002	−.013	n.a.
1880-1970	−.086	−.090	−.262	−.285	−.279	n.a.
1840*-1880*	−.100	−.057	+.067	+.048	n.a.	n.a.

Source: Table 1-18.

Table 1-18
Cost-of-Living Impact: Coefficient of Variation in State Income per Worker, Nominal and Real, 1840-1970

	Region					
	North		North and South		United States	
Year	Nominal	Real	Nominal	Real	Nominal	Real
1970	.087	.067	.119	.081	.117	.086
1950	.087	.074	.183	.156	.177	.156
1929	.178	.154	.325	.273	.312	.262
1920	.149	.138	.275	.265	.268	.259
1900	.140	.120	.380	.364	.383	n.a.
1880	.173	.157	.381	.366	.396	n.a.
1880*	.155	.135	.380	.365	n.a.	n.a.
1840*	.255	.192	.313	.317	n.a.	n.a.

Notes: See notes to table 1-7. The cost-of-living data used in the calculations can be found in table 1A-1 where New England = 100 in each year.

The first relates to cost-of-living differences. These have always been pronounced in the United States, so much so that nominal income and nominal wage comparisons often supply erroneous inferences on the operation of factor markets and measures of regional living standards. To the extent that regional policy is motivated by nominal indicators, those policies may well be misdirected. To repeat, there is no indication from the last century of U.S. experience that these cost-of-living differentials have declined in importance. Indeed, it seems likely, in the face of rising fuel costs and increased consumer interest in leisure-time outdoor activities and the quality of life, that regionally immobile "sunshine-amenity" effects will increasingly dominate worker location decisions. Furthermore, as retirement ages decline, these forces will become increasingly important for those still in productive ages but preparing for retirement.

Second, there is little evidence from U.S. history that interregional labor markets were very inefficient even during episodes of extraordinary stress. It appears that intraregional labor market "failure" has also been overdrawn in the literature. Part of the explanation for this failure in the literature appears to lie with inadequate attention to labor's heterogeneity. Furthermore, the key to factor immobility conditions in U.S. history seems to have been intraregional capital market imperfections and lags in adjustment. These imperfections seem to have departed from the U.S. scene between 1929 and 1950, at least as they relate to farm-nonfarm and corporate-noncorporate rate-of-return differentials.

Third, there is little consistency between regional and national inequality trends. While national inequality and regional inequality were both on the rise during the antebellum period and the 1920s, high and persistent inequality between 1880 and 1929 was consistent with convergence in regional incomes. While the national equality trends between 1929 and 1950 were consistent with regional convergence, continued regional convergence after 1950 was not accompanied by a continuation in national equality trends. The implication seems to be that between-state income variance has never been an important ingredient of total inequality trends in the United States, and within-state income variance has dominated. Policies directed toward distribution goals would never have been well served had they been directed along regional lines. The same holds for the 1970s.

Fourth, unbalanced sectoral rates of technological progress economywide seem to account for much of U.S. regional inequality experience for longer than a century. If the last third of the twentieth century were to produce similar degrees of technological imbalance, we might well expect serious regional implications. One obvious candidate as a source of future technological imbalance lies with the increasing scarcity of natural resources in general and fuels in particular. It is quite possible that the United States may be embarking on yet another episode of "technological shock" with regional divergence implications not unlike those of the antebellum period or of the first third of the present century. While the sources and key sectors may be quite different, the influence on "regional confrontation" may be very much the same.

References

[1] Abramovitz, M.S., and David, P.A. "Reinterpreting Economic Growth: Parables and Realities." *American Economic Review* 58 (May 1973):428-39.

[2] Bateman, F., and Weiss, T. *Industrialization and the Slave Economy*. Greenwich, Conn.: Johnson Associates, 1977.

[3] Borts, G., and Stein, J. *Economic Growth in a Free Market*. New York: Columbia University Press, 1964.

[4] Chiswick, B.R. *Income Inequality*. New York: National Bureau of Economic Research, 1974.

[5] Christensen, L.R. "Entrepreneurial Income: How Does It Measure Up?" *American Economic Review* 61, no. 4 (September 1971):575-85.

[6] Christensen, L.R., Cummings, D., and Jorgenson, D.W. *Economic Growth, 1947-1973: An International Comparison*. Discussion Paper No. 521, Harvard Institute for Economic Research (December 1976).

[7] Coelho, P., and Ghali, M. "The End of the North-South Wage Differential." *American Economic Review* 61, no. 5 (December 1971):932-37.

[8] Coelho, P., and Shepherd, J. "Differences in Regional Prices: The United States, 1851-1880." *Journal of Economic History* 35 (September 1974):551-91.

[9] Coelho, P., and Shepherd, J. "Regional Differences in Urban Prices and Wages: The United States in 1890." Paper presented to the Western Economic Association, San Diego, California (June 1975).

[10] Coelho, P., and Shepherd, J. "Regional Differences in Real Wages: The United States, 1851-1880." *Explorations in Economic History* 13, no. 2 (April 1976):203-230.

[11] Denison, E.F. *The Sources of Economic Growth in the United States and the Alternatives before Us*. New York: Committee for Economic Development, 1962.

[12] Denison, E.F. *Why Growth Rates Differ: Postwar Experience in Nine Western Countries*. Washington: The Brookings Institution, 1967.

[13] Denison, E.F. *Accounting for United States Economic Growth, 1929-1969*. Washington: The Brookings Institution, 1976.

[14] Denison, E.F., and Chung, W.K. *How Japan's Economy Grew So Fast*. Washington: The Brookings Institution, 1976.

[15] Easterlin, R.A. "State Income Estimates." In E.S. Lee et al., *Population Redistribution and Economic Growth, United States, 1870-1950*, vol. 1. Philadelphia: American Philosophical Society, 1957, pp. 703-759.

[16] Easterlin, E.A. "Interregional Differences in per Capita Income, Population, and Total Income, 1840-1950." In *Trends in the American Economy in the Nineteenth Century*. Princeton, N.J.: Princeton University Press, 1960.

[17] Easterlin, R.A. "Regional Income Trends, 1840-1950." In *American Economic History*, edited by S. Harris. New York: McGraw-Hill, 1961.

[18] Engerman, S.L. "Some Economic Factors in Southern Backwardness in the Nineteenth Century." In *Essays in Regional Economics*, edited by J.F. Kain and J.R. Meyer. Cambridge, Mass.: Harvard University Press, 1971.

[19] Fogel, R., and Engerman, S. *Time on the Cross*. Boston: Little, Brown, 1974.

[20] Gallman, R.E. "Commodity Output, 1839-1899." In *Trends in the American Economy in the Nineteenth Century*. Princeton, N.J.: Princeton University Press, 1960.

[21] Gilbert, M., and Associates. *Comparative National Products and Price Levels*. Paris: Organization for European Economic Cooperation, 1958.

[22] Gilbert, M., and Kravis, I.B. *An International Comparison of National Products and the Purchasing Power of Currencies*. Paris: Organization for European Economic Cooperation, 1954.

[23] Gunderson, G. "Southern Ante Bellum Income Reconsidered." *Explorations in Economic History* 10, 2 (Winter 1973):151-69.

[24] Hanna, F.A. *State Income Differentials, 1919-1954*. Durham, N.C.: Duke University Press, 1958.
[25] Hirschman, A.O. *The Strategy of Economic Development*. New Haven, Conn.: Yale University Press, 1958.
[26] Jones, R.W. "The Structure of Simple General Equilibrium Models." *Journal of Political Economy* 73 (December 1965):557-72.
[27] Kendrick, J.W. *Productivity Trends in the United States*. Princeton, N.J.: Princeton University Press, 1961.
[28] Kindleberger, C.P. *Europe's Postwar Growth: The Role of Labor Supply*. Cambridge, Mass.: Harvard University Press, 1967.
[29] Koffsky, N. "Farm and Urban Purchasing Power." In *Studies in Income and Wealth*, vol. 11. New York: National Bureau of Economic Research, 1949.
[30] Kuznets, S. "Economic Growth and Income Inequality." *American Economic Review* 45 (March 1955):1-28.
[31] Kuznets, S. *Modern Economic Growth: Rate, Structure and Spread*. New Haven, Conn.: Yale University Press, 1966.
[32] Lebergott, S.L. *Manpower in Economic Growth*. New York: McGraw-Hill, 1964.
[33] Lee, E.S., et al. *Population Redistribution and Economic Growth: United States, 1870-1950*, vol. 1. Philadelphia: American Philosophical Society, 1957.
[34] Lindert, P.H. *Fertility and Scarcity in America*. Princeton, N.J.: Princeton University Press, 1978.
[35] Lindert, P.H., and Williamson, J.G. "Three Centuries of American Inequality." In *Research in Economic History*, vol. 1, edited by P. Uselding. Greenwich, Conn.: Johnson Associates, 1976, pp. 69-123.
[36] McKee, D.L., Dean, R.D., and Leahy, W.H. *Regional Economics: Theory and Practice*. New York: The Free Press, 1970.
[37] McKinnon, R.I. *Money and Capital in Economic Development*. Washington: The Brookings Institution, 1973.
[38] Myrdal, G. *Economic Theory and Underdeveloped Regions*. London: Duckworth and Co., Ltd., 1957.
[39] Perloff, H.S., et al. *Regions, Resources, and Economic Growth*. Baltimore, Md.: Johns Hopkins Press, 1960.
[40] Perroux, F. "Note sur la notions de pole de croissance." *Cahiers de L'Institut de Science Economique Appliquée*, series D, no. 8 (1955).
[41] Richardson, H.W. *Regional Economics*. New York: Praeger Publishers, 1969.
[42] Romans, J.R. *Capital Exports and Growth among U.S. Regions*. Middletown, Conn.: Wesleyan University Press, 1975.
[43] Soltow, L. *Men and Wealth in the United States, 1850-1870*. New Haven, Conn.: Yale University Press, 1975.
[44] Stecker, M.L. "Inter-City Differences in Cost of Living in March

1935, 59 Cities." Works Progress Administration, Division of Social Research, *Research Monograph 12*. Washington: GPO, 1937.

[45] U.S. Department of Commerce. *Historical Statistics of the United States: Colonial Times to 1970*, pt. 1. Washington: GPO, 1975.

[46] U.S. Department of Labor, Bureau of Labor Statistics. *Handbook of Labor Statistics, 1970 and 1971*. Bulletin Nos. 1666 and 1705. Washington: GPO, 1970 and 1971.

[47] Williamson, J.G. "Regional Inequality and the Process of National Development." *Economic Development and Cultural Change* 13, no. 4, pt. II (July 1965), Supplement.

[48] Williamson, J.G. *Late Nineteenth Century American Development: A General Equilibrium History*. Cambridge: Cambridge University Press, 1974.

[49] Williamson, J.G. "The Sources of American Inequality, 1896-1948." *Review of Economics and Statistics* 58, no. 4 (November 1976):387-97.

[50] Williamson, J.G. "American Prices and Urban Inequality since 1820." *Journal of Economic History* 36 (June 1976):303-333.

[51] Williamson, J.G. "Inequality, Accumulation and Technological Imbalance: A Growth-Equity Conflict in American History?" Mimeo, 977.

[52] Williamson, J.G. "Strategic Wage Goods, Prices and Inequality." *American Economic Review* 67 (March 1977):29-41.

[53] Williamson, J.G., and Lindert, P.H. "A Macroeconomic History of American Inequality." Mimeo, 1976.

Appendix 1A:
Cost of Living Relatives

Table 1A-1
Koffsky-Adjusted Regional Cost-of-Living Relatives, 1840-1970
(New England = 100)

Region	1840	1880	1900	1920	1929	1950	1970
New England	*100.0*	*100.0*	*100.0*	*100.0*	*100.0*	*100.0*	*100.0*
Maine	97.0	95.2	96.3	99.2	97.2	95.9	96.7
New Hampshire	96.7	97.1	98.2	99.6	96.7	94.5	94.3
Vermont	95.1	92.1	94.7	98.6	93.5	94.5	94.3
Massachusetts	104.4	103.1	101.5	100.2	103.0	104.1	104.9
Rhode Island	104.6	103.3	101.8	100.2	97.8	104.1	104.9
Connecticut	99.8	100.5	100.3	100.0	101.2	102.7	103.3
Mid-Atlantic	*98.5*	*95.7*	*99.1*	*101.3*	*101.3*	*100.0*	*98.4*
New York	98.6	95.6	99.1	101.4	101.1	105.5	105.7
New Jersey	99.2	97.8	100.2	101.5	102.0	106.8	106.6
Pennsylvania	98.7	95.9	99.3	101.3	101.8	97.3	95.1
Delaware	96.7	92.1	96.2	100.7	99.1	94.5	91.8
Maryland	96.9	92.6	94.8	100.9	100.1	94.5	91.8
East North Central	*80.2*	*82.5*	*88.0*	*96.1*	*99.0*	*100.0*	*95.9*
Ohio	80.8	83.3	88.9	96.3	98.6	98.6	94.3
Indiana	79.4	81.0	86.7	95.7	90.6	98.6	94.3
Illinois	79.4	82.4	88.9	96.4	101.6	101.4	97.5
Michigan	79.1	82.9	87.6	96.1	101.3	98.6	95.1
Wisconsin	82.4	82.1	87.0	95.5	101.2	102.7	96.7
West North Central	*100.8*	*93.4*	*93.0*	*98.8*	*92.7*	*95.9*	*94.3*
Minnesota	—	94.0	94.0	99.0	102.6	102.7	100.0
Iowa	100.9	92.9	92.8	98.7	86.6	100.0	97.5
Missouri	100.6	94.1	93.8	99.1	96.7	97.3	95.1
Dakotas	—	95.3	90.7	97.5	90.3	90.4	89.3
Nebraska	—	92.5	92.1	98.4	91.2	90.4	89.3
Kansas	—	92.1	92.1	98.6	83.2	95.9	94.3
South Atlantic	*123.3*	*100.4*	*95.7*	*99.5*	*91.6*	*95.9*	*91.0*
Virginia	125.3	102.6	96.9	100.2	93.1	102.7	98.4
West Virginia	—	102.5	98.0	100.8	89.9	97.3	93.4
North Carolina	122.3	99.6	95.1	99.1	87.7	94.5	90.2
South Carolina	122.1	99.1	94.6	98.5	84.3	94.5	90.2
Georgia	121.7	99.6	94.9	99.0	90.8	91.8	87.7
Florida	123.9	101.4	97.4	100.6	91.7	90.4	86.9
East South Central	*83.5*	*80.6*	*90.6*	*97.2*	*84.0*	*94.5*	*88.5*
Kentucky	84.4	81.9	91.5	97.6	88.3	98.6	91.8
Tennessee	83.4	81.0	91.1	97.5	86.7	93.2	86.9
Alabama	82.9	80.0	90.5	97.1	81.7	91.8	87.7
Mississippi	82.6	79.4	89.3	96.3	78.2	91.8	87.7

Table 1A-1 *(cont.)*

Region	1840	1880	1900	1920	1929	1950	1970
West South Central	92.1	81.9	87.3	94.6	86.1	91.8	87.7
Arkansas	90.4	80.8	87.0	93.7	79.2	91.8	87.7
Louisiana	92.6	82.7	88.0	95.0	89.9	91.8	87.7
Texas and Oklahoma	—	81.9	87.1	94.7	86.1	91.8	86.9
Mountain				117.6	94.0	98.6	95.1
Montana				117.3	93.5	98.6	95.1
Idaho				116.9	92.7	98.6	95.1
Wyoming				117.7	94.2	98.6	95.1
Colorado				118.1	92.9	98.6	95.1
New Mexico				116.8	93.8	98.6	95.1
Arizona				118.1	96.8	98.6	95.1
Utah				117.9	93.2	98.6	95.1
Nevada				118.3	93.9	98.6	95.1
Pacific				103.1	97.7	100.0	97.5
Washington				103.1	94.1	101.4	98.4
Oregon				102.7	92.4	101.4	98.4
California				103.1	103.7	95.9	95.9

Sources and Notes:

1. The table lists the region and state cost-of-living relatives used for the regional inequality in the text. The data are expressed in terms of New England = 100 throughout. Thus, it is useful only for relative deviations in cost of living across space, not time.

2. Basic Sources:

Nathan Koffsky, "Farm and Urban Purchasing Power," in *Studies in Income and Wealth* (New York: National Bureau of Economic Research, 1949), pp. 151-78.

Jeffrey G. Williamson and Peter H. Lindert, "A Macroeconomic History of American Inequality," mimeo., 1976, chapter 5.

Philip Coelho and James Shepherd, "Differences in Regional Prices: The United States, 1851-1880," *Journal of Economic History*, September 1974, pp. 551-91.

Margaret L. Stecker, "Intercity Differences in Cost of Living in March, 1935, 59 Cities," Works Progress Administration WPA, Division of Social Research, Research Monograph XII (Washington: GPO, 1937).

U.S. Department of Labor, Bureau of Labor Statistics (BLS), *Handbook of Labor Statistics, 1970 and 1971*, Bulletin Nos. 1666 and 1705 (Washington: GPO, 1970 and 1971).

3. *1950-1970*. Data are taken directly from BLS *Handbooks*; for 1970, table 139, p. 326; for 1971, table 118, pp. 270-73. Both years refer to an "intermediate budget," four-person family, using (primarily) metropolitan areas. The regional figures are unweighted averages of the states themselves.

4. *1840-1929*. The procedure for these earlier years was to first collect estimates of urban price indices by state. These urban price indices were then adjusted to include the impact of cost-of-living differentials between urban and farm areas. We refer to this adjustment procedure as the "Koffsky adjustment."

5. *The Koffsky Adjustment*. Using 1935-1936 budget weights and 1941 prices, Koffsky estimated that the urban cost of living C_j^u was higher than the farm cost of living C_j^f, by 27 percent when farm weights were used and by 14 percent when city weights were used. In both cases, the budgets of the "lowest significant income levels" were used. In Williamson and Lindert, it is estimated that the ratios of these two costs of living were almost exactly the same in 1929 and

Table 1A-1 (cont.)

1941. Thus, we assume that the Koffsky differential was the same in 1929. Using the cost-of-living trends in Williamson and Lindert, we have estimated the Koffsky differential for the following years:

	Farm Weights	City Weights	(Z^{-1}) Unweighted Average
1891-1892	31.9%	18.3%	25.1%
1900	24.0	11.3	17.7
1919-1921	12.2	0.8	6.5
1929	27.0	14.0	20.5
1948	11.9	0.5	6.2

How can we use this information to derive state cost-of-living indices from data on urban price indices? The state cost-of-living index $\hat{C}_j$ can be estimated by the simple formula

$$\hat{C}_j = C_j \left[1 + (Z^{-1} - 1) \alpha_j^R \right] \qquad Z^{-1} = \frac{1}{Z} \text{ (see unweighted average)}$$

where α_j^R is a weight, the share of agricultural labor force in the total labor force of the given state j. Thus, the Koffsky adjustment supplies a means by which the state urban cost-of-living indices can be blown up to state coverage.

For the 1840 and 1880 calculations, we assume that the 1891-1892 unweighted average differential (approximating at 25 percent) applies. For the remaining years prior to 1950, we use the unweighted averages as indicated. No "Koffsky adjustments" were made in the 1950 and 1970 state price indices reported by the BLS since the Koffsky differentials are apparently quite small and, of course, the farm labor force was a very small share of total labor force in both years and in all states.

6. *The Urban State Relative Prices, 1840-1929.* These were derived from the sources above, with some adjustments. First, the 1840 relative prices are, in fact, for 1851. Second, Coelho and Shepherd (table 4, p. 571) were unable to directly observe price data for the West South Central region at all over their period (1851-1880), and only over 1866-1880 for the South Atlantic. We have estimated the missing values by interpolation. In effect, we have assumed that the cost-of-living ratio between the South Atlantic and the East South Central for 1866-1869 holds for the antebellum period as well, and that the ratio of the West South Central to the East South Central for 1867 and 1869 (Coelho and Shepherd, footnote 36, p. 572) holds for antebellum period as well. Third, 1880, 1900, and 1920 use Lindert (table G-1). Fourth, the 1929 figure is actually the 1935 estimate supplied by Stecker (tables 3 and 6, pp. 8, 162-63).

2

Current Demographic Change in Regions of the United States

Peter A. Morrison

Introduction

People with strong bonds to a region become aroused, understandably, when they perceive threats to the economic, social, and political interests of their region. The most basic kind of threat—and probably the most difficult to deal with—attaches to population change in a region, especially when the rate of change is accelerated. Fertility, mortality, and migration continuously alter the composition and size of any population, but so gradually in most places that few can realize how thorough and potent the process is. When for a host of reasons the rate of change speeds up, the more rapid alteration of the population disturbs social, economic, and political arrangements not only within regions but between them. The traditional power bases in the nation become reshuffled.

Signs of such realignment are today cropping up locally and especially nationally. In late 1975, for example, at the height of New York City's first skirmish with bankruptcy, the New York *Daily News* carried the stunning headline, "Ford to City: Drop Dead!" In 1976 a virtually unknown and modestly experienced politician from the Deep South was elected president of the United States.

The shifts in political power exposed in these separate events reflect the massive demographic transformations that have been quietly underway for a decade which, taken together, have destabilized the social and economic status quo in a large number of places.[1] A falloff in the birthrate, reversal of the historic trend in migration from rural to urban settings, and redirection of regional population movements have acted as a single force to alter the national landscape of growth and decline. Like wind drifting snow, they have created new features, obliterated old ones, and in some places exposed what lies beneath.

The falloff in the birthrate has throttled the rate of population growth nationally and, in the process, revealed migratory comings and goings as the principal determinant of local growth and decline in many places. Federal

---

This chapter draws on research supported by a grant from the Economic Development Administration, U.S. Department of Commerce. It was published in November 1977 as Rand P-6000 and is in the public domain. The author thanks Judith P. Wheeler, Will Harriss, Phyllis Ellickson, Ira S. Lowry, David Lyon, and Kevin F. McCarthy for helpful comments on earlier drafts.

statistics now register the beginnings of population decline in thirty-six of the nation's metropolitan areas, partly as a result of the low birthrate but mostly because of the excess of departing migrants over arriving ones. As the cities' magnetism has waned, population has stopped growing or begun to decline—a situation generally regarded as the prelude to economic stagnation—and has severely strained traditional mechanisms of municipal finance.

The counterpart of this trend toward urban decline is the "rural renaissance," the revival or acceleration of population growth in small cities and towns, even those that are remote from metropolitan areas. These small, once stable communities are ill equipped to deal with sudden population growth. Like new celebrities they lack the full array of legal and institutional structures for coping with unaccustomed attention.

The most complex facet of this demographic transformation, the redistribution of population among regions, has the most far-reaching implications. The changing directions of internal migration during the 1970s signal, and at the same time reinforce, new patterns in the regional distribution of economic vitality. These shifts are responsible for a variety of new regional conflicts of interest as well as for the new regional political coalitions whose power is just starting to be felt. Demographic change has been a catalyst for larger political issues that have been quick to mature in a period of economic depression and energy shortages. The widely publicized regional shifts of population in this decade have given rise to the Sunbelt and the Frostbelt (not to mention other regions that the media have dubbed the Welfare Belt, the Brain Belt, the Smog Belt, and so forth). Although the geographic boundaries of these regions are loosely defined at best, the interests and grievances they share are quite specific—and a grievance more often defines a region than a boundary.

This "tournament of the belts" is a kind of preliminary bout to the main event—the tension between growing and declining cities. Here the contending factions are made up of places, the Clevelands and Detroits, where growth has stopped, and the Tucsons and Boulders and Petalumas, which, far from enjoying their growth, often see themselves as victimized by the access that migration confers on places. The broad redirection of migration flows, as measured in regional statistics, is most palpable at the local scale, especially where an earlier trend has been completely reversed.

The severity of the problems confronting some areas has prompted demands for relief—no-growth policies have been adopted in a number of places, while others are clamoring for subsidies for population that is *not* there. One might suppose that migration rates had taken a strong upswing, but that is not the case. What has happened is that the population growth which was conferred more or less uniformly throughout the nation during the postwar baby boom decades has stopped, exposing the effects of the highly differential growth conferred by migration. Moreover, besides creating a quite different distribution pattern, these two means of growth—natural increase versus migration—create quite different political implications. Population gained by new births will not enter a voting booth for eighteen years. The bulk of the population gained (or lost) by migration, however, represents immediate poltical power. For a decade or so, unusually large numbers of people—the baby-boom generation—have been entering not only the labor and housing markets but also the prime ages for migrating.

The National Demographic Context[2]

Over the past decade, growth of the national population has slowed considerably, owing to a sharp decline in fertility with no offsetting change in mortality. The decline in fertility reflects an interplay between the widespread use of highly effective contraceptive methods and changing attitudes toward childbearing. While contraceptive practice was being modernized over the last ten years, fertility norms were shifting downward; young adults were shying away from large families and generally were postponing childbearing.

Estimating the impact of lower fertility now and in the future is complicated by the fact that while birthrates continue at an all-time low, household formation is at an all-time high. The average annual rate of increase in households rose from 1.8 percent during 1955-1965 to 2.3 percent during 1970-1976, an artifact largely of the post-World War II baby boom (which has now matured into a "young adult" boom). As figure 2-1 shows, the dissimilarity in population and household growth trajectories is projected to persist into the foreseeable future (thirteen years).

That dissimilarity blurs the growth-slowing effects of lower fertility, for "population" can be said to be contracting or expanding, depending on whether people or households is the unit of measurement. That is why in many cities where population is declining, South Bend, Indiana, for example, public institutions must nevertheless meet the needs of an expanding "population" of households.

Changed Patterns of Migration

Like fertility trends, migration trends have taken new directions whose effect generally is to intensify the impact of lower fertility in metropolitan areas and nullify it in nonmetropolitan areas. As is now widely known, national statistics show that more people in the United States are moving away

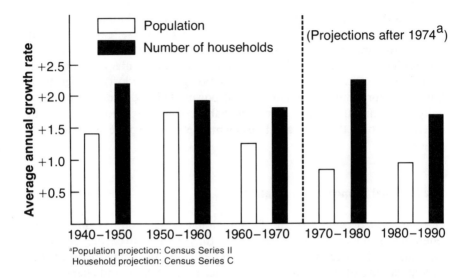

Source: U.S. Bureau of the Census, *Current Population Reports*, Series P-25, Nos. 601, 607.
Figure 2-1. Annual Growth Rates in Population and Number of Households, 1940-1974, and Projections to 1990

from metropolitan areas than are moving to them, reversing a long established urbanization trend.[3] Each year between 1970 and 1975, 131 people moved out of the metropolitan sector for every 100 who moved in. For the most recent period measured (1975-1976), the reversal continued.[4] As a result, nearly two-thirds of all nonmetropolitan counties are gaining migrants, compared with only one-quarter in the 1960s and one-tenth in the 1950s.

The underlying demographic structure of these shifts is suggested by data comparing the direction of population change in the 1960s with that of 1970-1975 (table 2-1). The reversal of net migration is particularly noticeable for counties with little commuting and those that are entirely rural and not adjacent to a standard metropolitan statistical area (SMSA). Clearly, the migration reversal cannot be explained away as just more metropolitan sprawl or "spillover," for it is affecting distinctly remote and totally rural nonmetropolitan areas as well as those adjacent to metropolitan centers.[5] The resulting net migration to these places may be due to less "push," that is, greater retention of native residents who no longer feel compelled by economic pressures to migrate, or to more "pull," that is, increased inflow of urbanites who are drawn there for a variety of reasons.

In any case, nonmetropolitan areas have become distinctly more attrac-

Table 2-1
Components of Population Change for Groups of Metropolitan and Nonmetropolitan Counties: 1960-1970 and 1970-1975

Population Category	Provisional 1975 Population (000s)	Annual Population Change Rate		Annual Natural Increase Rate		Annual Net Migration Rate[a]	
		1960-1970	1970-1975	1960-1970	1970-1975	1960-1970	1970-1975
United States Total	213,051	1.3	0.9	1.1	0.7	0.2	0.2
Metropolitan							
Total, all SMSAs[b]	156,098	1.6	0.8	1.1	0.7	0.5	0.1
>1.0 million	94,537	1.6	0.5	1.1	0.6	0.6	-0.2
0.5-1.0 million	23,782	1.5	1.0	1.2	0.8	0.4	0.3
0.25-0.5 million	19,554	1.4	1.3	1.2	0.8	0.2	0.5
<0.25 million	18,225	1.4	1.5	1.2	0.8	0.2	0.7
Nonmetropolitan							
Total, all nonmetropolitan counties	56,954	0.4	1.2	0.9	0.6	-0.5	0.6
In counties from which:							
≥20% commute to SMSAs	4,407	0.9	1.8	0.8	0.5	0.1	1.3
10%-19% commute to SMSAs	10,011	0.7	1.3	0.8	0.5	-0.1	0.8
3%-9% commute to SMSAs	14,338	0.5	1.2	0.9	0.6	-0.4	0.6
<3% commute to SMSAs	28,197	0.2	1.1	1.0	0.6	-0.8	0.5
Entirely rural counties[c] not adjacent to an SMSA	4,661	-0.4	1.3	0.8	0.4	-1.2	0.9

Source: Unpublished preliminary statistics furnished by Richard L. Forstall, Population Division, U.S. Bureau of the Census; and Calvin L. Beale, Economic Research Service, U.S. Department of Agriculture.

[a]Includes net immigration from abroad, which contributes newcomers to the United States as a whole and to the metropolitan sector, thereby producing positive net migration rates for both.

[b]Population inside SMSAs or, where defined, standard consolidated statistical areas (SCSAs). In New England, New England county metropolitan areas (NECMAs) are used.

[c]"Entirely rural" means the counties contain no town of 2,500 or more inhabitants.

tive both to their residents and to outsiders, whereas metropolitan areas have become less so. Paradoxically, though, migrants to the nonmetropolitan sector no longer clearly favor the high-income areas within it—an indication, perhaps, that scenery has begun to compete with salary.[6]

If the nonmetropolitan sector has become known in the 1970s for the revival of population growth, the metropolitan sector has become notorious for population loss.[7] Where only one of the nation's twenty-five largest SMSAs lost population during the 1960s (Pittsburgh), fully nine were declining as of 1975: New York, Los Angeles-Long Beach, Philadelphia, St. Louis, Pittsburgh, Newark, Cleveland, Seattle-Everett, and Cincinnati. Approximately one-sixth of the nation's 259 metropolitan areas were losing population at mid-decade, and one-third of metropolitan residents lived in areas of population decline.

It would be misleading to project a picture of widespread abandonment of urban and suburban territory in these declining metropolitan centers. On the contrary, most SMSAs are still growing, and even those with declining population overall contain communities which are continuing or even accelerating their growth, particularly on the metropolitan fringe. But migration is clearly differentiating among individual metropolitan areas, and its influence has become all the more apparent as the receding tide of natural increase has exposed the places that people are leaving.[8] Migration, then, determines *where* and how soon the manifestations of slowdown in overall population growth will first appear (for example, New York State); and, conversely, it is evident that even approaching zero poulation growth will not eliminate the impact of rapid growth *somewhere* (for example, Houston or Fort Lauderdale).

Outlook for the Future

Judging from current fertility expectations and experience, a gradual transition to population stability—zero population growth—appears to be underway. For now, the most reasonable conditional forecast in light of current birth expectations is the Census Bureau's projection series II (associated with an ultimate, completed cohort fertility rate of 2.1 births per woman).[9] That projection indicates a slight rise from present levels to annual increases of 2.1 million during most of the 1980s. Thereafter, the annual increase falls in a range of 1.4 to 2.0 million during the 1990s and drifts lower in later years (figure 2-2).

Events can upset forecasts, of course. Future fertility, the principal determinant of population growth, is uncertain; future growth rates may be higher or lower than those in the series II projection (which I regard as most reasonable). For that reason, the Census Bureau also compiled projection

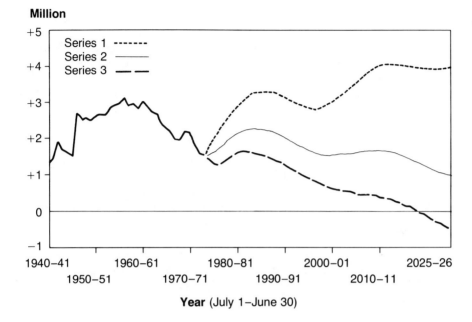

Figure 2-2. Estimates and Projections of Annual Population Change: 1940 to 2025

series I (higher fertility, associated with an assumed 2.7 births per woman) and projection series III (lower fertility, associated with an assumed 1.7 births per woman). These projections, also shown in figure 2-2, bracket what most demographers would regard as a reasonable range of future possibilities.[10]

Trends in migration are even more difficult to forecast, because the changing spatial distribution of economic opportunities to which migrants respond lies largely beyond our predictive reach. The changing fortunes of geographic regions result from a complex accumulation of forces, whose future course is as much a matter of accident as of intent.[11] It is unclear whether the 1970s reversal of migration between metropolitan and nonmetropolitan sectors will prove to be temporary or long-term, for the shift coincided with and may be due in part to the economic recession of the past several years. If so, a resumption of metropolis-bound migration would be expected with improvement in the economy (although while the latter has occurred, the former has not). At the same time, the taste for rural living may persist, so that metropolitan and nonmetropolitan growth will coexist in the future. Whatever its nature and likely longevity, though, the migration reversal has manifested itself diversely in various geographic regions of the United States.

New Patterns of Regional Growth and Decline

While national population growth is slowing, the fortunes of different regions present a complex picture of growth and decline with quite marked breaks with the past. I have restricted my discussion to the post-1950 period and to just two of the many possible systems of regions with which demographic trends could be comprehended.[12] These are (1) the nine Census divisions (and the four Census regions in which they are nested) and (2) the twenty-six economic subregions, formulated and applied by Calvin Beale and his associates at the U.S. Department of Agriculture. The two types (figures 2-3 and 2-4) each have distinctive advantages for analytical purposes.[13] Data for the first of these scales enables me to contrast developments in the Northeast and the South and to draw attention to the first of the three principal points I want to make:

1. The shift in migration, together with moderating natural increase, is determining where the symptoms of national decline first appear.

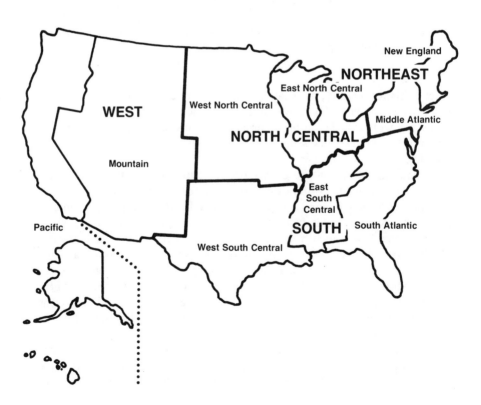

Figure 2-3. Census Regions and Geographic Divisions of the United States

Data at the subregional level allow me to demonstrate my second and third points:

2. Migration continues to support metropolitan growth in the 1970s, although the locus of this growth has shifted away from the Northeast.
3. Nonmetropolitan settlement patterns everywhere have evolved beyond the point where proximity to metropolitan activity is a prerequisite for local growth.

Regional Migration Shifts

Roughly since 1970, underlying patterns of regional migration have transformed population growth at the regional scale (see figure 2-5). The South's 5.1 million population increase in the first five years of this decade is a sharp departure from other recent five-year periods. The Northeast, in

1. Northern New England—St. Lawrence
2. Northeastern Metropolitan Belt
3. Mohawk Valley and New York—Pennsylvania Border
4. Northern Appalachian Coal Fields
5. Lower Great Lakes Industrial
6. Upper Great Lakes
7. Dairy Belt
8. Central Corn Belt
9. Southern Corn Belt
10. Southern Interior Uplands
11. Southern Appalachian Coal Fields
12. Blue Ridge, Great Smokies, and Great Valley
13. Southern Piedmont
14. Coastal Plain Tobacco and Peanut Belt
15. Old Coastal Plain Cotton Belt
16. Mississippi Delta
17. Gulf of Mexico and South Atlantic Coast
18. Florida Peninsula
19. East Texas and Adjoining Coastal Plain
20. Ozark—Ouachita Uplands
21. Rio Grande
22. Southern Great Plains
23. Northern Great Plains
24. Rocky Mountains, Mormon Valleys, and Columbia Basin
25. North Pacific Coast (including Alaska)
26. The Southwest (including Hawaii)

Figure 2-4. Twenty-six Economic Subregions

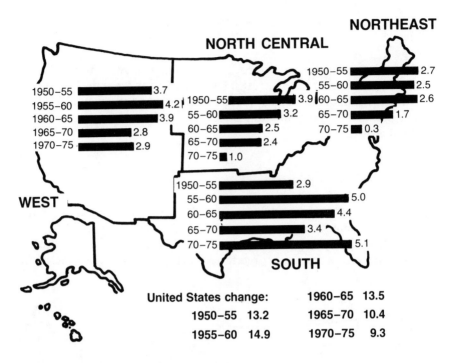

Periods beginning July 1. Change expressed in millions.

Figure 2-5. Population Change for Five-Year Periods by Regions: 1950 to 1975

contrast, has entered an era of virtual stability, and the North Central region's population increase has slowed considerably.

The current situation of the Northeast illustrates how a regional shift in net migration, combined with the overall drop in the birthrate, can halt population growth and weaken the political influence of a historically dominant region. Natural increase has diminished everywhere (although somewhat more so in the Northeast than in other regions) since 1960 (table 2-2). The Northeast gained 2.3 million people through natural increase between 1960 and 1965, but only 1.0 million between 1970 and 1975. Net migration, however, has changed very differently in the Northeast than in the South. Before 1970, the Northeast had nominal five-year gains of several hundred thousand from migration; thereafter the region began to suffer a sizable loss of 700,000 between 1970 and 1975. This loss has intensified since 1972, and the Northeast now registers net population decline since outmigration exceeds the population's natural increase.

These losses have been more severe in the three Mid-Atlantic states

Table 2-2
Population Change by Component for Each Region: Five-Year Periods, 1950 to 1975
(in millions; periods begin July 1)

Period	Natural Increase					Net Migration				
	United States	Region				United States	Region			
		North-east	North Central	South	West		North-east	North Central	South	West
1950-1955	12.1	2.3	3.5	4.5	1.9	1.0	0.4	0.4	-1.6	1.9
1955-1960	13.2	2.6	3.9	4.7	2.2	1.7	0.0	-0.7	0.3	2.0
1960-1965	12.0	2.3	3.3	4.2	2.2	1.5	0.3	-0.8	0.3	1.7
1965-1970	8.7	1.6	2.3	3.0	1.7	1.7	0.1	0.1	0.4	1.1
1970-1975	6.8	1.0	1.8	2.5	1.5	2.5	-0.7	-0.8	2.6	1.4

Source: U.S. Bureau of the Census, *Current Population Reports*, series P-25, no. 640, November 1976, table B.

(New York, New Jersey, and Pennsylvania) than in the five New England states which make up the remainder of the Northeast.[14]

The sharply increased streams of migrants out of some of the nation's colder regions have been gravitating to the South and West. Florida, for example, added 22 percent to its population between 1970 and 1976. More important, though, migratory growth appears to be diffusing throughout the South and West in this decade (see figure 2-6). Prior to 1970, the large number of inmigrants to Florida had offset what was, in fact, a migratory loss for the rest of the South (table 2-3). But since 1970, Florida's ballooning net inmigration has been matched by an equally large inmigration to the rest of the region. This migration into the other Southern states is noteworthy since it foreshadows a future wave of growth throughout the region.

The nation's other fast-growing region, the West, shows a comparable diffusion of migratory growth. California's migratory experience no longer dominates the West's regional migration growth as it did prior to 1965 (table 2-3). Between 1970 and 1975, the other twelve states in the West gained 1.0 million through net inmigration (more than twice California's

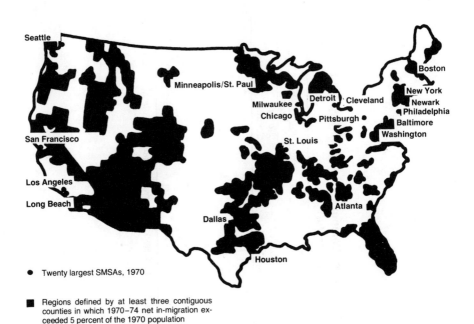

• Twenty largest SMSAs, 1970

■ Regions defined by at least three contiguous counties in which 1970–74 net in-migration exceeded 5 percent of the 1970 population

Source: Curtis C. Roseman, *Changing Migration Patterns within the United States*, Resource Papers for College Geography, Washington, 1977, figure 10.

Figure 2-6. Regions of Relative Growth due to Migration, 1970-1974

Table 2-3
Net Migration for Selected Areas: Five-Year Periods, 1950 to 1975

Period	California	Rest of West	Florida	Rest of South
In millions				
1950-1955	1.6	0.3	0.7	−2.3
1955-1960	1.7	0.4	0.9	−0.6
1960-1965	1.5	0.2	0.6	−0.4
1965-1970	0.5	0.6	0.7	−0.3
1970-1975	0.4	1.0	1.4	1.2
As percentage of beginning population				
1950-1955	14.7	3.0	24.8	−5.3
1955-1960	12.3	4.4	24.9	−1.3
1960-1965	9.7	1.8	12.8	−0.8
1965-1970	2.7	4.1	11.7	−0.5
1970-1975	2.1	6.7	19.9	2.1

Source: U.S. Bureau of the Census, *Current Population Reports*, series P-25, no. 640, November 1976, table C.

share), compared with only 0.2 million (less than one-seventh of California's share) between 1960 and 1965. Gains in the Mountain states were especially impressive: In the first six years of the 1970s, these eight states gained 913,000 through net inmigration, compared with 307,000 during the entire decade of the 1960s.

Overall, these regional trends are departures from the past, made all the more apparent locally by the low rate of natural increase. The course these migration trends will follow in the future is difficult to forecast, however, because their coincidence with an economic depression makes this an atypical period. As the sharply contrasting developments in the Northeast and the South clearly demonstrate, however, shifting migration, together with moderating natural increase, is determining where and when the symptoms of national population slowdown first appear.

Subregional Patterns

As we have seen, the attractiveness of metropolitan areas nationally has weakened since 1970, and a growing number of individual areas have begun to decline. At the same time, there has been a strong revival of population growth in nonmetropolitan areas. Are these national shifts pervasive, or are they more prevalent in some parts of the nation than in others?

Recent unpublished data kindly furnished by Calvin Beale show the net migration into or out of the individual counties that make up each of the twenty-six economic subregions shown in figure 2-4. They show the rate at which *counties* in a subregion (not in the entire subregion) are gaining or losing population through migration. That gain or loss, of course, comes about through an unknown combination of migration streams among and within subregions. The moves producing these changes may be intraregional (among counties within a subregion), or they may be interregional. Because we cannot distinguish between the two, we can gauge not gains or losses for a subregion itself, only the average experience of counties within it, classified by type. Even so, there are clear and diverse patterns among the twenty-six subregions which do not show up in the aggregate national statistics showing a heightened outmigration from the metropolitan sector since 1970.

The Metropolitan Sector. The recent slowdown in metropolitan growth has been measured nationally, but its occurrence is far from uniform in all sections of the country. Metropolitan population growth has been halted in some sections of the country but accelerated in others by subregional population movements. In restricting our focus to the metropolitan sector, it is evident that the locus of migratory attraction is shifting away from the metropolitan counties in four subregions in the North and East for the most part and toward eight subregions in the South and West. Simply put, migrants are moving to metropolitan areas in certain subregions and away from them in others. National metropolitan growth, then, continues and is even accelerating in certain areas, but they are not the same areas as before.

This pattern is illustrated in figure 2-7, which shows subregions where the metropolitan sector is losing migrants (vertical stripes) and gaining migrants (horizontal stripes) in the 1970s. The bolder patterns indicate that outflow or inflow has been initiated or intensified between this decade and the previous one; for example, heavy vertical stripes signify a higher outflow rate during the 1970s than the 1960s or a shift to net outmigration following net inmigration during the 1960s.

There are seven subregions (confined mostly to the North and East) where net migration is withdrawing population from the metropolitan sector. Of these seven subregions, four (heavy vertical stripes) are experiencing new or more intense net outmigration during the 1970s. Metropolitan population in the Northeastern metropolitan belt (number 2), for example, recorded net inmigration at 0.3 percent annually during the 1960s, but net outmigration at 0.4 percent annually afterward. The overall result has been a decisive halt to metropolitan growth through a combination of outmigration and declining natural increase (as was seen for the Northeast region as a whole). For the remaining three subregions where there is net outflow from the metropolitan sector, the post-1970 outmigration is less intense than it was earlier; their metropolitan areas are still losing population

Current Demographic Change

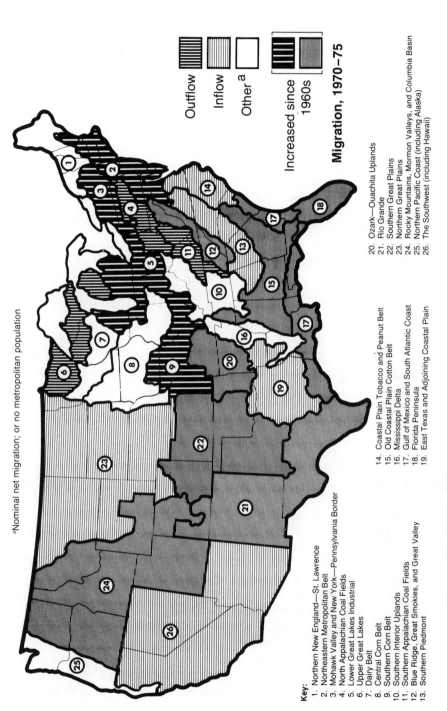

Figure 2-7. Metropolitan Counties: Changing Locus of Migratory Growth

through migration, but diminishingly so, and their populations have been less affected.

Most of the Southern and Western subregions show the mirror image of this pattern. There, net migration is adding to the metropolitan population in thirteen subregions. Eight of these have experienced net migration at an accelerated rate since 1970, in several instances reversing pre-1970 outmigration. An impressive illustration of such a reversal is seen in the Rio Grande subregion (number 21), where annual net migration for metropolitan counties changed from -0.8 percent during the 1960s to $+0.8$ percent thereafter.

Overall, metropolitan population growth has been halted in some sections of the country but accelerated in others by population movement at the subregional scale. The geographic locus of migratory attraction (a barometer of economic opportunity) has shifted noticeably in the 1970s, and will to an extent perpetuate population and employment growth throughout much of the South and West.

The Nonmetropolitan Sector. The strong revival of nonmetropolitan population growth has been produced by a single change—reversal of the historic outmigration from the nonmetropolitan sector—but the reasons for this "rural renaissance" are multifaceted. Migration toward nonmetropolitan areas in the 1970s reflects a number of significant (often mutually reinforcing) influences.[15] These areas are evolving along three basic dimensions.

The first is that of *ease of access to the national metropolitan economy*, a constraint which has been relaxed considerably in recent years by advances in transportation. Such advances have enabled more people to settle in locales which are separated from large urban centers without being isolated from them. The sharply accelerated migration into nonmetropolitan areas is partly a manifestation of metropolitan expansion along transportation routes—an evolution of metropolitan spatial form that gives rise to new urban nodes. Small, self-contained urban centers spring up in nonmetropolitan areas adjacent to existing metropolitan centers because business can now be efficiently transacted at a distance.

Industrial trends have also relaxed some of the traditional constraints on settlement patterns. Two noteworthy industrial trends are (1) decentralization of manufacturing in response to reduced transportation costs, inexpensive land, and low wage rates in nonmetropolitan areas, and (2) the revival or expansion of energy extraction, and highly localized large-scale energy-related industrial development in the Rocky Mountain states.

What can loosely be termed *changes in the U.S. life-style* comprise a third dimension of evolution. Included here are (1) the trend toward earlier retirement and semiretirement, which has multiplied the ranks of retirees and lengthened the interval during later life when a person is no longer tied to a specific place by a job, (2) new soures of income, such as pensions and other payments which either were earned elsewhere in younger years or are a transfer of public funds from taxes paid elsewhere[16] (these new kinds of income have expanded retirees' roles as consumers, whose presence in an increasingly service-oriented society creates jobs wherever they go), and (3) an increased orientation at all ages toward leisure activities, caused in part by rising per capita incomes and centered on amenity-rich areas outside the daily range of metropolitan commuting.[17]

A common feature of these several changes is that they have laid the foundation for expansion of nonmetropolitan employment. Servicing the arriving migrants and temporary residents provides opportunities which induce existing residents to stay and which also entice more newcomers. This synergistic development has appeared in locales as diverse as northern New England and the upper Great Lakes (popular as recreation areas), the Ozark-Ouachita Uplands (favored by retirees), and the northern Great Plains (the site of energy-related industrial development).[18] Although particular circumstances vary from place to place, the outcomes are much the same: initial base employment opportunities, however created, furnish the jobs that retain existing residents and draw opportunity-seeking migrants from elsewhere. The resulting population, larger and more affluent, enlarges local demand for goods and services, thereby creating new jobs for more migrants.

The changing economic fortunes of nonmetropolitan areas have manifested themselves diversely. Some subregions stand out because preexisting growth has accelerated sharply. What is more often noteworthy, though, is simply the reversal of previous decline, seen in the comparison of a given subregion's nonmetropolitan population trends in the 1970s with those in the 1960s.

Adjacent counties. Accessibility is central to the idea of functional integration into the national metropolitan economy. Freeways and other means of transportation that have fostered the phenomenon of long-distance commuting afford metropolitan access to areas that lie considerable distances from centers of economic activity. If metropolitan size makes for economies of scale, then nonmetropolitan areas are increasingly able to "borrow" that size through daily access.[19]

The federal metropolitan-nonmetropolitan distinction was designed to reflect the presence or absence of social and economic integration into city life that is conferred by residence in a particular location.[20] Since metropolitan and nonmetropolitan areas are defined in terms of whole counties, which are often geographically large and socioeconomically heterogeneous, the approximation to a functional definition is crude at best. Moreover, many persons who are classified as "nonmetropolitan" reside in counties that are within easy commuting range of metropolitan centers. A considerable proportion of these adjacent nonmetropolitan residents are functionally metropolitan: their life-styles are more like those of the SMSA residents than those of people who live in remote rural areas.

This functionally metropolitan population consists mostly of persons residing in nonmetropolitan counties adjacent to presently defined SMSAs. These counties are the ones most likely to benefit from the outward dispersal of population from a neighboring metropolitan center.

The data in figure 2-8, which refer to this "nonmetropolitan adjacent" sector, enable us to distinguish this "disguised metropolitan growth" within each subregion. Subregions are grouped according to several patterns of demographic change in this sector: The first pattern (represented in vertical stripes) shows subregions that experienced migration outflow in the 1960s which ended in the 1970s. In some of these subregions (the heavier vertical stripes) that outflow had been so severe as to result in absolute population decrease. (All that decrease, of course, has now ended.) The second (represented in horizontal stripes) shows subregions that experienced net inmigration during 1960-1970, accelerating afterward.

In these areas adjacent to the nation's metropolitan centers, the pervasiveness of the "rural renaissance" is evident. Fully twenty of the twenty-six subregions registered net migration loss in the "nonmetropolitan adjacent" sector during the 1960s; in five, that loss was severe enough to produce absolute population decline despite the moderately high birthrates in that decade. Yet, in the 1970s, net migration has become distinctly more positive: In seventeen of these twenty, net migration reversed from negative to positive, and despite lower birthrates in this decade, all subregions experienced population increase in this sector.

This cessation of previous, often severe outmigration from the nonmetropolitan adjacent sector suggests that metropolitan growth continues, although perhaps not always within the arbitrary boundaries of SMSAs. The true picture undoubtedly is more complex than these data can reveal and does not lend itself to simple generalizations. Judging from the impressive growth trends indicated, however, it is reasonable to infer that the nonmetropolitan adjacent sector has fallen much more heavily under the sway of metropolitan influence in the 1970s than before—and in virtually every section of the country. Thus, the new locus of migratory attraction includes the nonmetropolitan adjacent sector in *all* subregions, as well as the metropolitan sector in many of those Southern and Western ones noted earlier.

Nonadjacent counties. In the past, metropolitan integration offered a generic explanation of why some nonmetropolitan areas grew but others did not. But the nonmetropolitan growth of the 1970s, by many indications, also reflects specialized activities occurring in areas located beyond the

Current Demographic Change

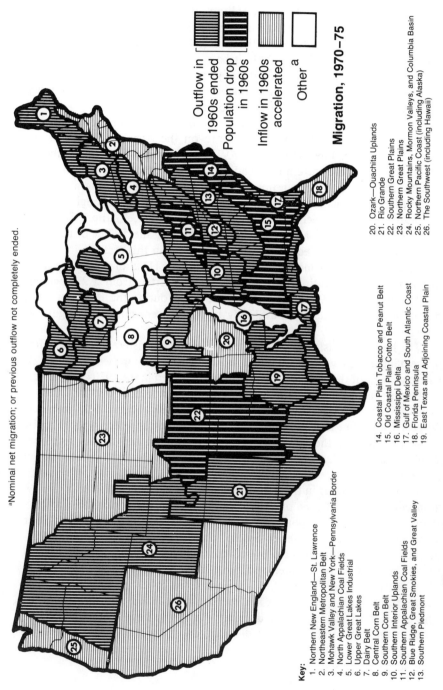

Figure 2-8. Nonmetropolitan Counties (Adjacent): The Reversal of Population Loss

immediate sphere of daily metropolitan life. Population trends in the "nonadjacent" sector reflect developments in these areas, which presumably are less closely tied to the national metropolitan economy. Such counties by no means lack sizable urban centers; but by definition such centers are below the minimum 50,000 population threshold that qualifies an urban county as a metropolitan one. In all cases, however, these smaller cities and towns are not near a metropolitan area.

The data in figure 2-9 refer to the "nonmetropolitan nonadjacent" sector. Subregions are grouped as before on the basis of change in net migration between 1960-1970 and 1970-1975. The pattern of change shown closely resembles the pattern in the adjacent sector, and often with more intensity. The nonmetropolitan nonadjacent sector was losing migrants in all but a few subregions in the 1960s, and that loss was severe enough to incur absolute population decline in fully ten of them. Yet, by the 1970s, that outflow had ended virtually everywhere, eradicating the decline of the past. Only one subregion failed to register any growth in this sector.

The southern Appalachian coal fields (number 11) is an exemplary case: The annual net migration rate for this sector shifted from a 2.6 percent outflow during the 1960s to a 0.8 percent infow during the 1970s; and the population, which had been declining by 1.3 percent annually, began increasing by about 1.6 percent.

The heavily industrialized subregions also have registered noticeable improvement in this sector. The Northeastern metropolitan belt (number 2), where growth of the *metropolitan* population halted in the 1970s, is registering better than a 1.7 percent annual increase in the nonmetropolitan nonadjacent sector—considerably above that for the 1960s. Much of this increase reflects formation and growth of small communities which appear to be prospering despite the larger trend of metropolitan no-growth. Regional differences in the laws and regulations governing annexation also play a part.[21]

Population trends in the nonadjacent sector closely parallel those in the adjacent one. The degree of new attractiveness in the former and its pervasiveness (despite enormous variations in the overall fortunes of particular regions) are especially noteworthy, for they suggest that metropolitan expansion is being supplemented by self-contained local urbanization, even in remote reaches of the nonmetropolitan United States.

Clearly, the pattern of U.S. settlement has evolved beyond the point where nearness to a metropolis is a prerequisite to local migratory growth. The cultures of city slicker and county bumpkin have merged—with an assist from television and the federal highway program—and Safeways, Sears, and Sizzlers have diffused down the urban hierarchy to serve even small and remote settlements.

Current Demographic Change

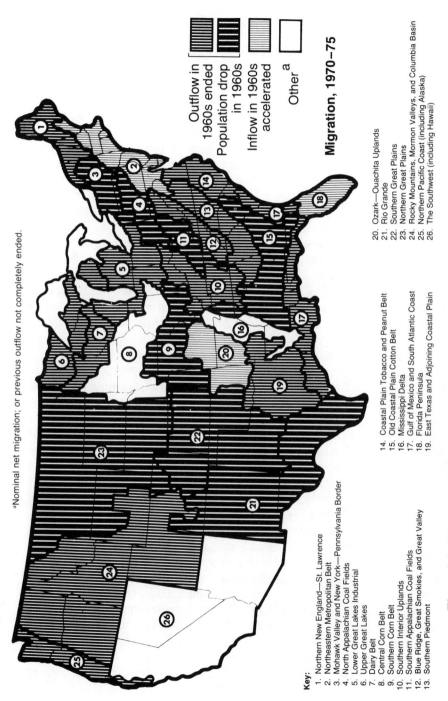

Figure 2-9. Nonmetropolitan Counties (Nonadjacent): The Reversal of Population Loss

Public Concerns and Regional Population Shifts

Population is continuously replaced and recomposed, enlarged, or reduced over time through two types of mechanisms. Population is replaced and may grow naturally or decline as people die and babies are born. When there are more births than deaths, the population grows by natural increase. Since the death rate is virtually constant over time in the United States, the fluctuating birthrate governs population change by this natural mechanism.

Population can also grow, shrink, or be replaced and recomposed through the movement of migrants to or from a region. Population growth or decline through migration is measured in a given area by net migration—the numerical difference between incoming and outgoing migrants—but that figure registers only the surface ripples of much larger cross-flows of migrants going in opposite directions. The total number of people moving (gross migration) is far larger than the net migration figure that results. For example, between 1965 and 1970 Texas recorded a net migration gain of 306,000, which happened to be the difference between the 1,148,000 people who moved there and the 842,000 who left. It could as easily have been the difference between 306,001 arrivals and 1 departure, which shows the extent to which net migration statistics fail to indicate the volume of migration; in this case, roughly 2 million acts of interstate migration were involved in a net population gain of less than 0.33 million.

In almost every place in the United States, the population is continuously recomposed by a gradual procession of migrants coming and going, for the most part deliberately. That element of deliberate choice in most moves sharply differentiates population growth by net migration from growth by natural increase; and more important, it differentiates the *places* that are being affected. A place that grows by net migration of 1,000 has gained 1,000 people who are there because they *want* to be *there*, and a place that loses 1,000 migrants has lost 1,000 people who do *not* want to be there. Natural increase does not contribute deliberate residents; it only adds to population by lottery.[22]

The purposefulness of migration makes it a largely autonomous process and one that, at least in the United States, is indicative of opportunity seeking. The view that personal success is achievable as readily outside as inside one's native region is a distinctive and deeply ingrained element of U.S. culture. It is the product of the persistent pull of economic opportunities in other places, which enables individuals alert to opportunity to exploit newly developed resources or knowledge quickly. The U.S. economy benefits from the readiness of its population to migrate: Without a tradition of migration, which moves people from areas where jobs are dwindling to places where workers are needed, U.S. economic development would be sluggish and less efficient than it actually has been.

As this U.S. predilection to migrate fosters rearrangement of population in space to answer the changing needs of the national economy, it also restructures regional political interests. These political effects are far more evident at the local and regional levels, where growth often comes as much from net migration as from natural increase. Equally important, natural increase has no direct electoral consequences for at least eighteen years, whereas growth (or decline) through migration has immediate effects. For example, the current flow of net migration toward the South and West and toward nonmetropolitan areas is altering the status quo along a broad front. Specific concerns can be grouped in several major categories:

1. *Head-count concerns.* Formulas whereby federal largesse (for example, revenue sharing, Law Enforcement Assistance Administration funding to states, and the like) is distributed among localities and regions typically give weight to the number of people the area claims as its inhabitants. Regions unable to boast more bodies (or, worse, to establish even as many as they had last year) will lose funds, even though—perhaps for that very reason—they may merit more federal assistance. This point has not been lost on cities and regions with shrinking populations, and they are pressing now for revisions of these formulas that will direct federal dollars to areas losing population. But if a formula that compensates areas for population growth is politically unpalatable, one that gives dollar credits for population loss may be equally offensive. The image arises of an urban "black hole," a once-great city into which an unending stream of federal dollars disappears as its population vanishes.

2. *Human capital concerns.* In recomposing a region's population, net migration may alter its labor pool, replenishing or depleting its stock of human capital. Skilled workers may depart, to be replaced by less skilled inmigrants; young adults educated at one region's expense may move themselves, and that investment, to another.

3. *Dependency concerns.* Some segments of the population (say, persons claiming public assistance) are recognizably a public burden, and others (for example, uneducated rural-to-urban migrants) are thought to be. Their migration can scarcely be a matter of local indifference because of the real or perceived costs they impose. Certain popular notions persist, though, in the face of contrary research evidence, for example, that low-income migrants move to large cities like New York to gain access to generous public assistance allowances.[23]

4. *Concern with undocumented aliens.* This issue figures prominently in California, Texas, and a few other states to which many aliens make their way. Illegal aliens formerly stayed close to the border, often in rural areas, but as substantial numbers have scattered throughout the nation, the issue of their local impact has come to figure prominently in many parts of the country. There is much disagreement about what effects they have: some

insist that alien Mexican workers take jobs which could otherwise be filled by unemployed citizens or legal migrants and that illegal aliens overburden social services of all kinds, taking more in the form of social welfare services than they contribute in taxes.[24] Since evidence is scarce, these allegations and the thorny issues they pose have gained wide currency.[25]

5. *Sociopolitical concerns.* More generally, newly declining areas discover that shrinkage cannot be accomplished merely by reversing the process of expansion within an existing organizational setting; and newly growing ones resent the problems that arriving migrants cause and the unwelcome access this migration confers on destination areas. The impact of growth through migration versus natural increase can be suggested by this hypothetical situation: The population of a town composed exclusively of 300 childless couples is doubled overnight when every couple produces twins on the same day. Compare the impact on the physical capacity, economy, and the "feel" of the town of this doubling of population with one that arises from the arrival one day of 600 adult newcomers, all of whom must be housed, transported to work, and supplied with municipal services.

These and other concerns surface at different geographic scales and travel under different guises, but they are bound to persist. In an era of slowed population growth at the national level, it is likely that the ebb and flow of migration will figure prominently in both the fortunes of particular regions and the public debate about regional change.[26] It is, for example, now fairly apparent that migratory gains in the Sunbelt South are occurring largely at the expense of Northern industrial states. This recognition has been one factor prompting the debut of such regional alliances as the Coalition of Northeast Governors, the Northeast-Midwest Economic Advancement Coalition, and the Southern Growth Policies Board. In the course of only a decade, migration's pivotal role in determining which areas of the nation grow and which do not has become apparent. Consequently, population movements have taken on a political weight beyond their demographic and economic import.

The data examined in this section show that migration is changing the patterns of growth for many regions defined at several different scales. Some regions exhibit a clean break with an earlier trend, with the results of the break showing up most clearly at the local level. Cincinnati, Los Angeles, and Philadelphia are only three of thirty-six metropolitan areas that now enjoy local zero population growth (ZPG) along with its unavoidable "shrinking pains."[27]

Metropolitan growth, we have seen, has gravitated away from the North and East and toward the South and West. Regions where metropolitan no-growth is becoming common will continue to face problems of excess capacity and a depreciated stock of public facilities—most notably, vacant and abandoned housing and vandalized, underused schools. Selective outmigration will add still more to the burden of dependency that elderly and low-income citizens impose on a locality by lingering on after younger and more mobile people have left.

Inmigration can be beneficial, but new kinds of concerns can also crop up in newly growing areas. Around the turn of the century, Oregonians were said to offer an occasional prayer:

> We thank the goodness and the grace
> That brought us to this lovely place;
> And now with all our hearts we pray
> That other folks will stay away.

Today, residents of small communities in growing regions are equally concerned with the ailments they associate with "newcomers." Some of those ailments can be perceived readily only by examining tables of statistical data; consequently, the sheer visibility of migrants seems to have made them a ready focus of territorial issues concerning rights of access to places. Submerged beneath this concern are more profound legal and political questions, such as Who gets to live where? and Who is to decide that question, and by what criteria?

Migration enables people to live where they believe their opportunities for personal satisfaction are greatest. The benefits accrue to both the individual migrant and society as a whole, as the exodus of blacks from the rural South and their subsequent return to its flourishing metropolitan sector today amply demonstrate. Despite its long-term benefits, though, many people in the United States (including migrants themselves) still harbor a faint distrust of both the process and its participants when their day-to-day lives are affected.

Migration, then, has come to be viewed as a powerful and unpredictable force in the newly declining and growing areas alike, but it is not clear what role, if any, public policy should assume vis-à-vis the issues it generates. We need a framework for dealing with the issues themselves. The following paragraphs do not presume to prescribe policy; rather, they discuss possible directions that policy may take and what consequences may ensue.

Migration's public policy consequences arise from two distinct sources: people's inability to make well-informed assessments of their long-run opportunities in an existing setting or a potential new one and the public effects of their migration on others whom they leave behind at origin and join at destination.[28] In the former case, government agencies can try to provide information to potential migrants to improve the choices they make. In the latter case, the externalities (either costs or benefits) are transferred by instruments of public finance to the migrant or to a higher level of social organization encompassing the affected populations.

This perspective suggests three possible approaches for dealing with the issues that migration is now posing in our society:

1. *Control or redirect migration flows* if the patterns of contemporary regional growth and decline are judged to be unsatisfactory. Migration policy might intervene to alter these patterns to achieve specified broad purposes, thereby addressing the factionalism engendered by regional economic change. For example, policies might be designed to blunt the forces that prompt migrants to leave the Northeast, or to steer Sunbelt in-migrants away from places where existing residents feel they are being inundated by newcomers. Job information and relocation assistance might be provided to certain target groups (say, unemployed youth and other "economic refugees") whose job prospects may be far brighter in job-surplus regions of the country.[29] That can be done, and has been; witness the federal government's efforts to relocate Cuban and Vietnamese refugees. The idea of having migration policies surely is not immune to criticism, and the purposes themselves are controversial. Trying to tinker with so vast and complex a system as the one which continuously redistributes people, their productive capacity, and their votes would be politically complex.

Perhaps the existing migration streams are only part of some long-run equilibrium process designed to rectify the original mistake the British settlers made when they landed in the upper right-hand corner of the map instead of proceeding directly to Houston. In any event, if the government does not adopt official migration policies, we can be sure that other forces will fill the vacuum, and perhaps in ways not to our liking. Critics and advocates of migration policy alike acknowledge that "hidden" migration policies are already at work. When government agencies build highways, regulate energy prices, and impose environmental restrictions or regulations, they simultaneously redistribute employment growth and alter incentives for private investment. Although such government activities are nominally unrelated, they exert a powerful undirected influence on migration patterns.

2. *Ameliorate the visible consequences of migration.* This approach, in contrast to the first, sees migration shifts as mostly inevitable. It focuses on helping localities find workable solutions to the common problems imposed by regionwide growth and decline, thereby responding to the factionalism engendered by the highly visible consequences of local population change. Many of the large urban centers in the Northeastern metropolitan belt, for example, face the prospect of continued outmigration and its ensuing shrinking pains—a withering tax base without a corresponding reduction in demands for public services, and the need to manage excess housing and underused facilities. These common concerns that follow in the wake of urban ZPG might be addressed by policies that promote an orderly thinning out and reduction of excess capacity.

In the newly growing nonmetropolitan sector, population growth will increase the demand for public services. The kinds of services demanded will depend on the age composition of inmigrants: Younger couples will create demands for more schools; older inmigrants will enlarge the demand for health services, now or in future years.

3. *Federalize the redistribution of resources to complement the pattern of migratory redistribution.* From the national perspective, internal migration is largely a zero-sum game. Outmigration may relieve Mississippi of its poor, but these same outmigrants will appear as poor inmigrants somewhere else (although their poverty may cease there). Likewise, physicians trained at Boston's expense may remove themselves (and Massachusetts' considerable human capital investment) to Florida. Migration's inherently selective nature makes it inevitable that some regions will gain and others will lose in terms of the distinctive kinds of people that accumulate in places. Acknowledging these facts of life, the third policy approach would be to transform such national problems as welfare dependency (which are disproportionately accumulated in some parts of the country) into national responsibilities.

These approaches do not define policy; instead, they express ways to view regional change and to devise organizational responses to the problems that change engenders. The choice of which policy stance should be taken depends on one's convictions about the proper role of the public sector, on what one believes policy has in its power to do, and on the extent to which these processes of change themselves can be harnessed toward deliberate ends.

Notes

1. See George Sternlieb and James W. Hughes, "New Regional and Metropolitan Realities of America," *Journal of the American Institute of Planners* 43, no. 3 (July 1977):227-41.

2. This section is drawn from two earlier papers prepared by the author: "New York State's Transition to Stability: The Demographic Outlook," The Rand Corporation, P-5794, January 1977; and *Emerging Public Concerns over U.S. Population Movements in an Era of Slowing Growth*, The Rand Corporation, P-5873, October 1977. Also see Thomas J. Espenshade and William J. Serow (eds.), *The Economic Consequences of Slowing Population Growth* (New York: Academic Press, forthcoming).

3. See Calvin L. Beale, *The Revival of Population Growth in Nonmetropolitan America*, ERS-605, Economic Development Division, Economic Research Service, U.S. Department of Agriculture, June 1975; Curtis C. Roseman, *Changing Migration Patterns within the United States*, Resource Papers for College Geography, No. 77-2 (Washington: Associa-

tion of American Geographers, 1977); John M. Wardwell, "Equilibrium and Change in Nonmetropolitan Growth," *Rural Sociology* 42, no. 2 (Summer 1977):156-79; C. Jack Tucker, "Changing Patterns of Migration between Metropolitan and Nonmetropolitan Areas in the United States: Recent Evidence," *Demography* 13, no. 4 (November 1976):435-43; Peter A. Morrison, "Rural Renaissance in America? The Revival of Population Growth in Remote Areas," *Population Bulletin* 31, no. 3 (1976).

4. Nonmetropolitan areas have continued gaining migrants after the economic depression earlier in the decade. Between 1975 and 1976, migrants to the metropolitan sector were outnumbered by those moving out by a ratio of 6 to 5. U.S. Bureau of the Census, *Current Population Reports*, series P-20, no. 305, January 1977.

5. Related to this point, see D.R. Vining, Jr., and A. Strauss, "A Demonstration that the Current Deconcentration of Population in the United States Is a Clean Break with the Past," *Environment and Planning* 9 (1977):751-58.

6. Calvin L. Beale, "Current Status of the Shift of U.S. Population to Smaller Communities" (Paper presented at the annual meeting of the Population Association of America, April 1977, St. Louis).

7. The post-1970 halt in metropolitan population growth has been observed in other countries as well. See, for example, Daniel R. Vining, Jr., and Thomas Kontuly, *Population Dispersal from Major Metropolitan Regions: An International Comparison*, Discussion Paper No. 100, Regional Science Research Institute, Philadelphia, September 1977; Norman Glickman, *Growth and Change in the Japanese Urban System: The Experience of the 1970s*, International Institute for Applied Systems Analysis, Laxenburg, Austria, August 1977.

8. William Alonso, "The Current Halt in the Metropolitan Phenomenon" (Paper prepared for the Symposium on Challenges and Opportunities in the Mature Metropolis, St. Louis, June 6-8, 1977).

9. In 1976 wives 18 to 24 years old expected a lifetime average of 2.1 births. When adjusted to include those women who have not yet married or who will remain single, the birth expectations data suggest, as of this time, that the average lifetime fertility of all women 18 to 24 will be about 2.0 births. U.S. Bureau of the Census, *Current Population Reports*, series P-20, no. 300, November 1976.

10. A substantial body of evidence exists on which an informed judgment about future fertility trends can be based. In my judgment, the long-term trend of future fertility (and hence population growth through natural increase) is unlikely to rise above the level assumed in series I; it seems plausible, on the other hand, that fertility could fall below the level assumed in series III.

11. For further detail, see Roger J. Vaughan, *The Urban Impacts of Federal Policies*, vol. 2: *Economic Development*, R-2028-KF/RC, The

Rand Corporation, June 1977; U.S. Senate, Committee on Appropriations, *Patterns of Regional Change—The Changes, the Federal Role, and the Federal Response: Selected Essays*, Committee Print (Washington: GPO, October 1977).

12. Regions may be defined as (1) areas of relative homogeneity, with borders determined by discontinuities (*homogeneous* regions), (2) heterogeneous areas, whose different parts complement and support one another and typically form a functional hierarchy (*polarized* regions), or (3) a space whose various parts depend on the same set of decisions (*planning* regions). For further discussion of the rationale for various regionalizations, see Niles M. Hansen, *A Critique of Economic Regionalizations of the United States*, Research Report RR-75-32, International Institute for Applied Systems Analysis, Laxenburg, Austria, September 1975.

13. *Census divisions and regions* are convenient mostly as territorial common denominators for which data are available, especially when our analyses include earlier decades of this century. Also, because they are composed of whole states, they are regions with administrative and political meaning.

Economic subregions divide the nation into twenty-six economically and culturally distinct groupings of counties. Blind to the often artificial boundaries that separate states, they reflect the administratively untidy economic and cultural geography of the nation: differences in regional resource endowment, in economic activity, in the evolution and present form of human settlement, and so forth. The analytical strengths of using subregions are straightforward: (1) They are entities that have had historically persistent demographic characteristics which people seem to recognize and understand, and (2) they reflect the considered judgment of knowledgeable persons, not narrowly conceived administrative criteria.

14. Between 1970 and 1976, for example, 851,000 more persons migrated away from these three states than migrated to them, resulting in a 2.3 percent loss of the Middle Atlantic states' 1970 population. In conjunction with comparatively low natural increase, this loss has virtually halted population growth.

15. See Beale, *The Revival of Population Growth in Nonmetropolitan America*; idem, "Rural Development: Population and Settlement Prospects," *Journal of Soil and Water Conservation* 29 (1974):23-27; "A Further Look at Nonmetropolitan Population Growth since 1970," *American Journal of Agricultural Economics* 58, no. 5 (1976):953-58. The following recent studies, which focus on the combinations of influences at work in specific regions of the nation, are especially useful: James P. Allen, "Population Changes in the Nonmetropolitan West, 1970-1975," *The Great Plains-Rocky Mountain Geographical Journal*, forthcoming; E. Evan Brunson and Thomas D. Bever, "Southern Growth Trends: 1970-

1976," Research report published by the Southern Growth Policies Board, Research Triangle Park, N.C., June 1977; C. Shannon Stokes, "Population Trends in Pennsylvania and the Northeast," *Farm Economics*, periodical published by Pennsylvania State University in cooperation with the Cooperative Extension Service, U.S. Department of Agriculture, University Park, Pa., July 1976; Glenn V. Fuguitt and Calvin L. Beale, "Post-1970 Shifts in the Pattern of Population Change in the North Central Region," CDE Working Paper 76-17, Center for Demography and Ecology, University of Wisconsin, Madison, May 1976; John A. Kuehn and Curtis Braschler, *New Manufacturing Plants in the Nonmetro Ozarks Region*, Agricultural Economics Report No. 384, Economic Research Service, U.S. Department of Agriculture, September 1977; Richard L. Morrill, "What's behind the Rural Recovery? Population Trends in the Pacific Northwest, 1970-1975" (Paper presented at the annual meeting, Association of American Geographers, Salt Lake City, April 1977); Wilbur Zelinsky et al., "Population Change and Redistribution in Nonmetropolitan Pennsylvania, 1940-1970," Report submitted by The Pennsylvania State University's Population Issues Research Office to the Center for Population Research, National Institute of Child Health and Human Development, HEW, November 1974.

16. See Vernon Renshaw and Howard L. Friedenberg, "Transfer Payments: Regional Patterns, 1965-1975," *Survey of Current Business* 57, no. 5 (May 1977):15-19.

17. See Richard Lamb, "Metropolitan Impacts on Rural America," Research paper no. 162, Department of Geography, University of Chicago, 1975; and "Intra-Regional Growth in Non-Metropolitan America: Change in the Pattern of Change" (Paper presented at the annual meeting, Association of American Geographers, Salt Lake City, 1977). The latter study demonstrates that whereas much of the migration to nonmetropolitan areas before 1970 could be attributed to proximity to urban areas interacting with amenity appeal, migration in the 1970s to amenity-rich areas appears to have been freed from traditional distance constraints.

18. These developments are widespread and often highly localized, occurring in specific counties within these subregions. Useful regionally focused studies are Robert W. Marans et al., *Waterfront Living: A Report on Permanent and Seasonal Residents in Northern Michigan* (Ann Arbor, Mich.: Institute for Social Research, University of Michigan, 1976); Stephen J. Tordella, "Recreational Housing in Wisconsin Counties, 1970," *Population Notes* no. 4 (August 1977), published by the Applied Population Laboratory, Department of Rural Sociology, University of Wisconsin, Madison.

19. William Alonso, "Urban Zero Population Growth," *Daedalus* 102, no. 4 (Fall 1973):191-206.

20. For many purposes, that area in and around a city where activities form an integrated economic and social system is considered a unit. The standard metropolitan statistical area (SMSA) was developed in 1960 as such a unit, for which many general-purpose statistics are now tabulated. All persons who reside outside of SMSAs are defined as nonmetropolitan.

21. The propensity of places to annex population, and the amount of population acquired per annexing place, varies among regions. In the Northeast, especially, legal factors inhibiting annexation prevent cities from acquiring much of the suburban growth that occurs beyond their boundaries. For further elaboration see Glenn V. Fuguitt and Calvin L. Beale, "Recent Trends in City Population Growth and Distribution" (Paper prepared for the 1976 Public Policy Forum, Problems of Small Cities, Washington, December 1976).

22. Not surprisingly, places that grow primarily through migration possess a distinctive regional character that echoes common motives and expectations behind the original settlement and is perpetuated by an ongoing process of selective migration. Well-known examples are Yankee New England and Mormon Utah. See Raymond D. Gastil, *Cultural Regions of the United States* (Seattle: University of Washington Press, 1975); Wilbur Zelinsky, *The Cultural Geography of the United States* (Englewood Cliffs, N.J.: Prentice-Hall, 1973); Dorothy O. Johansen, " A Working Hypothesis for the Study of Migrations," *Pacific Historical Review* 36 (1967):1-12.

23. For recent evidence on this point, see David M. DeFerranti et al., *The Welfare and Nonwelfare Poor in New York City*, R-1381-NYC, The Rand Corporation, June 1974; Miriam Ostow and Anna B. Dutka, *Work and Welfare in New York City* (Baltimore, Md.: Johns Hopkins University Press, 1975), p. 76; Larry H. Long, "Poverty Status and Receipt of Welfare among Migrants and Nonmigrants in Large Cities," *American Sociological Review* 39 (February 1974):46-56; Gene B. Peterson et al., *Southern Newcomers to Northern Cities* (New York: Praeger, 1977).

24. Leon F. Bouvier et al., "International Migration: Yesterday, Today and Tomorrow," *Population Bulletin* 32, no. 4 (1977).

25. A summary of recent evidence is given in Wayne A. Cornelius, *Illegal Mexican Migration to the United States: Recent Research Findings, Policy Implications and Research Priorities*, M.I.T. Monograph Series on Migration and Development, no. C/77-11, May 1977, Cambridge, Mass.; Jorge Bustamente, "Undocumented Immigration from Mexico: Research Report," *International Migration Review* 11, no. 2 (Summer 1977):149-77; U.S. House of Representatives, Committee on the Judiciary, *Illegal Aliens: Analysis and Background*, Committee Print, June 1977.

26. Carol L. Jusenius and Larry C. Ledebur, *Federal and Regional Responses to the Economic Decline of the Northern Industrial Tier*, Office

of Economic Research, Economic Development Administration, Department of Commerce, March 1977; idem, *A Myth in the Making: The Southern Economic Challenge and Northern Economic Decline*, November 1976.

27. See Gurney Breckenfeld, "It's Up to the Cities to Save Themselves," *Fortune*, March 1977, pp. 194-206.

28. I have drawn heavily here on ideas propounded by my colleague T. Paul Schultz.

29. A proposal recently made by Congressman Reuss. See U.S. House of Representatives, Subcommittee on the City, Committee on Banking, Finance, and Urban Affairs, *To Save a City*, Committee Print, September 1977, p. 32.

**Part II
Regional Economic Shifts
and Differences**

3 Regional Shifts in Economic Base and Structure in the United States since 1940

William H. Miernyk

Regional Growth and Regional Structure

When economists talk about structure, they generally mean the industrial distribution of employment. This distribution has been changing since Colonial days, at times almost imperceptibly, at other times fairly rapidly. There are major differences in economic structure among regions, although they are not as pronounced now as they once were.

The best and most comprehensive study of the regional structure of the U.S. economy is the historical analysis made by Perloff et al. at Resources for the Future.[1] Although their latest data are now almost a quarter of a century old, it provides a useful background for discussion of more recent changes.

In 1840, according to Easterlin's estimates, the ratio of agricultural to nonagricultural workers in the United States was 3.8. New England and the Middle Atlantic states had ratios of 1.6 and 2.1, respectively. At the other extreme, the three southern regions—South Atlantic, East South Central, and West South Central (which then consisted of Arkansas and Louisiana only)—had ratios of 8.8, 11.6, and 6.1. The West South Central region reported the highest per capita income in 1840, followed by New England and the Middle Atlantic states.[2]

By 1860, manufacturing had become an important economic activity in the United States. The five largest industries, in terms of value added, produced cloth, lumber, boots and shoes, men's clothing, and flour (including meal). The iron and machinery industries ranked sixth and seventh, followed by leather, woolen goods, and liquors. More than 71 percent of all manufacturing employment in 1860 was in New England and the Middle Atlantic states. When the Great Lakes states are included, the Northeast accounted for almost 84 percent of all manufacturing employment in 1860.[3]

A century later Perloff et al. wrote that "The most striking feature in the history of American manufacturing is the enduring strength of the Northeast...even today the great industrial belt in the Northeast continues to dominate the regional structure of American manufactures much as it did at the beginning of this century."[4] But their analyses of regional changes in value added by manufacture, and in industrial employment, showed a steady drift from the Northeast to the South and West. Between 1870 and

1910, the drift proceeded at a glacial pace. But between 1910 and 1954, the last year of their study, the rate of industrial migration was accelerating.[5]

During the latter period the industrial structure of the national economy also changed more rapidly than it had in the past as a growing number of workers shifted from agriculture to manufacturing and construction, and subsequently to trade and service occupations.

All regions have grown, and all have undergone major changes in industrial structure. But some things have remained constant. The Northeast and the Far West have maintained their preeminence in personal income per capita.[6] Although Arkansas and Louisiana had enjoyed above average per capita income in the early days of the last century, this was not to last. Despite major changes in economic structure in recent decades, the Southwest and Southeast have remained at the bottom of per capita income tables.

Regional Growth Theory and Regional Development Policy

The literature of regional growth theory is fairly sparse if one excludes the burgeoning list of growth pole-growth center studies.[7] Richardson's recent book includes references to most of the important English-language works on regional growth published through the early 1970s, as well as numerous references to publications which are only obliquely related to regional growth. The monumental study by Perloff et al. was conducted largely without benefit of theory, although they briefly discussed "export base" and "sector" growth theories. An attempt to develop a general theory of regional growth was made by Borts and Stein in the early 1960s.[8] Essentially, they added a regional dimension to conventional neoclassical growth theory. The focus of their theory is on the relationship between factor mobility and differential factor returns.

An appropriate regional policy for a nation, according to Borts, is one which maximizes real gross *national* product.[9] As Siebert has pointed out, however, there is at least "the possibility of a goal conflict not only between growth policies of different regions but also between regional and national growth goals."[10]

In their 1962 paper, Borts and Stein attempted to relate regional growth to changes in economic structure while remaining faithful to their basic neoclassical model.[11] The regional component of their model is a simple export-base concept. Employment shifts among sectors, they assert, have an effect on wage differentials. And "autonomous shifts of labor between sectors of a region are capable of setting a growth process into motion."[12] They do not, however, go into the causes of these autonomous shifts.

An outspoken critic of neoclassical regional growth theory has been

Harry W. Richardson. He pointed out the incompatibility between the neoclassical assumption of full employment and the existence of regional labor markets with different degrees of labor utilizatiion. And as Lösch had done many years earlier, Richardson pointed out that explicit recognition of space and the existence of transportation costs vitiates the neoclassical assumption of perfect competition.[13]

Richardson briefly discusses export base, econometric, input-output, and multisector development planning models. It is questionable whether these analytical techniques can be considered as "theories" per se. He does, however, discuss the Myrdal-Kaldor cumulative causation hypothesis, which comes closer to the mark of a legitimate growth theory.

The cumulative causation hypothesis is of interest to regional economists because it yields predictions diametrically opposed to those of neoclassical growth theory. Neoclassical regional theory concludes that growth will lead to convergence in regional per capita incomes. The cumulative causation hypothesis suggests, however, that there will be divergence in regional per capita incomes as a national economy grows and trade among regions increases.

Richardson has developed a "theoretical approach to regional growth that incorporates space both between and within regions." It places emphasis "on spatial agglomeration economies and locational preferences as opposed to the traditional neoclassical variables of wage and capital yield differentials." He feels it is essential to consider urbanization explicitly and to analyze the ways in which spatial structure influences resource mobility. Finally, he states that the "doctrine of balanced growth both within and between regions does not make good economic sense. . . ."[14] Richardson is somewhat apologetic about his inability to provide a complete empirical test of his theory.[15]

Both neoclassical and Richardsonian growth theory are demand-oriented. Factor supplies are implicitly assumed to be infinitely elastic in both models. Thus whether the objective is to maximize real gross national product, as neoclassical scholars maintain, or to minimize regional disparities in per capita incomes and unemployment rates, the modus operandi is the same; it is to stimulate demand.[16]

Neoclassical regional economists, like their macroeconomic counterparts, want to minimize government intervention. Supporters of the cumulative causation hypothesis are less concerned about direct intervention because when trade between regions is considered, "neither the principles of comparative advantage nor classical mechanisms of adjustment work. Instead, increasing returns favour the rich regions and inhibit development in the poor; because of scale effects the rich regions gain a virtual monopoly of industrial production. Also, since competition in industry is imperfect while near perfect competition prevails in agriculture, movements in the terms of trade favour the rich regions."[17] Indeed, Richardson states that to Kaldor the "principle of cumulative causation is nothing more or less than the existence of increasing returns to scale [in the widest sense, that is, including external and agglomeration economies] in manufacturing...."[18] Increasing returns will generate rapid increases in productivity in the regions which experience the most rapid growth rates in output. And this is why "the relatively fast-growing regions tend to acquire cumulative advantages over the slow-growing ones."[19]

The focus on demand in contemporary regional growth theory is total, or nearly so. This is not surprising since most contemporary economics is still strongly wedded to Keynesian macroeconomic principles. Supply constraints—of natural resources, and particularly of energy—are conspicuously absent in discussions of regional growth theory. Again, this is not surprising since with a few exceptions economists have tended to shy away from the difficult problems raised by the explicit introduction of supply constraints.[20]

The trends discussed by Kaldor in his 1970 article might have accurately described past regional growth patterns in some countries. But it is questionable whether they apply to recent changes in the United States. During the present decade several resource-based state economies have exhibited robust growth in employment and per capita incomes, while some of the older industrial states, which presumably are beneficiaries of external economies, have experienced slow rates of growth or even decline. The evidence is not definitive, but there are straws in the wind which suggest that there have been real income transfers from industrial to resource-based states.[21]

If the events of the 1970s signify changes in trend, rather than short-term fluctuations, there is a clear need for a complete rethinking of regional growth theory. Regional growth policy also must be adapted to new conditions where supply constraints may be among the most important determinants of regional growth rates.

In his list of regional growth theories Richardson does not include what Perloff has called "the sector theory."[22] Elsewhere he dismisses it summarily because "the sector theory is inadequate...it offers no insights into the causes of growth itself."[23] In *Regional Growth Theory* he does not discuss structural change at all, except as it is subsumed under the concept of agglomeration. The terms *agglomeration* and *structure* should not be considered synonymous, however, since the former deals largely with urban phenomena while agriculture and other "primary" activities play a significant role in the "sector" hypothesis of regional growth.

The "sector theory" of regional development, also called the Clark-Fisher hypothesis, after Colin Clark and A.G.B. Fisher who developed it independently, has been discussed in detail elsewhere, and only the bare rudiments are given here.[24] Clark and Fisher hypothesized a functional relationship between per capita income and the stage of industrialization in national economies. Lewis Bean extended this hypothesis to regions. According to Bean, the stage of industrialization is "the proportion of a country's working population engaged in primary occupations—agriculture, forestry, and fishing." He also defined the "pattern of industrialization" as "the relative importance of secondary occupations—manufacturing, mining, and construction; and of tertiary occupations—trades and services."[25]

The major disadvantages of the sector theory, according to Perloff et al., are that it is partial in scope and too highly aggregated.[26] As noted, Richardson was critical because he feels it does not deal with causality. An assiduous critic could also point out that empirical versions of the Clark-Fisher model are exercises in comparative statics rather than dynamics. In spite of these criticisms one can argue that it is a useful heuristic device for studying the process of regional structural change.[27]

The Changing Regional Structure of the U.S. Economy, 1940-1975

The thirty-five years discussed in this chapter cover a period of rapid economic growth in the United States. Population rose from 132 million to 213 million, an increase of 61 percent. Gross national product went from $100 million to $1.5 trillion, a fifteenfold increase. In constant (1972) dollars, the nation's output increased less, going from $344 billion to $1.2 trillion. This is still an impressive increase of more than 346 percent. On a per capita basis, real gross national product more than doubled, rising from $2,600 in 1940 to $5,580 in 1975. There were major increases in productivity. Between 1947 (the earliest year for which comparable data are available) and 1975, an index of the output per hour of all employed persons rose from 52.3 to 111.5, an increase of 113 percent.[28]

Increasing productivity has been a major cause of structural change in the U.S. economy. Reductions in labor requirements in some sectors have released labor which was then employed in others. Rising real income has stimulated the demand for services, and thus the derived demand for labor to provide those services. In 1940, for example, 11 million persons were employed on farms. This number had dropped to about 4.4 million by 1975. Meanwhile, the index of agricultural output had increased from 62 to 102, and the index of agricultural output per hour of farm work had increased from 20 to 110.[29] There were smaller, but still substantial, gains in the nonfarm sector where the index of productivity increased from 58.7 in 1947 to 109.4 in 1975, a gain of 86 percent.[30]

Economic growth between 1940 and 1975 was not uniformly distributed on a geographical basis. The center of population continued the westward drift which has been occurring since the initial settlement of the East Coast.[31] By 1976, according to estimates prepared by the Metropolitan Life

Insurance Company, the South and the West contained more than half the nation's population for the first time.[32] Increases in population do not by themselves signify improvement in economic well-being, as Perloff et al. pointed out when they distinguished between "volume" growth and "welfare" growth.[33] But the South and the West have been gaining in economic well-being, if per capita personal income can be used as a rough approximation of the latter.[34]

Per capita personal income by state and changes between 1940 and 1975 are given in table 3-1. The states are arranged by Census regions. Because frequent reference is made to them later in this chapter, the Sunbelt states have been italicized in this table.

There has been considerable convergence in per capita income during the period under review. This is indicated by the two columns which show each state's per capita income as a percentage of the U.S. average in 1940 and in 1975. In 1940, the range of relative per capita income was from 36.6 percent in Mississippi to 196.6 percent in Washington, D.C., or, if one wishes to eliminate the nation's capital as a "special case," to 168.7 percent in Delaware. By 1975 there was a much flatter distribution. Mississippi remained at the bottom of the list, at 68.7 percent of the U.S. average, while Alaska's per capita income was 160.1 percent of the average. If both Alaska and Washington are considered as "special cases," the next highest relative per capita income was in Delaware, at 113.3 percent of the national average.

Table 3-1 supports the convergence conclusion of neoclassical regional growth theory, while contradicting the Myrdal-Kaldor cumulative causation hypothesis. Let me hasten to add, however, that while the evidence rules out the cumulative causation hypothesis as an explanation of differential regional growth rates in the United States, it does not provide proof of the correctness of neoclassical regional growth theory.

One possible alternative explanation is that agglomeration and locational preferences are behind the growth of relatively low-income states, as Richardson has suggested. But while I personally prefer Richardson's model to neoclassical theory, I feel it does not explain differential rates of growth in regional income satisfactorily. I doubt, for example, that either theory explains the increase in per capita income in North Dakota from 59 to 97 percent of the U.S. average or the slightly less spectacular increases recorded by, say, South Dakota and Kansas during this period. Table 3-1 shows that most states which have recorded substantial improvement in relative per capita income position have economies that are heavily based on agriculture or other natural resources, notably coal, oil, and natural gas. This is brought out by the distribution of location quotients for selected sectors given in table 3-2. The states with a higher relative per capita income in 1975 than they had in 1940 are italicized in this table; an asterisk indicates that the state had above average per capita income in 1975.[35]

Table 3-1
Per Capita Income by State, 1940 and 1975

Percent	1940		1975		Changes 1940-1975	
	Dollars	Percent of U.S.	Dollars	Percent of U.S.	Dollars Difference	Percent
New England						
Maine	523	87.9	4,786	81.1	4,263	815.10
New Hampshire	579	97.3	5,315	90.1	4,736	817.96
Vermont	507	85.2	4,960	84.0	4,453	878.30
Massachusetts	784	131.8	6,114	103.6	5,330	679.84
Rhode Island	743	124.9	5,814	99.0	5,071	682.50
Connecticut	917	154.1	6,973	118.1	6,056	660.41
Middle Atlantic						
New York	870	146.2	6,564	112.1	5,694	654.48
New Jersey	822	138.2	6,722	113.9	5,900	717.76
Pennsylvania	648	108.9	5,943	100.7	5,295	817.12
East North Central						
Ohio	665	111.8	5,810	98.4	5,145	773.68
Indiana	553	92.9	5,653	95.8	5,100	922.24
Illinois	754	126.7	6,789	111.5	6,035	800.39
Michigan	679	114.1	6,173	104.6	5,494	809.13
Wisconsin	554	93.1	5,669	96.1	5,115	923.28
West North Central						
Minnesota	526	88.4	5,807	98.4	5,281	1,003.99
Iowa	501	84.2	6,077	103.0	5,576	1,112.97
Missouri	524	88.1	5,510	93.4	4,986	951.52
North Dakota	350	58.8	5,737	97.2	5,387	1,539.14
South Dakota	359	60.3	4,924	83.4	4,565	1,271.58
Nebraska	439	73.8	6,087	103.1	5,648	1,286.56
Kansas	426	71.6	6,023	102.1	5,597	1,313.84

Table 3-1 *(cont.)*

	1940		1975		Changes 1940-1975	
Percent	Dollars	Percent of U.S.	Dollars	Percent of U.S.	Dollars Difference	Percent
South Atlantic						
Delaware	1,004	168.7	6,748	114.3	5,744	572.11
Maryland	712	119.7	6,474	109.7	5,762	809.26
District of Columbia	1,170	196.6	7,742	131.2	6,572	561.70
Virginia	466	78.3	5,785	98.0	5,319	1,141.41
West Virginia	407	68.4	4,918	83.3	4,511	1,108.35
North Carolina	328	55.1	4,952	83.9	4,624	1,409.75
South Carolina	307	51.6	4,618	78.2	4,311	1,404.23
Georgia	340	57.1	5,086	86.2	4,746	1,395.88
Florida	513	86.2	5,638	95.5	5,125	999.02
East South Central						
Kentucky	320	53.8	4,871	82.5	4,551	1,422.18
Tennessee	339	57.0	4,895	82.9	4,556	1,343.95
Alabama	282	47.4	4,643	78.7	4,361	1,546.45
Mississippi	218	36.6	4,052	68.7	3,834	1,758.71
West South Central						
Arkansas	256	43.0	4,620	78.3	4,364	1,704.68
Louisiana	363	61.0	4,904	83.1	4,541	1,250.96
Oklahoma	373	62.7	5,250	89.0	4,877	1,307.50
Texas	432	72.6	5,631	95.4	5,199	1,203.47
Mountain						
Montana	570	95.8	5,422	91.9	4,852	851.22
Idaho	464	78.0	5,159	87.4	4,695	1,011.85
Wyoming	608	102.2	6,131	103.9	5,523	908.38
Colorado	546	91.8	5,985	101.4	5,439	996.15
New Mexico	375	63.0	4,775	80.9	4,400	1,173.33
Arizona	497	83.5	5,355	90.7	4,858	977.46

Utah	487	81.8	4,923	83.4	4,436	910.88
Nevada	876	147.2	6,647	112.6	5,771	658.78
Pacific						
Washington	662	111.3	6,247	105.8	5,585	843.65
Oregon	623	104.7	5,769	97.7	5,146	826.00
California	840	141.2	6,593	111.7	5,753	684.88
Alaska	n.a.	n.a.	9,448	160.1	n.a.	N/A
Hawaii	577	97.0	6,658	112.8	6,081	1,053.89
United States	595		5,902		5,307	891.93

Source: U.S. Bureau of the Census *Statistical Abstract of the U.S.*, Washington: GPO 1965, table 458, p. 334; U.S. Department of Commerce, Bureau of Economic Analysis, *Survey of Current Business* (Washington: USGPO, August 1976), table 2, p. 13.

One shortcoming of the conventional location quotient is that it fails to take into account differences in labor and capital intensity among regions. For example, in 1975, West Virginia had a location quotient (LQ) of 10.14 in mining while Kentucky's LQ was only 4.19. This would suggest, ceteris paribus, that West Virginia produced more coal than Kentucky. In fact, it did not—it produced less. Most West Virginia coal is produced by relatively labor-intensive deep mines while most of Kentucky's coal is the output of capital-intensive surface mines.[36] Thus while location quotients tell us something about the industrial structure of states and regions, they must be interpreted with caution.

To supplement the traditional location quotients of table 3-2, the same formula was used to calculate relative sources of personal income by region. The results are given in table 3-3. These two tables are not directly comparable, and should be examined independently.[37] Table 3-3 covers too short a period to reveal the kinds of structural change that are visible in table 3-2. Even during this short period, however, a few changes stand out, such as the sharp drop in the relative importance of military spending in New England between 1969 and 1975, with smaller relative declines in other regions. The increasing importance of state and local government spending in all regions is also indicated by table 3-3.

Kaldor argued that one cause of cumulative causation is the existence of imperfect competition and stable prices in industrial regions with concomitant downward pressure on prices exerted by the more nearly perfect competition found in agricultural regions.[38] Such differences in market structure would, he believed, progressively alter the terms of trade in favor of industrial regions. Tables 3-1 and 3-2 deny the applicability of this line of reasoning to the United States. The data in these tables suggest that the terms of trade have been turning in favor of agricultural and other resource-based regions. Kaldor's model, as is true of every other theory of regional growth with which I am familiar, simply ignores supply constraints and the effects that such constraints can have on the growth of incomes in resource-based regions.[39]

What about the Clark-Fisher-Bean hypothesis? What light can it shed on the process of convergence in per capita incomes among the states? This hypothesis is best examined by means of simple regression analysis. In each case, relative per capita income is the dependent variable while the percentage of total state employment in each of the three broad sectors defined by Bean is the independent variable. It was not possible to obtain data on employment in forestry and fisheries in each of the states, so the analysis of "primary" employment which follows is limited to the relationship between relative state per capita income and employment in agriculture. The results for 1940 are illustrated graphically in figure 3-1a.

Table 3-2
Selected Location Quotients, 1940 and 1975

State	Agriculture 1940	Agriculture 1975	Mining 1940	Mining 1975	Manufacturing 1940	Manufacturing 1975	Contract Construction 1940	Contract Construction 1975	Services 1940	Services 1975
New England	0.27	0.29	0.08	0.15	1.65	1.22	1.03	0.84	1.05	1.15
Maine	0.76	0.81	0.10	0	1.41	1.15	0.95	1.08	1.02	0.93
New Hampshire	0.49	0.36	0.09	0	1.69	1.27	1.13	1.03	1.03	1.12
Vermont	1.32	1.11	0.58	4.39	0.94	0.98	0.99	0.93	1.07	1.16
Massachusetts	0.14	0.18	0.05	0	1.57	1.13	0.98	0.79	1.08	1.27
Rhode Island	0.11	0.16	0.03	0	1.95	1.41	1.08	0.69	0.88	1.10
Connecticut	0.21	0.22	0.04	0	1.86	1.41	1.08	0.84	0.96	1.05
Middle Atlantic	0.24	0.34	1.23	0.44	1.33	1.10	1.06	0.80	1.09	1.12
New York	0.23	0.27	0.09	0.11	1.18	0.91	1.08	0.69	1.20	1.22
New Jersey	0.17	0.13	0.11	0.12	1.57	1.22	1.11	0.80	0.99	1.02
Pennsylvania	0.32	0.56	3.45	1.14	1.42	1.31	0.97	0.97	0.93	1.03
East North Central	0.92	0.90	0.58	0.52	1.28	1.29	0.86	0.82	0.87	0.94
Ohio	0.59	0.68	0.68	0.71	1.43	1.35	0.95	0.83	0.93	0.98
Indiana	0.95	1.25	0.58	0.42	1.28	1.39	0.96	0.90	0.87	0.76
Illinois	0.53	0.61	0.84	0.60	1.22	1.19	0.89	0.92	0.96	1.00
Michigan	0.63	0.63	0.43	0.48	1.64	1.35	0.88	0.70	0.85	0.94
Wisconsin	1.38	1.88	0.12	0.11	1.09	1.21	0.82	0.71	0.87	0.92
West North Central	1.62	2.60	0.51	0.69	0.53	0.79	0.89	0.90	0.97	0.91
Minnesota	1.62	2.19	0.42	0.92	0.53	0.83	0.89	0.87	1.00	0.97
Iowa*	1.91	3.03	0.37	0.28	0.49	0.86	0.92	0.93	0.92	0.86
Missouri	1.26	1.18	0.47	0.52	0.81	0.93	0.96	0.83	0.96	0.94
North Dakota	2.84	4.26	0.24	0.82	0.11	0.27	0.45	1.23	0.86	0.87
South Dakota	2.56	4.82	0.69	1.17	0.19	0.31	0.67	0.74	0.90	0.92
Nebraska*	1.99	3.17	0.07	0.36	0.29	0.56	0.86	0.95	0.97	0.87
Kansas*	1.67	2.14	1.29	1.34	0.39	0.79	0.94	1.01	0.98	0.86
South Atlantic	1.29	0.91	1.23	0.93	0.88	0.93	1.04	1.28	1.03	0.96
Delaware	0.75	0.53	0.54	0	1.25	1.28	1.51	1.52	1.02	0.93
Maryland	0.56	0.42	0.29	0.14	1.12	0.70	1.26	1.49	1.05	1.14
District of Columbia	0.01	0	0.02	0	0.31	0.12	1.40	0.82	1.43	1.48

Table 3-2 (cont.)

State	Agriculture		Mining		Manufacturing		Contract Construction		Services	
	1940	1975	1940	1975	1940	1975	1940	1975	1940	1975
Virginia	1.31	0.94	1.28	1.20	0.86	0.88	1.14	1.34	0.93	0.92
West Virginia	0.82	1.17	10.80	10.14	0.75	0.79	0.87	0.91	0.83	0.69
North Carolina	1.80	1.49	0.12	0.20	1.15	1.51	0.85	1.26	0.86	0.71
South Carolina	2.10	1.13	0.10	0.21	0.97	1.44	0.74	1.38	0.89	0.69
Georgia	1.86	0.97	0.18	0.42	0.79	1.06	0.82	1.04	1.02	0.81
Florida	1.01	0.60	0.19	0.42	0.50	0.52	1.41	1.43	1.41	1.20
East South Central										
Kentucky	2.07	1.78	1.58	1.64	0.62	1.13	0.86	1.09	0.87	0.80
Tennessee	1.94	2.24	3.53	4.19	0.51	0.95	0.92	0.93	0.80	0.81
Alabama	1.77	1.47	0.76	0.61	0.78	1.24	1.00	1.04	0.94	0.85
Mississippi	2.12	1.27	1.66	1.07	0.74	1.16	0.71	1.30	0.88	0.79
	3.08	2.41	0.13	0.86	0.39	1.15	0.73	1.10	0.76	0.66
West South Central										
Arkansas	1.73	1.28	1.43	3.29	0.44	0.76	1.03	1.36	1.04	0.91
Louisiana	2.75	2.46	0.50	0.61	0.42	1.09	0.65	1.02	0.76	0.73
Oklahoma	1.76	1.04	0.95	4.88	0.55	0.64	0.99	1.67	1.07	0.89
Texas	1.77	1.98	2.63	4.43	0.33	0.67	0.90	1.01	1.02	0.84
	1.59	0.99	1.41	3.07	0.42	0.77	1.13	1.42	1.06	0.96
Mountain										
Montana	1.33	1.14	3.03	3.29	0.39	0.51	1.23	1.30	1.02	1.11
Idaho	1.71	2.44	3.60	2.79	0.32	0.35	1.04	1.02	0.87	0.95
Wyoming*	1.98	2.57	2.10	1.41	0.34	0.67	1.00	1.21	0.86	0.86
Colorado*	1.57	1.83	3.58	12.54	0.23	0.22	1.05	2.08	0.89	0.77
New Mexico	1.13	1.14	2.25	1.97	0.44	0.60	1.13	1.33	1.12	1.06
Arizona	1.72	1.00	3.12	5.74	0.27	0.31	1.31	1.47	1.03	1.03
Utah	1.16	0.54	4.22	3.56	0.36	0.59	1.29	1.33	1.12	1.06
Nevada	1.03	0.84	3.36	3.12	0.47	0.65	1.19	1.18	0.96	0.96
	0.82	0.33	7.51	1.66	0.19	0.20	1.63	1.06	1.03	2.40

Pacific	0.62	0.79	0.77	0.40	0.77	0.84	1.32	0.93	1.16	1.10
Washington	0.75	1.13	0.42	0.17	0.89	0.84	1.31	1.00	1.11	0.99
Oregon	1.00	1.09	0.38	0.24	0.89	0.92	1.17	0.93	0.99	0.96
California	0.57	0.63	0.89	0.43	0.70	0.87	1.32	0.83	1.21	1.14
Alaska		4.74		0.64		0.22		1.94		
Hawaii										0.84

Note: Italics indicate an increase in relative per capita income. Asterisks mean that the state had above average per capita income in 1975.

Table 3-3
Industrial Sources of Personal Income by Region, Expressed as Location Quotients, 1969 and 1975

	New England		Middle Atlantic		East North Central		West North Central	
	1969	1975	1969	1975	1969	1975	1969	1975
Metal mining			0.200	0.200	0.400	0.467	2.733	2.867
Coal mining			1.043	1.022	0.826	0.761	0.174	0.152
Crude petroleum and natural gas		0.015	0.043	0.119	0.174	0.239	0.522	0.537
Nonmetal mining, except fuels	0.353	0.444	0.588	0.556	1.000	0.944	1.118	1.333
Contract construction	1.028	0.893	0.869	0.783	1.033	0.903	1.084	1.046
Food and kindred products	0.644	0.573	0.895	0.849	1.046	1.096	1.933	1.872
Textile mills	2.524	1.363	0.874	0.875	0.136	0.100	0.039	0.075
Apparel and other accessories	0.882	0.907	1.795	1.804	0.394	0.485	0.512	0.536
Lumber and furniture	0.663	0.748	0.445	0.450	0.789	0.775	0.563	1.081
Paper and allied products	1.670	1.621	0.991	0.968	1.160	1.137	1.123	1.211
Printing and publishing	1.117	1.172	1.417	1.400	1.209	1.207	1.086	1.193
Chemicals and allied products	0.577	0.717	1.291	1.321	0.974	0.995	0.714	0.630
Petroleum refining	0.109	0.220	1.109	1.180	0.848	0.840	0.304	0.480
Primary metals	0.681	0.594	1.332	1.370	1.933	1.995	0.340	0.361
Fabricated metals	1.279	1.385	0.813	0.933	1.607	1.759	0.787	0.846
Machinery, except electrical	1.333	1.314	0.988	0.967	1.930	1.861	1.145	1.118
Electric machinery and supplies	1.478	1.710	1.177	1.117	1.328	1.234	0.826	0.734
Motor vehicles	0.153	0.136	0.419	0.414	3.355	3.500	0.759	0.803
Transportation equipment	1.783	2.020	0.557	0.584	0.552	0.617	1.015	0.644
Other manufacturing	1.317	1.825	1.388	1.278	1.023	1.088	0.467	0.927
Transportation, commerce and other utilities	0.794	0.794	1.092	1.070	0.901	0.898	1.193	1.206
Wholesale and retail trade	0.955	0.951	0.991	0.959	0.958	0.949	1.187	1.184
Fire	1.110	1.154	1.298	1.283	0.734	0.828	1.013	0.987
Services	1.133	1.147	1.104	1.118	0.856	0.862	0.994	0.923
Federal civilian	0.721	0.698	0.651	0.691	0.547	0.549	0.867	0.862
State and local government	0.913	0.934	0.996	1.054	0.883	0.904	1.110	1.035
Federal military	0.708	0.587	0.333	0.318	0.309	0.341	0.847	0.857

Source: Calculated from data in U.S. Department of Commerce, Regional Economic Analysis Division, "State Projections of Income, Employment and Prosecutions to 1990," in *Survey of Current Business* 54, no. 4 (Washington: USGPO, April, 1974) pp. 19-45, and U.S. Department of Commerce, State Personal Income 1974-75," in *Survey of Current Business* 56, no. 8 (Washington: USGPO, August, 1976), pp. 14-27.

Shifts in Economic Base and Structure

South Atlantic		East South Central		West South Central		Mountain		Pacific	
1969	1975	1969	1975	1969	1975	1969	1975	1969	1975
	0.067	0.400	0.333	0.133	0.067	16.467	14.667	0.267	0.200
2.826	2.239	4.435	5.043	0.043	0.239	0.739	0.935		0.022
0.087	0.149	0.522	0.463	8.783	6.776	2.457	2.328	0.783	0.642
1.176	1.222	1.294	1.167	1.353	1.278	2.059	2.778	0.765	0.667
1.020	1.100	1.111	1.048	1.148	1.263	1.170	1.360	1.005	1.212
0.816	0.679	1.188	1.087	0.971	0.963	0.916	0.931	1.050	1.032
4.117	4.025	1.874	1.638	0.165	0.188	0.010	0.025	0.117	0.150
1.189	1.258	2.315	2.381	0.709	0.887	0.197	0.320	0.449	0.588
1.336	1.414	2.180	1.802	0.977	0.928	0.938	1.108	1.742	1.793
0.976	1.000	1.387	1.389	0.783	0.800	0.085	0.200	0.717	0.726
0.669	0.724	0.742	0.690	0.613	0.648	0.620	0.648	0.767	0.752
1.359	1.326	1.963	1.598	1.238	1.293	0.190	0.310	0.429	0.408
0.152	0.180	0.413	0.580	3.370	3.620	0.717	0.660	1.326	1.080
0.517	0.429	1.366	1.164	0.412	0.475	0.445	0.589	0.446	0.461
0.360	0.621	0.651	0.928	0.632	0.903	0.375	0.390	0.993	0.718
0.330	0.363	0.617	0.701	0.571	0.734	0.441	0.498	0.614	0.634
0.462	0.504	0.806	0.794	0.525	0.589	0.485	0.581	1.184	1.198
0.261	0.273	0.394	0.505	0.192	0.212	0.069	0.096	0.330	0.273
0.778	0.597	0.695	0.886	1.409	1.221	0.557	0.577	2.310	2.027
0.552	0.711	0.691	0.865	0.535	0.474	0.380	0.807	0.524	0.734
0.965	0.977	0.976	1.165	1.128	0.746	1.106	1.111	1.072	1.008
0.958	0.986	1.042	0.971	1.113	1.160	1.048	1.073	1.065	1.010
1.338	0.969	0.842	0.771	0.883	0.993	0.941	0.964	0.314	0.976
0.958	0.999	0.629	0.853	0.975	0.966	1.118	0.993	1.136	1.078
1.989	1.924	1.333	1.203	1.142	1.037	1.705	1.551	1.185	1.084
0.947	0.986	1.071	0.905	0.984	0.930	1.234	1.172	1.250	1.146
2.080	1.951	1.521	1.278	1.743	1.632	2.483	1.709	1.576	1.480

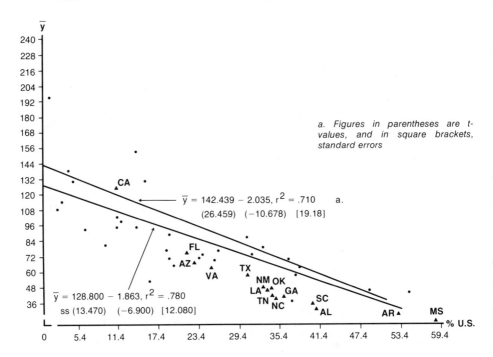

Figure 3-1a. Relationship between per Capita Income as Percentage of U.S. Total and Relative Share of Agricultural Employment, by State, 1904

The fit of the regression line for all states is moderately good. The coefficient of determination (r^2) is .71, and the t values for the constant term and slope are quite large—in both cases they are statistically significant well above the 1 percent level.

The relationship is the one postulated by Clark, Fisher, and Bean, that is, a downward-sloping regression line. The scatter diagram shows clearly that as dependence on employment in agriculture increases, relative per capita income declines. On average, if agricultural employment in a state amounted to more that 20 percent of total employment in 1940, per capita income was below the U.S. mean.

A second regression was run for the Sunbelt states alone.[40] The equation for the Sunbelt states fits slightly better than that for all states. The coefficient of determination rose to .78, and the standard error dropped from about 19 to 12. Once again, the t values are large enough to indicate high significance levels. The slope of the Sunbelt equation is slightly less than that for all states, but this is no surprise since all the Sunbelt states, except California, had per capita incomes well below the national average.

On purely statistical grounds one might argue that California should have been excluded from the Sunbelt analysis.[41] The most important argument for excluding California would be that only the southern half of the state is included in the generally accepted definition of the Sunbelt.[42] Comparable data on per capita income and employment are not readily available for the southern part of the state, however. And to leave California out of the analysis would have been to exclude from the Sunbelt an area with a population about equal to that of Texas.[43]

A second regression for the Sunbelt states expressed the variables as logarithms. The scatter diagram is shown as figure 3-1b.

The logarithmic fit is excellent with an r^2 of .97 and a standard error of estimate much smaller than the arithmetic relationship. The t values are large, indicating high levels of significance. The primary reason for running the logarithmic regression was the fairly obvious nonlinearity exhibited by the Sunbelt states in figure 3-1a.

Figure 3-1a and 3-1b supports the Clark-Fisher-Bean hypothesis of a negative relationship between relative per capita income and dependence on agricultural employment. Has this been a stable relationship? Figure 3-1c shows that it has not. When all states are considered, the scatter diagram shows a poor relationship between income and agricultural employment. The r^2 is only .15—about 85 percent of the variation is *not* "explained" by the regression equation. The slope is negative, but it is not steep. Its t value shows that this coefficient is barely significant.

The equation for the Sunbelt states is better. It "explains" about 42 percent of the variation, and the slope coefficient is slightly larger than 2. The latter is also statistically significant. But one would not want to make predictions on the basis of this regression equation.[44]

A number of states which depend more on agricultural employment than most of the Sunbelt states had substantially higher per capita incomes than the latter in 1975. South Dakota, the state with the largest relative dependence on agricultural employment, had a per capita income greater than that of half the Sunbelt states. And North Dakota, which ranked second in terms of agricultural dependence, had a higher per capita income than all the Sunbelt states except California in 1975. Clearly, something had happened to weaken the relationship between relatively low per capita income and a high degree of dependence on agricultural employment between 1940 and 1975.

Consider next the relationship between relative per capita income and "secondary" employment.[45] Figure 3-2a is the scatter diagram for 1940. The regression equation has the proper signs. On average, relative per capita income increases as the proportion of total employment in the secondary sector increases. Statistically, however, the relationship is not good. The coefficient of determination is only .19, and the standard error is

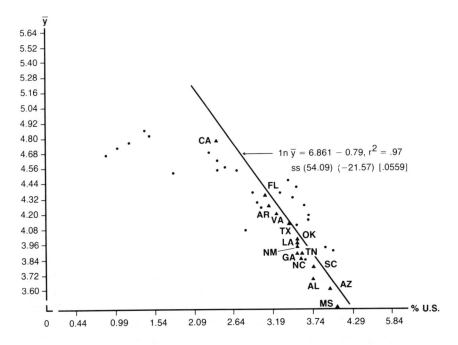

Figure 3-1b. Logarithimic Regression of Relative per Capita Income and Relative Share of Agricultural Employment, Sunbelt States, 1940

quite large. The t values for the constant term and slope are relatively small, although they are significant at the 1 percent level. Figure 3-2a does not deny the Clark-Fisher-Bean hypothesis, that relative per capita income increases with increasing dependence on secondary employment, but the statistical support is weak.

Much the same is true of the regression for the Sunbelt states. The r^2 is considerably larger (.60), but the standard error is also quite large, and neither of the t statistics is significant at the 5 percent level. As in the general case, the Sunbelt regression does not deny the Clark-Fisher-Bean hypothesis, but lends it only weak support.

One problem is that the high degree of aggregation conceals the heterogeneity of the broad sectors defined by Clark and Fisher. For example, North Carolina was the most industrialized of the Sunbelt states in 1940, yet its per capita income was only 55 percent of the national average. Meanwhile, Florida, which has much less industry than most of the Sunbelt states, had a per capita income only 15 percent less than the national average. Income in North Carolina was strongly influenced by the

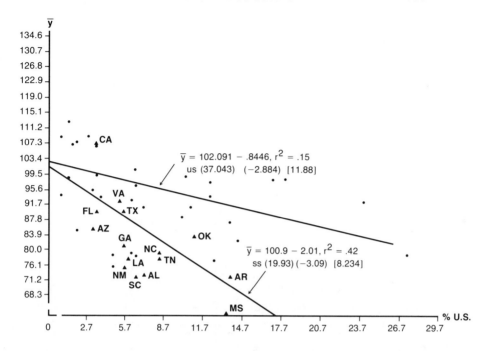

Figure 3-1c. Relationship between per Capita Income as Percentage of U.S. Total and Relative Share of Agricultural Employment, by State, 1975

preponderance of relatively low-wage manufacturing activities in the state—primarily textile mills and related activities—while Florida's income was favorably affected by its tourist industry which catered to relatively high-income groups.

What about the situation in 1975? Was it better or worse than that for 1940? Figure 3-2b shows at a glance tht it was much worse for the national case. The coefficient of determination was considerably higher than that for 1940 (.62), and the standard error was lower. Unfortunately, the slope coefficient has the wrong sign—it is negative when it should be positive. Also, it is not statistically significant at the 5 percent level. The same is true when the Sunbelt states are considered. The r^2 is quite low (.23); the regression equation is negatively sloped, and again it is not statistically significant at the 5 percent level. One belabors the obvious by stressing that figure 3-2b offers no support to the Clark-Fisher-Bean hypothesis.

Consider, finally, the relationship between relative per capita income and employment in the trades and services, or more accurately all other activities not covered in figures 3-1 and 3-2. Clark and Fisher postulated a positive relationship, and figure 3-3a shows that this relationship existed in

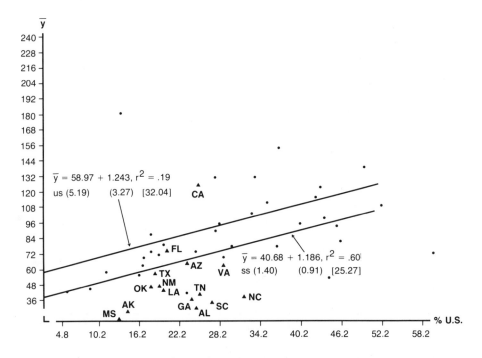

Figure 3-2a. Relationship between per Capita Income as Percentage of U.S. Total and Relative Share of Employment in Manufacturing, Mining, and Construction, by State, 1940

1940. The regression for the Sunbelt states is fairly good. About 76 percent of the variation is explained by the equation; the standard error is not unduly large, the slope term has the right sign, and it is statistically significant well above the 1 percent level. The Sunbelt states as a group lend fairly strong support to this aspect of the Clark-Fisher-Bean hypothesis. But when all states are considered, the statistical relationship is weaker. The r^2 drops to .45, and the standard error is more than twice as large as that for the Sunbelt states. If a regression had been run for all states except the Sunbelt states, there would have been virtually no correlation.

By 1975, as figure 3-3b shows, the relationship had weakened for the Sunbelt states and had improved only slightly for the nation as a whole. The slopes of both regression lines have the proper signs, but the national equation explains only 52 percent of the variation compared with 40 percent for the Sunbelt states. Again, one would hesitate to make projections based on either of these regressions.

What can be said, in summary, about the usefulness of a Clark-Fisher-Bean analysis? I think it is worth doing because it shows that the process of

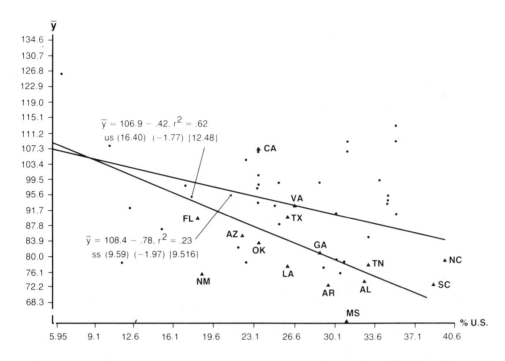

Figure 3-2b. Relationship between per Capita Income as Percentage of U.S. Total and Relative Share of Employment in Manufacturing, Mining, and Construction, by State, 1975

regional economic growth is far more complicated than simply one of moving from one "stage" of development to another. The experience of the Southeastern states, in particular, indicates that industrialization per se does not bring with it immediate advancement up the relative per capita income ladder. Also, above average dependence on agriculture does not mean that a state is destined to remain in the low per capita income category. While there might be occasional overproduction of specific crops, which will cause the prices of those crops to fall, the era of chronic agricultural surpluses in the United States, which extended from about the early 1920s until the early 1970s, appears to have ended. The rising worldwide demand for food and fiber is likely to ensure expanding markets for all we can produce.

Major changes in energy markets are also likely to have profound and lasting effects on the incomes of energy-producing states in this country. The ten states which specialize most in the production of energy belong to that group whose per capita incomes rose more rapidly than the national average between 1940 and 1975.[46] The impacts of rising energy prices on per

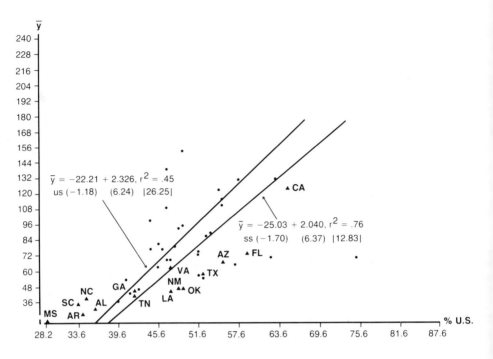

Figure 3-3a. Relationship between per Capita Income as Percentage of U.S. Total and Relative Share of All Other Employment, by State, 1940

capita incomes came, of course, during the last five years of this period. Since energy prices are likely to rise more rapidly than the general price level in the immediate future, per capita incomes in energy-producing states should continue to rise more rapidly than the national average. Several states with expanding energy sectors, such as Wyoming, North Dakota, Montana, and Kansas, are still heavily agricultural. They are likely to benefit from supply constraints, and rising prices, in both sectors.

When I first became interested in the process of regional growth, shortly after the end of World War II, I was convinced that regional development was a simple process. The Clark-Fisher-Bean hypothesis, which was quite new at that time, made sense. If a region wanted to prosper, it should industrialize. After an appropriate export base had been established, the region could afford the luxury of an expanding tertiary sector. The older regions of the East had already gone through this process and had achieved high levels of per capita income. All that remained was for the Southern and Western regions to catch up with their Eastern neighbors, and the economic millenium would soon be at hand.

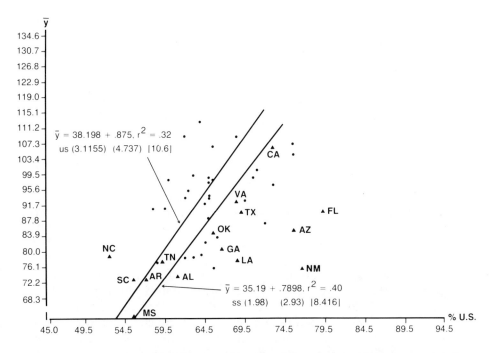

Figure 3-3b. Relationship between per Capita Income as Percentage of U.S. Total and Relative Share of Agricultural Employment, by State, 1975

All of us at that time had been caught up in the excitement of Keynes's grand new design for permanent prosperity everywhere. We had learned earlier about the scarcity of resources, but somehow this was forgotten. The problem at hand was finding employment for *surplus* human resources—or that had been the problem until World War II started. So in the 1940s, economists did not talk about supply constraints. Perhaps we should have paid more attention to some of the things Keynes wrote after the *General Theory* was published, however. In an article which appeared after his death, Keynes warned that we should not throw out all classical economic doctrine.[47] And the focus of classical economics was on the allocation of *scarce* resources.

Today, a number of economists are emphasizing scarcity. The optimists among them, like Herman Daly, have argued that we should prepare for a steady-state economy.[48] The steady-state economists evidently feel that resource constraints will not be a major problem for "a very long time." Others, such as Georgescu-Roegen, are less sanguine. They believe that we are already feeling the effects of some resource constraints and that these will become increasingly severe unless rigorous conservation measures are enacted and enforced.

Unfortunately, regional shifts in income, investment, and employment create problems. In the past, the problem was how to deal with areas which national economic growth had bypassed, or those which had been hit suddenly by an outmigration of industry. These problems are fairly easy to deal with, at least conceptually, within the context of a growing economy. This was the situation when area redevelopment legislation was passed in the 1960s.

The regional problems of today are quite different from those of the past. The national growth rate has slowed since 1970; in real terms it has been quite low. And the brunt of the slowdown has been borne by the nation's older industrial areas which have to pay more for the resources they consume or fabricate as well as the energy to fabricate them. In many markets the terms of trade have turned against the industrial states, and this condition is likely to persist.

Coalitions of governors and other state officials are being formed, particularly in the Northeast and the Midwest, to deal with their problems.[49] The leaders of older industrial areas will do what they can to protect the status quo. Meanwhile, political leaders in the South and West will tend to resist policies that could impair the comparative advantages that are in large part responsible for increased growth rates in these areas.

There is no simple or easy solution to this problem, such as that advocated by Keynes more than forty years ago. Available regional growth theory provides few useful policy guidelines. The problem today is not simply one of maintaining adequate aggregate demand. It is the much more difficult one of learning to live with supply constraints which are becoming an increasingly important fact of everyday economic life.

There are, of course, the perennial optimists who believe that all our present problems are transitory, that science and technology will "solve" even such difficult problems as dwindling supplies of fossil fuels. They are the heirs of the Victorians who developed the idea of progress, an idea which some of today's optimists believe it is fatuous to question. Let us hope they are right. But even the optimists are not talking about the immediate future. Solar energy and other technical fixes are seen to be "practical" and "cheap" only on the other side of a long-run horizon (usually some time in the next century). Meanwhile there are difficult problems of regional adjustment that will have to be dealt with in terms of present resources and present technology. These adjustments will require more than the vague hope that we can somehow step up the national growth rate so that growth will become geographically ubiquitous, and all of us can look forward to living happily ever after.

Notes

1. Harvey S. Perloff et al., *Regions, Resources, and Economic Growth* (Baltimore, Md.: Johns Hopkins Press, 1960).
2. R.A. Easterlin, "Interregional Differences in per Capita Income, Population and Total Income, U.S., 1840-1950," cited by Perloff et al., ibid., p. 114. Easterlin made separate calculations of per capita income from commodity production and commerce, from commodity production alone, and from agriculture. Each of these measures showed the West South Central region in first place in 1840.
3. Perloff et al., *Regions, Resources, and Economic Growth*, p. 120.
4. Ibid., p. 151.
5. Ibid., pp. 153-59.
6. Ibid., p. 276.
7. See Harry W. Richardson, *Regional Growth Theory* (London: Macmillan, 1973), pp. 237-53. For a discussion of the growth pole-growth center literature see David F. Darwent, "Growth Poles and Growth Centers in Regional Planning—A Review," *Environment and Planning* 1 (1969):5-32.
8. G.H. Borts and J.L. Stein, *Economic Growth in a Free Market* (New York: Columbia University Press, 1964). See also Borts, "Criteria for the Evaluation of Regional Development Programs," in *Regional Accounting for Policy Decisions* (Baltimore: Johns Hopkins Press, 1966), pp. 183-218; and Borts and Stein, "Regional Growth and Maturity in the United States," *Schweizerische Zeitschrift für Volkswirtschaft und Statistik*, 3 (1962):290-321.
9. Borts, "Criteria for the Evaluation of Regional Development Programs," p. 184.
10. Horst Siebert, *Regional Economic Growth: Theory and Policy* (Scranton, Pa.: International Textbook Company, 1969) p. 171. Siebert does not, incidentally limit his discussion of regional growth to a neoclassical model. When considering a regional policy, in particular, he introduces regional and interregional input-output models as potentially useful instruments for the implementation of regional growth policy. See pp. 176-85.
11. Borts and Stein, "Regional Growth and Maturity," see especially pp. 297-8.
12. Ibid., p. 311. Earlier (p. 300), they had defined "the process of economic growth [as] movements of resources to superior employments." Thus regional growth is at least partly a function of structural change.
13. Richardson, *Regional Growth Theory*, p. 22. Compare August Losch, *The Economics of Location* (New Haven, Conn.: Yale University Press, 1954), pp. 105-108. This idea was first advanced by Losch in "The

Nature of Economic Regions," *Southern Economic Journal* 5 (1938):71-78.

14. Richardson, *Regional Growth Theory*, pp. vii-viii.

15. For a generally laudatory review of Richardson's work, see Edwin von Boventer, *Urban Studies* 12 (February 1975):1-29. Borts, incidentally, feels that Richardson's regional growth theory is more of an extension of neoclassical regional theory than an indictment of the latter. See his review of Richardson's book in *Journal of Economic Literature* 12 (June 1974):546-47.

16. The term *demand* is used broadly here to include the demand for factors of production as well as final demand. One form of demand stimulus, as Borts suggests, would be labor subsidies to depressed areas to improve the mobility of labor. See G.H. Borts, "Criteria," pp. 205-206.

17. Richardson, *Regional Growth Theory*, p. 30.

18. Ibid.; compare Nicholas Kaldor, "The Case for Regional Policies," *Scottish Journal of Political Economy* 17 (1970):337-47.

19. Richardson, *Regional Growth Theory*, p. 31. This alleged differential rate of change in productivity among regions has been referred to as "Verdoorn's law" by Kaldor and Richardson.

20. An outstanding exception is Nicholas Georgescu-Roegen. For an excellent summary of his views see *Energy and Economic Myths* (New York: Pergamon Press, 1976), Chapter 1, pp. 3-36.

21. See William H. Miernyk et al., *The Regional Impacts of Rising Energy Prices*, Report to the Economic Development Administration (Morgantown, W.Va.: Regional Research Institute, 1977).

22. Perloff et al., *Regions, Resources, and Economic Growth*, pp. 58-60.

23. Harry W. Richardson, *Regional Economics* (New York: Praeger Publishers, 1969), p. 341.

24. See William H. Miernyk, *The Changing Structure of the Southern Economy*, Southern Growth Policies Board Occasional Paper no. 2 (Research Triangle Park, N.C.: The Southern Growth Policies Board, January 1977), pp. 2-8. See also Perloff et al., *Regions, Resources, and Economic Growth*, pp. 58-62.

25. Lewis H. Bean, "International Industrialization and per Capita Income," pt. V, *Studies in Income and Wealth* (New York: National Bureau of Economic Research, 1946), p. 121.

26. Perloff et al., *Regions, Resources, and Economic Growth*, p. 60. The ideal vehicle for analyzing structural change would be a set of multistate input-output tables covering a period sufficiently long to show changes in technical relationships and trade patterns. Data collected for such tables could be aggregated to provide regional and national tables.

27. One need not, however, accept the Clark-Fisher-Bean conclusion that progressive shifts from primary to secondary and then to tertiary ac-

tivities always mean concomitant improvement in economic well-being. On this, see William H. Miernyk, "Labor Mobility and Regional Growth," *Economic Geography* 31 (October 1955):321-23. See also Seymour E. Harris, *The Economics of New England* (Cambridge, Mass.: Harvard University Press, 1952), p. 286; and Jean Fouristè, *le Grand Espoit du XXe Siecle* (Paris: Presses Universitaires de France, 1950), pp. 92-93.

28. *Economic Report of the President*, Government Printing Office (GPO), Washington, D.C., January 1977, pp. 188-189, 217, and 228.

29. Ibid., p. 292.

30. Ibid., p. 228.

31. In 1790 the center of population was close to Baltimore, Maryland. By 1970, it was near the Illinois-Missouri border. See Bureau of the Census, U.S. Department of Commerce, *Statistical Abstract*, Government Printing Office (GPO), Washington, D.C., 1975, p. 9.

32. *Washington Post*, May 29, 1977.

33. Perloff et al., *Regions, Resources, and Economic Growth*, pp. 3-4.

34. A major weakness of per capita personal income data is that they are available only in money terms. The limited evidence available on geographic differences in the cost of living, or on family budgets, suggests that increases in money income in the South understate increases in real income, while the reverse is true of the North. Unfortunately, available data are limited to urban areas, and the budgets that have been compiled by the Bureau of Labor Statistics are normative rather than average budgets, so that accurate comparisons of real income differentials among regions cannot be made. For what they are worth, the most recent data on urban budget differentials are summarized in an article by Jill Goldman, "It Matters Where You Live," *MBA*, June 1977, p. 32. Such data are published on an irregular basis, generally in the form of press releases, by the Bureau of Labor Statistics.

35. One state—Wyoming—had above average per capita income in both 1940 and 1975. The location quotient, for those not familiar with this statistic which is widely used by regional economists, measures the relative importance of an economic activity in a state (or other area for which the LQ might be calculated) compared to the importance of that activity in the nation. For further details see Miernyk, *The Changing Structure of the Southern Economy*, pp. 13-14.

36. Location quotients are often used to "adjust" the technical coefficients of national input-output tables to state or local levels when shortcut methods are employed to construct state and local input-output tables. The example in the text, which could be repeated for other sectors where production data are available, shows how misleading the results of such estimating techniques can be.

37. The employment LQs are for selected sectors only, and the two

tables cover different periods because income data in this disaggregated form have been available only since 1969.

38. Kaldor, "The Case for Regional Policies," p. 341.

39. It is interesting that all the agricultural states in the East North Central and West North Central regions had higher agricultural location quotients in 1975 than in 1940. This is also true in Oklahoma, Montana, Idaho, Kentucky, and Wyoming. But agricultural states in the South, such as Alabama, Mississippi, Arkansas, and Louisiana, registered marked declines in agricultural location quotients. Increases in location quotients do not mean, however, that the agricultural sectors of those states have expanded. Location quotients are *relative* measures, and the increases noted reflect in part the relative contraction of agricultural sectors in other states.

40. Washington, D.C., was excluded because it has no employment in agriculture.

41. In many ways California can be considered as an "independent" region. It stands out in size, and the state economy is highly diversified. California's population is exceeded by only thirty-four independent countries, including the United States, while it is larger than 124 countries.

42. See Gurney Breckenfeld, "Business Loves the Sunbelt (and Vice Versa)," *Fortune*, June 1977, pp. 133-46.

43. One can only hope that variances within California are lower than those between California and other Sunbelt states.

44. Both the U.S. and Sunbelt regressions for 1975 were run in logarithmic terms although the results are not shown graphically. There was some improvement with tightening of the scatter diagram, particularly for the Sunbelt states. But the coefficient of determination for the forty-eight states was only .29, and the regression line had a slope of only .08. The logarithmic regression for the Sunbelt states was somewhat better, with an r^2 of .49 and a standard error of .09. But the scatter diagram still shows a weak relationship.

45. Employment in the secondary sector, following Bean's definition, includes manufacturing, mining, and construction.

46. See William H. Miernyk, "Regional Economic Consequences of High Energy Prices in the United States," *Journal of Energy and Development* One (Spring, 1976):323-33.

47. Keynes, "The Balance of Payments of the United States," *The Economic Journal* June 1946, pp. 172-87.

48. See Herman E. Daly, *Toward a Steady-State Economy* (San Francisco: W.H. Freeman, 1973), pp. 1-27.

49. See Robert W. Rafuse, Jr., *The New Regional Debate: A National Overview* (Washington: National Governors' Conference, Center for Policy Research and Analysis, April 1977).

4

Regional Differences in Factor Costs: Labor, Land, Capital, and Transportation

David L. Birch

Introduction

Factor costs have long been considered important determinants of the location of economic activity. The basic notion is that the business executive looks at the firm's income statement; notes the importance of such expense items (or factors) as payroll, property depreciation, shipping costs, the cost of borrowed money, and, for some, energy costs; and locates the firm's activities so as to minimize such costs and thereby maximize profit. The purposes of this chapter are, first, to determine whether regional differences exist in four of these expense items—labor, land, capital, and transportation—and, second, to see, whenever possible, whether any relationship exists between factor cost differences and growth patterns.

For all the mention of factor costs in economic theory, one would expect to find a rich literature describing factor cost differentials and their role in location analysis. To our surprise, and to the surprise of others who have searched the literature lately, with a few notable exceptions, no such body of writings exists. Very few researchers have compiled detailed, empirically based regional comparisons of factor costs and analyzed their role in influencing location decisions. After pursuing the subject for a while, we see the reason for this paucity of analysis: factor costs have been poorly measured in the past. With the major exception of wage rates, data on factor costs have not been systematically compiled for a comparable set of places and time periods. Many such data have only recently been available in any useful form. Those who have attempted to understand regional differences in cost elements have been forced, therefore, to piece together scraps of largely uncomparable data to reach any conclusions at all.

Before we plunge into our own analysis, it is important to state precisely what is and is not being explored. A first, and frequently unasked, question is, Factor cost to whom? The economist usually assumes that factor costs influence the location of economic activity (often just manufacturing) and that it is the costs to firms that should be considered. Any student of migration and the interregional shifts of jobs and people quickly learns, however, that the redistribution of people and jobs is hopelessly intertwined, and factor costs to people may be just as important as factor costs to firms. Commonly labeled "cost of living," such costs include the cost to the individual (or standardized family) of housing, food, transportation, entertainment, and so forth. Since this chapter attempts nothing so ambitious as an integrated explanation of regional change, these factor costs to individuals are ignored. Those taking a broader view, though, must treat them at least as seriously as the costs to firms.

Within the firm population we find great variations in the relative importance of different expense items for different kinds of firms. The only solid data we can obtain across industrial sectors come from the Census Bureau's *Enterprize Statistics*, which reports payroll and sales for various groups of firms:[1]

Sector	Payroll as a Percentage of Sales
Manufacturing	22%
Wholesale trade	5
Retail trade	12
Selected services	29
Construction	24
Mining	14
Central administrative offices	47
All firms	16

The variation in the importance of labor costs is obvious, and we can postulate similar variations in the roles played by land, capital, and transportation as we think about the needs and functions of firms in these different sectors. We should, therefore, expect to find different classes of firms responding differently to the same pattern of factor cost differentials.

Another difficult issue is causality. The pure economic theory postulates a profit-maximizing executive locating to minimize factor costs. Such an executive should relocate a firm every time the combined discounted value of factor cost savings is greater than the cost of moving to the area with the lower costs, assuming constant prices for the firm's products. The limited interview and directory data available suggest a far more complex phenomenon. First, virtually no firms relocate their operations across metropolitan or rural area boundaries. Of a sample of 5.6 million firms analyzed by the author,[2] on the average only 0.03 percent relocate across such boundaries in any particular year. This is in comparison to 8 to 10 percent per year shifts in the business population as a result of firms forming and expanding. Furthermore, for any particular area, the very small flow out tends to very nearly equal the flow in, leaving a truly negligible net change attributable to migration. This finding is true in all sections of the country for different periods in the 1970s.

This evidence suggests either that discounted cost savings are smaller than moving costs or that the simplistic model of the cost-minimizing executive needs closer examination, or both. Shortly we examine factor cost differences to see if they are large enough to make a significant difference. Before doing so, however, we should consider carefully our model of executive behavior.

Looking more closely at how employment relocation does take place in our sample of 5.6 million firms, we notice that underlying relatively gradual net change are rather large gross flows, which can be broken into births, expansions, and inmigrations on the gain side and deaths, contractions, and outmigrations on the loss side. We know already that in- and outmigration are negligible. Interestingly, the other two loss components (contractions and deaths) in combination are more or less the same across all states regardless of how fast states are growing or declining. In other words, the processes by which firms go out of business or contract are not particularly sensitive to where the firm is located. They have more to do with management skill. This finding holds below the state level and for several different industrial sectors.

The explanation of net gain or loss in employment, therefore, lies almost entirely in the rate at which new firms and expanding firms make up for the relatively fixed losses occurring everywhere. As can be seen in table 4-1, the great majority of employment gains are attributable to independent, unaffiliated firms, not to the branches or subsidiaries of large firms of which the rhetoric of regional disparities is so filled. Roughly two-thirds of all new jobs created are created by independent entrepreneurs who have started up their own companies.[3] We do note, to be sure, that Southern expansions have a significantly greater tendency than expansions in other parts of the country to take place in branches of firms headquartered out of state, which is consistent with the model of differential branch expansion in the South by Northern and Western firms. This differential expansion, however, amounts to only 11 percent of the total growth in the South and explains only a small part of the economic prosperity of that section of the country. This pattern is remarkably invariant over time, with a slight tendency at the margin for entrepreneurial activity to increase rather than decrease in importance. If factor costs are to play a role, therefore, they do so not by inducing firms to migrate or establish branches, but mostly by influencing where independent entrepreneurs decide to locate and expand facilities. This being the case, the attitudes and motivations of such entrepreneurs and the role played by factor costs in their calculations become crucial.

Evidence on entrepreneurial motivations is very thin indeed. The data assembled in this chapter suggest that, in some cases, a direct and strong relationship exists between low factor costs and the location of new and expanding firms, that in other cases no strong relationship exists, and that in

Table 4-1
Status of Firms versus Employment Gains by Region, 1969-1972, 1972-1974, 1974-1976

| | | Births | | | | | |
| | | | Percent Employment Gains in Firms that Are: | | | | |
Region	Period	Inde-pendent	Head-quarters	Subsi-diary	Branch/HQ in State	Branch/HQ out of State
Northeast	1969-1972	77.4	12.2	10.2	0.1	0.1
	1972-1974	81.6	9.5	8.9	0.0	0.0
	1974-1976	87.3	7.4	5.2	0.0	0.1
North Central	1969-1972	79.9	12.7	7.2	0.1	0.1
	1972-1974	83.4	9.6	6.9	0.1	0.0
	1974-1976	88.8	6.4	4.7	0.1	0.0
South	1969-1972	78.3	11.8	9.7	0.1	0.1
	1972-1974	83.7	9.2	6.9	0.1	0.1
	1974-1976	88.9	5.8	5.1	0.1	0.1
West	1969-1972	80.7	11.0	8.1	0.1	0.1
	1972-1974	87.1	8.0	4.9	0.0	0.0
	1974-1976	89.5	6.4	4.0	0.1	0.0
		Expansions				
Northeast	1969-1972	62.9	16.7	4.3	4.4	11.7
	1972-1974	55.9	20.6	5.9	5.7	11.9
	1974-1976	57.9	21.4	6.8	4.2	9.7
North Central	1969-1972	58.2	15.4	3.0	8.1	15.3
	1972-1974	55.4	20.8	4.7	6.0	13.1
	1974-1976	54.3	21.2	5.0	6.3	13.2
South	1969-1972	59.2	13.3	4.8	4.2	18.4
	1972-1974	56.0	16.0	5.0	3.7	19.3
	1974-1976	54.2	17.4	5.8	4.6	18.0
West	1969-1972	60.3	15.8	3.1	7.5	13.3
	1972-1974	58.1	21.1	3.8	6.0	11.0
	1974-1976	56.7	22.3	4.7	5.3	11.0

Regional Differences in Factor Costs 129

still other cases high factor costs and high employment growth go hand in hand. In these latter situations a far more complex model must be employed to explain firm location, allowing for quality-of-life considerations and costs not considered in this chapter (for example, energy costs and taxes). Since our purpose here is not a coherent explanation of growth but rather the analysis of four very specific factor costs, the complexities of how other aspects of location enter the picture are not considered in detail. They are raised here only to warn the reader about the hazards of relying on simpleminded correlations between factor costs and growth in trying to explain that growth.

Recent Shifts in Population and Jobs

The differential growth of various parts of the United States has been well documented by the Census Bureau and consumers of its products. For comparative purposes we thus simply summarize a few of the salient facts. The first, and most dominant, regional features are the emergence of the West and more recently the South (see table 4-2). While the South's growth has been consistently high since the mid-1950s, its dominance in recent years is clear from the table.

A more recently noted shift is concerned not so much with regions as with the scale of places lived in. Since the Industrial Revolution began, we have, as a nation, flocked to our cities as the centers of economic activity. Now, for all the reasons noted, these centers offer no comparative advantages for some firms and households and are considered congested and obsolete and crime-ridden by many. In response, people and jobs are leaving, or not settling there in the first place (particularly in the case of firms), causing a significant redistribution. This phenomenon has revealed itself in data series mostly since 1970, as these comparisons of Census data show:

Type of Area[4]	Net Migration (Millions of People)	
	1960-1970	1970-1974
Large metropolitan	4.2	−1.2
Small and medium metropolitan	2.2	1.7
Nonmetropolitan	−3.2	1.5

These shifts in net migration have influenced aggregate levels of population in places of different sizes. To probe these shifts in greater detail, we divided the United States into 315 areas[5] and grouped them into five categories: (1) supercities, (2) regional centers, (3) large metropolitan, (4) small metropolitan, and (5) rural.[6] Only in the South have cities of any great size continued to grow at their pre-1970 levels (see table 4-3). In all

Table 4-2
Net Change in Population by Region, 1950-1975

Five-Year Period	Northeast	North Central	South	West
1950-1955	2.7	3.9	2.9	3.7
1955-1960	2.5	3.2	5.0	4.2
1960-1965	2.6	2.5	4.4	3.9
1965-1970	1.7	2.4	3.4	2.7
1970-1975	0.3	0.9	5.1	2.9
Total (1965-1975)	9.8	12.9	20.8	17.4

other regions, the trends in the aggregates reflect the reversal of migration flows as people move toward smaller-scale environments.

Paralleling shifts in population have been changes in the location of employment. Recent employment data are more difficult to obtain for the entire firm population. Our estimates based on an 82 percent sample, however, reveal a pattern similar to the population movements (see table

Table 4-3
Population Changes by Region and Area Type, 1960, 1970, and 1974

Region/Area	Population (Millions)			Annual Percentage Change	
	1960	1970	1974	1960-1970	1970-1974
Northeast					
Supercity	15.6	17.3	17.1	1.0	-0.3
Regional center	12.5	13.7	13.9	0.9	0.4
Large metropolitan	10.0	11.1	11.4	1.0	0.7
Small metropolitan	1.1	1.1	1.1	0.0	0.0
Rural	4.8	5.0	5.5	0.4	1.5
North Central					
Supercity	6.6	7.5	7.5	1.3	0.0
Regional center	14.3	16.2	16.2	1.3	0.0
Large metropolitan	9.3	10.5	10.8	1.2	0.7
Small metropolitan	4.1	4.5	4.6	0.9	0.6
Rural	15.9	16.2	16.8	0.2	0.9
South					
Supercity	—	—	—	—	—
Regional center	9.8	13.3	14.5	3.1	2.2
Large metropolitan	14.4	17.1	18.7	1.7	2.3
Small metropolitan	6.0	6.8	7.3	1.3	1.8
Rural	22.0	22.3	23.6	0.1	1.4
West					
Supercity	10.5	13.2	13.6	2.3	0.7
Regional center	3.1	4.1	4.4	2.8	1.8
Large metropolitan	5.1	6.8	7.6	2.9	2.8
Small metropolitan	1.8	2.2	2.5	2.0	3.2
Rural	5.0	5.5	6.1	1.0	2.6

Note: These population figures do not agree exactly with Census Bureau totals because people in the military or in group quarters have not been included.

4-4). Rural areas are consistently experiencing greater employment gains, particularly in the trade and service sectors, to support growth in population.

Having noted these shifts in jobs and people, and having forewarned ourselves about the complexities and dangers inherent in relating factor costs to growth, let us now proceed to an analysis of the costs themselves. Wage differences are treated first, followed by land, capital, and transportation.

Cost of Labor

As we have seen already, labor cost is one of the major items in the income statements of many firms, particularly for firms in the service sector and in office complexes generally. That is probably a major reason why it is by far the most thoroughly measured factor cost. There are still holes, however. Recent wage data, for example, are complete only for manufacturing. Also, no account is taken for variations in efficiency or employer-contributed benefits. But, by and large, the coverage during the past several decades is quite good.

One benefit of the Bureau of Labor Statistics' (BLS) extensive reporting system is the ability it gives us to compare wages for the same small areas that businesses choose among in making location decisions. We are not limited to the much more aggregate regions for which most factor-cost data are provided. We have capitalized on this detail by utilizing our 315 small areas. These areas may be sorted by size and growth rate as well as by region, permitting much finer resolution. We can thus explore the distribution of opportunities facing a firm as a firm sees them.

The first thing we notice is the lower wage rates in the South. Table 4-5 presents the average hourly wage by industrial sector by region. For every sector of the economy studied, the South has the lowest average wages, of-

Table 4-4
Net Percentage Change in Employment for Metropolitan and Rural Areas by Economic Sector

Industry	Net Percentage Change[a], 1970-1972	
	Metropolitan	Rural
Agriculture	−3.6	−1.7
Manufacturing	−10.0	−5.5
Other industry	−2.7	0.2
Trade	2.7	14.0
Service	1.6	8.3
Total	−3.9	1.5

[a]These net change figures have not been adjusted for a known underreporting of new firm formation. They are nevertheless consistent, although low, across the board.

Table 4-5
Average Hourly Wage by Industrial Sector by Region in 1967

Sector	Northeast	Midwest	South	West
Retail trade	$1.84	$1.74	$1.63	$1.93
Wholesale trade	2.94	2.86	2.55	2.97
Manufacturing	3.00	3.05	2.54	3.12
Service	2.00	1.75	1.63	1.92

fering a firm looking for inexpensive labor many places to choose among. Wage rates in the West, in contrast, appear to be the farthest above the U.S. average, but this difference is not nearly so dominant, in most cases, as the lower rates in the South.

Since the figures in table 4-5 are now ten years out of date, an obvious question is, What is happening to this gap? One way to gain some feeling for the trends over time is to look at changes during the 1960s. We computed the compound annual growth rate of wages by economic sector by region (see table 4-6). We see that, in general, wages in Southern areas tend to be growing faster, but not always. The big gap in manufacturing appears to be closing somewhat, for example. There is some tendency also for Southern workers in wholesale trade and services to be gaining on their counterparts elsewhere. The West appears to be moving in just the opposite direction, compensating for its tendency to pay higher than average wage rates.

For manufacturing wage rates, where more recent data are available at the state level, we can see how these differences are working themselves out in the present. The average hourly rates for each region (the best we can do without small-area data) in 1975 are as follows:

Region	Average Rate
Northeast	$4.41
North Central	5.07
South	4.23
West	4.95

Southern rates are obviously converging on Northern ones, although some difference still exists, and the West has dropped below the North Central area, bringing its rates more in line.

Regional breakdowns, while the focus of this chapter, are somewhat arbitrary and tend to mask several distributional phenomena at work. One of the most interesting trends in recent years has been the shift from higher to lower densities and from larger- to smaller-scale environments. Table 4-7

Regional Differences in Factor Costs

Table 4-6
Average Annual Percentage Change in Wages by Industrial Sector by Region, 1962-1967

Sector	Northeast	Midwest	South	West
Retail trade	4.8%	4.1%	3.8%	3.3%
Wholesale trade	4.0	4.1	4.9	4.5
Manufacturing	3.5	2.6	3.3	1.5
Service	5.3	5.9	6.5	3.7

presents wage rate comparisons along our five-point "urbanness scale" in the same format as table 4-5. There is little doubt about the outcome. The larger the scale of the place, the higher its wages. This is true for all sectors and all points along the urban-scale dimension. Apparently workers in larger labor markets are able to organize and bargain with greater effect. Conversely, workers in thinner markets have fewer options and are forced to settle for less. The South, with its predominance of rural areas and population, falls mostly on the low end of the scale, and it is not surprising that its rates are lower.

To see whether the redistribution of population and jobs to less dense, smaller-scale environments is narrowing the gap, we looked at rates of change (see table 4-8). The pattern is, at best, mixed. There is no clear tendency for rates in one type of area to be gaining more or less rapidly than in others. The period being examined (1962-1967) may explain part of this ambiguity, since the trend toward more rural areas gained much of its momentum in the late 1960s and early 1970s, but the figures are comparable in time to those in table 4-7, suggesting that many of the regional shifts in wage rates in the future will be caused by movements of areas up and down the urbanness scale, as well as the spread of organized bargaining activity, rather than the area's relationship to the Mason-Dixon line or the Mississippi River.

It is a commonly held notion that employers seek out areas with lower

Table 4-7
Average Wage by Industrial Sector by Area Type, 1967

	Area Type				
Sector	Super-city	Regional Center	Large Metropolitan	Small Metropolitan	Rural
Retail trade	$2.12	$1.88	$1.80	$1.77	$1.66
Wholesale trade	3.67	3.42	3.07	2.83	2.47
Manufacturing	3.65	3.48	3.17	2.87	2.58
Services	2.63	2.15	1.89	1.75	1.62

Table 4-8
Average Annual Percentage Change in Wages by Industrial Sector by Area Type, 1962-1967

Sector	Area Type				
	Super-city	Regional Center	Large Metropolitan	Small Metropolitan	Rural
Retail trade	4.3%	3.9%	4.1%	3.9%	3.9%
Wholesale trade	3.1	3.6	4.2	3.7	5.1
Manufacturing	3.9	3.0	3.9	.1	3.7
Services	4.1	5.2	5.3	5.4	6.2

wage rates and that these areas consequently grow. We wished to test this notion, so we sorted areas according to how rapidly their aggregate employment grew between 1960 and 1970 and examined corresponding wage rates (see table 4-9). The results certainly do not support the model of employers seeking cheap labor. With great regularity, faster growth is taking place in areas with higher rather than lower average wages. Furthermore, the simple explanation that these more rapidly growing places are out of equilibrium, and that the tightness of their markets causes their wage rates to rise faster, is not consistent with the evidence. As table 4-10 shows, faster-growing places tend to have average, or even less than average, rates of wage rate inflation for three different wage levels.

These findings so jar our basic notion of firms minimizing (rather than maximizing) factor costs that we demand a more detailed explanation. Averages, of course, can mask a great deal of clustering and thereby lead to false inferences. The first step we took in unraveling this puzzle was to identify places with unusually high or low wage rates (relative to the U.S. mode) and unusually high or low rates of growth. We did this for two quite different industrial sectors—manufacturing and service. Of the eight possible categories (high-low wages by high-low change for service and manufacturing) only five contained more than a few areas. In other words, there is considerable clustering. The five heavily populated cells are:

1. Low manufacturing wages and low growth
2. Low service wages and low growth
3. Low manufacturing wages and high growth
4. High manufacturing wages and high growth
5. High service wages and high growth

The first two groups overlap a great deal, and are predominantly rural, mostly Southern areas. The third group is the stereotype that we had expected would dominate the averages. It is the group of places (again mostly

Table 4-9
Average Wage by Industrial Sector by Rate of Aggregate Employment Change, 1962-1967

Sector	Employment Change, 1962-1967				
	−12%−−1%	0%-10%	11%-20%	21%-35%	36%-100%
Retail trade	$1.64	$1.65	$1.72	$1.78	$1.83
Wholesale trade	2.31	2.53	2.71	2.92	3.05
Manufacturing	2.52	2.60	2.88	2.92	3.11
Service	1.54	1.59	1.74	1.81	2.00

Southern) that are attracting manufacturing (and related) growth by offering inexpensive labor.

The fourth and fifth groups are the ones that do not fit the offer-low-wages model. Included are places like Orlando, Atlanta, Houston, Denver, Dallas, Seattle, Phoenix, and San Diego, all of which are growing rapidly and pay significantly higher than average wages. Because of their size and importance, and because of the large number of low-wage areas that are growing slowly, the high-wage, high-growth areas dominate the averages.

We return to these five groups shortly. It is worth noting first, however, that virtually no areas fall into the cells defined by high manufacturing wages and low change (apparently high wages are not by themselves a deterrent to growth), low service wages and high change (service firms apparently do not seek out low-wage areas the way some manufacturers do), and high service wages and low change (again, high wages do not appear to act as an explicit deterrent).

Table 4-10
Average Annual Percentage Change in Manufacturing and Service Wages by Wage Level in 1967 and Rate of Aggregate Employment Change, 1962-1967

1967 Average Manufacturing Wage Level	Area Employment Change, 1962-1967				
	−12%−−1%	0%-10%	11%-20%	21%-35%	36%-100%
0-3.00	4.1%	4.4%	4.2%	4.8%	3.5%
3.01-3.50	6.0	4.0	4.2	4.0	3.2
3.51+	—	2.8	3.6	3.8	3.9
1967 Average Service Wage Level					
0-1.50	6.3%	5.6%	6.1%	5.1%	2.4%
1.51-2.00	6.2	5.8	6.3	5.4	4.4
2.01+	—	6.7	5.3	5.8	5.9

A possible explanation for the high-wage, high-growth syndrome lies in the mix of industries on which the growth is based. Many of these places have attracted sophisticated, high-technology firms in industries such as aerospace, computer, electronics, or, in the service sector, in higher education or finance. To test this postulate, we turned to the Census Bureau's Public Use Sample Tape, which contains a 1 percent sample of the U.S. population (over 2 million individuals) that can be traced geographically to areas closely related (and in most cases identical) to our 315 subareas. Using this data file, we explored the mix of industries and occupations in each of the groups of areas that emerged from our search for clusters.

The first step was to streamline the clusters. Groups 1 and 2, which overlapped considerably, were merged, retaining those areas in the sample whose wages were low in both manufacturing and service and that grew slowly. Some overlap between groups 4 and 5 was resolved by placing the few rapidly growing areas with both high manufacturing and high service wages into the manufacturing group. Finally, the few ambivalent areas that paid high wages in one industrial sector and low wages in the other were dropped from the sample. The resulting four groups of areas are:

1. Low wages generally, low growth
2. Low manufacturing wages, high growth
3. High manufacturing wages, high growth
4. High service wages, high growth

The areas included in each category are presented in appendix 4A.

The next step was to group industries and occupations in a way that would most effectively reveal differences across groups. Industries were divided into:

1. Sophisticated manufacturing
2. Sophisticated services
3. Other manufacturing
4. Other services
5. Other [including agriculture, mining, construction, transportation (except air), and utilities]

The list of manufacturing and service industries designated as "sophisticated" because of their high technology or level of required skills is presented in appendix 4B. Occupations are grouped along a more traditional dimension reflecting degree of skill and/or education involved:

1. Professional
2. Managerial

3. Sales and clerical
4. Craftsman
5. Operatives
6. Other (including laborers and service workers)

The first thing we notice from the nationwide sample is that, as we suspected, the high-wages, high growth areas are more heavily dependent on the more sophisticated industries—by a factor of 3 or 4 in the case of manufacturing (see table 4-11). We would thus expect a better educated, higher-skilled, and higher-paid workforce, and, as can be seen in table 4-12, this is exactly what is found. The high-wage, high-growth areas have a substantially greater proportion of their workforce classified as "professional," by a factor of about 2, and a corresponding lower proportion on the low end of the scale. We are not surprised, therefore, to find that such places pay higher than average wages.

A key question remains: Are wages in high-wage, high-growth areas higher simply because their mix of occupations favors the high end of the scale, or, in addition, do they pay more for any given occupational group? Said another way, Are sophisticated manufacturers actually minimizing costs by choosing places with lower-paid engineers (even though engineers as a group are highly paid), or are they growing in places inhabited by more expensive than average engineers? The striking thing (see table 4-13) is that they are paying more—considerably more—for each occupational group than the lower-wage areas, and more than the U.S. average for these groups. In other words, growth in these high-wage areas is taking place at a rapid pace despite the high wages.

One final explanation for this phenomenon is that, under the broad occupational groupings used in table 4-13, we have buried diverse groups. The group labeled Professional, for example, includes school teachers (who are notoriously underpaid) and scientists. The high-technology places could

Table 4-11
Percentage of Labor Force by Industry by Class of Place

	Class of Place			
	Low Wages Generally/ Low Change	Low Manufacturing Wages/ High Change	High Manufacturing Wages/ High Change	High Service Wages/ High Change
Sophisticated manufacturing	1.7%	2.5%	9.2%	6.3%
Sophisticated services	11.4	16.1	17.8	17.2
Other manufacturing	23.6	21.5	20.0	13.1
Other services	38.5	40.6	38.6	48.9
Other	24.8	19.4	14.3	14.4

Table 4-12
Percentage of Labor Force by Occupation by Class of Place

	Class of Place			
Occupation	Low Wages Generally/ Low Change	Low Manufacturing Wages/ High Change	High Manufacturing Wages/ High Change	High Service Wages/ High Change
Professional	11.0%	13.1%	19.4%	22.0%
Managerial	10.6	12.0	12.0	13.4
Sales and clerical	15.7	20.1	22.6	23.2
Craftsmen	19.2	20.8	18.4	16.0
Operative	21.7	18.0	14.6	11.5
Other	21.8	16.1	13.0	13.9
Total	100.0	100.0	100.0	100.0

have a higher proportion of scientists in their professional category, and that might explain the variation. To check on this proposition, we repeated the analysis in table 4-13 for five particular occupational groups: computer specialists, engineers, scientists, scientific technicians, and office and nonretail sales managers. The results (see table 4-14) exhibit a pattern very similar to that in table 4-13. Even after controlling for very particular occupations, the high-wage, high-growth areas pay more per occupation than other areas.

We must thus address the question, Why should firms be flourishing in areas whose wages are higher across the board than the U.S. average and much higher than some low-wage areas that are also growing? We have no way of answering the question definitively without interviewing a large sample of the firms involved. We can observe, however, that the high-wage

Table 4-13
Average Annual Wage by Occupation by Class of Place

	Class of Place				
Occupation	Low Wages Generally/ Low Change	Low Manufacturing Wages/ High Change	High Manufacturing Wages/ High Change	High Service Wages/ Low Change	U.S. Average
Professional	$8,642	$10,199	$12,208	$12,622	$11,233
Managerial	10,088	12,014	14,030	14,109	13,129
Sales and clerical	6,525	7,374	8,343	8,273	8,088
Craftsmen	6,895	7,605	9,603	9,049	8,865
Operative	5,796	5,857	7,964	7,291	7,163
Other	3,982	4,649	6,443	5,819	5,644

Table 4-14
Average Annual Wage by Special Occupation by Class of Place

		Class of Place			
Occupation	Low Wages Generally/ Low Change	Low Manufacturing Wages/ High Change	High Manufacturing Wages/ High Change	High Service Wages/ Low Change	U.S. Average
Computer Specialist	$8,026	$10,537	$12,215	$11,962	$11,373
Engineer	11,774	12,400	14,726	15,158	13,782
Scientist	11,449	12,389	13,444	14,706	12,666
Technician	7,267	8,655	10,162	9,899	9,394
Office manager	10,590	12,856	15,152	14,460	14,179

places are typically viewed as especially attractive places to live. Orlando, Houston, Denver, Seattle, Phoenix, San Diego, Los Angeles, San Francisco, Atlanta, Dallas, and Santa Barbara all boast about their attractive climates, outdoor recreational opportunities, and/or sophisticated urban settings. Perhaps firms are locating there to experience these advantages, or perhaps they feel that, by being there, they can attract a higher-caliber person for less money than that same person would demand in a less attractive setting, and thereby offset the higher wages with higher efficiency. Whatever the explanation, we certainly cannot ignore the high-wage, high-growth phenomenon because of its magnitude and the challenge it poses to our traditional notions of minimizing factor costs. Until we get much better measures of efficiency and productivity, we will not be able to sort out whether aesthetics buys greater output per dollar or whether people are simply willing to absorb higher costs per unit of output in order to live in a place that they find physically attractive. Resolving this issue should be high on our research agenda.

Land Cost

As we move from an agricultural, and more recently a manufacturing, economy into an era dominated by service industries, raw land becomes much less important as a factor cost. Cost per square foot of the office or store space occupied by service firms, not cost per acre of land, becomes the major property-related expense item. The cost of such office space depends critically on its precise location, the architectural style of the building, its landscaping, and so forth in addition to the raw land cost.

Land costs still enter the picture for some forms of industry that require large single-story plants and parking lots, however, and it certainly is a major factor in the cost of housing, and the corresponding cost of living. Indirectly, then, it may affect labor costs as well as property costs and taxes. We should not, therefore, ignore it.

Unfortunately, like so many factor costs, the cost of land is not easily measured on a region-by-region basis over time. Few reporting systems converge on a central location that can, and will, compile statistics. The three major exceptions are the U.S. Census Bureau, the housing programs of the U.S. Department of Housing and Urban Development, and the U.S. Department of Agriculture. They collect and publish the cost of residential and raw farmland. Residential land costs may well serve as an indication of the cost of land to new firms building structures in the growth sectors of metropolitan areas, and the cost of farmland could indicate the cost to manufacturers seeking larger, more remote sites for their plants. But both measures are, at best, surrogates and not the real thing we are after—the cost of land to the firm.

Residential Land

It has only recently been possible to obtain consistent measures of residential land cost per square foot used in new construction. Earlier data supplied by the Census Bureau did not control for lot size, and, as the data below indicate, average lot size varies considerably from region to region:

Region	Average Lot Size in 1975 (sq. ft.)
Northeast	15,481
North Central	11,147
South	11,971
West	8,124

Reports by the Department of Housing and Urban Development are restricted to certain kinds of construction encompassed in one or another of its housing programs, also biasing the results to an unknown degree. In 1974 and 1975, however, the Census Bureau systematically reported on land costs, controlling for lot size, and thereby gave us a valid basis comparison.[7]

The results of the Census Bureau's gathering efforts are as follows:

Region	Cost per Square Foot 1974	1975	Percentage Change
Northeast	$0.61	$0.67	9.8%
North Central	0.62	0.66	6.5
South	0.52	0.56	7.7
West	0.85	0.95	11.8

The South maintains a clear advantage over other areas, and its rate of increase is the second lowest, giving it a good chance of retaining that advantage. The West, in contrast, values its residential land most highly and shows the greatest tendency to increase its margin over other parts of the country. The Northeast and North Central areas fall in the middle.

Farmland

Farmland is not a perfect measure of raw land since it includes only land in agricultural production. Such land already has minimum improvements and is mostly cleared. Also, much of it is too remote for any commercial, industrial, or residential uses. Anyone familiar with land prices realizes that the price of land at the fringe of development behaves quite differently from the average price of land in a state or region. States whose farmland is practically all within 100 miles of a metropolitan area can expect, for example, to experience greater price fluctuations than states with no metropolitan concentrations, and to have higher value in general as a result. Keeping in mind these locational imperfections, however, farmland prices do give us some feeling for the value of raw land to a firm in different parts of the country.

The Economic Research Service of the Department of Agriculture has done an admirable job of keeping track of farmland prices. No doubt leaning on its well-established extension service, the Agriculture Department has traced average farmland prices by state as far back as 1850. Looking at recent experience first (see table 4-15), we note a marked variation in land costs from region to region. Once again, the South tends to be low and the North high, sometimes by a factor of 2. Farmland prices in the West, however, show quite a different pattern from the residential counterparts. They tend to be considerably below the U.S. average and are appreciating less rapidly. This is something of a puzzle, given the controversy in the West

Table 4-15
Average Cost per Acre for Farmland by Region, 1972-1975

Region	Cost in 1972	Cost in 1975	Percentage Change
New England	$619	$1,104	78%
Mid-Atlantic	660	1,214	84
East North Central	409	1,024	150
West North Central	207	565	173
South Atlantic	404	788	95
East South Central	270	502	86
West South Central	262	447	71
Mountain	97	188	94
Pacific	308	491	59

over the usurping of farmland by developers (presumably driving up the price), and the problem does not go away when individual states are taken one at a time. California's land prices, which are higher than those of other Western states, still averaged only $686 per acre in 1977, which is slightly above average for the United States but well below price levels in the East, Midwest, and even South Atlantic region. Either supply and demand are not operating properly in California, or the controversy is overstated, or farmland at the margin of development is priced much higher but amounts to such a small percentage of the total that it affects the average very little.

A striking feature of table 4-15 is the extent of appreciation in land prices generally. Over a five-year period, farmland is certainly gaining in value at a rate far greater than the rate of inflation. Thanks to the USDA efforts to retain past data, we can place recent trends in a broader perspective. Table 4-16 presents the longer term trends in farm real estate values (including buildings as well as land) since 1900. The land-only values are included in the last part of the table on an annualized basis. While not strictly comparable because they do not include buildings, all recent data on housing costs in the 1970s suggest that inclusion of buildings would only increase the rate of price growth and hence exaggerate the upward spiral.[8]

The figures show a relatively slow and even growth for the first fifty years of the century, with regional differences in 1900 being remarkably similar to those in 1950 and 1970. The Midwest has slipped slightly, and the South and West have gained a bit, but the pattern is still far more consistent than it is disparate over this rather long interval. The growth rates are approximately at, or slightly above, the rate of inflation during the first seventy years. Only recently have farmland prices climbed so rapidly, particularly in the Midwest, and only since 1960 have the Pacific states experienced considerably lower rates of appreciation.

Summary

Land costs are clearly lower in the South, whether residential or agricultural. The surge in residential land costs in the West has not been paralleled by nearly so rapid a rise in the cost of farmland. For most of this century, land in the Northeast and East North Central areas has been well above the national average, with the gap narrowing only slightly over a seventy-year period.

Capital Costs

More sophisticated students of corporate finance are quick to point out that the cost of capital to a firm is far more complicated than the cost of bor-

Table 4-16
Average Cost per Acre of Farm Real Estate by Region, 1900-1970

	Cost per Acre			
Region	1900	1950	1960	1970
New England	$33	$142	$235	$474
Mid-Atlantic	47	134	287	579
East North Central	41	127	231	371
West North Central	22	70	117	190
South Atlantic	16	81	173	331
East South Central	11	66	117	239
West South Central	9	62	114	226
Mountain	8	28	48	79
Pacific	17	100	194	283
	Ratio of Region to Average of All Regions			
New England	1.5	1.6	1.4	1.5
Mid-Atlantic	2.1	1.5	1.7	1.9
East North Central	1.8	1.4	1.4	1.2
West North Central	1.0	0.8	0.7	0.6
South Atlantic	0.7	0.9	1.0	1.1
East South Central	0.5	0.7	0.7	0.8
West South Central	0.4	0.7	0.7	0.7
Mountain	0.4	0.3	0.3	0.3
Pacific	0.7	1.1	1.2	0.9
	Average Annual Percentage Increase			
	1900-1950	1950-1960	1960-1970	1972-1975[a]
New England	3.0	5.2	7.3	12.3
Mid-Atlantic	2.1	7.9	7.3	13.0
East North Central	2.3	6.2	4.9	20.1
West North Central	2.3	5.3	5.0	22.2
South Atlantic	3.3	7.9	6.7	14.3
East South Central	3.6	5.9	7.4	13.2
West South Central	3.9	6.3	7.1	11.3
Mountain	2.5	5.5	5.1	14.2
Pacific	3.6	6.9	3.8	9.8

[a]The figures for 1972-1975 do not include buildings and hence are not strictly comparable.

rowed money alone, and relates to the relative mixes of debt and equity in the capital structure and the rate of return of potential opportunities available to the firm.[9] We cannot measure such sophisticated concepts on a regionwide basis and are restricted, therefore, to one component of the true cost of capital to the firm—the cost of its debt. We can and do, however, distinguish between short-term cyclical needs and longer-term borrowings.

Probably the best source of information on business borrowing is the Federal Reserve Board, which since 1967 has published in its *Bulletin* regional differences in interest rates charged by a sample of banks in thirty-five major centers across the country. Table 4-17 summarizes these data for 1967, 1972, and 1977. Unfortunately the 1967 data are not broken down by

Table 4-17
Interest Rates Charged by Banks on Business Loans, by Region and Type of Loan for 1967, 1972, 1977

	Year[a]		
Type of Loan	1967[b]	1972	1977
Short-term			
U.S. average	—	5.5	7.3
New York City	—	5.4	6.9
Other Northeast	—	5.7	7.6
North Central	—	5.4	7.3
Southeast	—	5.9	7.5
Southwest	—	5.8	7.3
West Coast	—	5.4	7.5
Revolving Credit			
U.S. average	6.1	5.2	7.2
New York City	5.9	5.1	7.2
Other Northeast	6.5	5.4	6.9
North Central	6.1	5.7	7.5
Southeast	6.1	5.8	7.1
Southwest	6.2	5.9	7.5
West Coast	6.3	5.1	7.1
Long-term			
U.S. average	—	5.6	7.5
New York City	—	5.4	7.4
Other Northeast	—	6.0	6.6
North Central	—	5.4	7.7
Southeast	—	7.1	7.6
Southwest	—	6.2	7.7
West Cost	—	5.8	8.0

[a]The three years are reported in May of each year by the Federal Reserve *Bulletin*. The actual dates of the rates are February 1967, February 1972, and November 1976.
[b]For 1967, only a weighted average of all types was presented. It is included here under the heading Revolving Credit.

type of loan. The *Bulletin* also breaks out its data by size of loan, which are not included here because they merely illustrate the phenomena of decreasing rates for larger loans, with little or no systematic, regional variation.

The most obvious conclusion to be drawn from table 4-17 is the similarity of rates. There is little systematic regional variation in rates, the main differences being attributable to the year and the type of loan. This finding should not surprise us. Capital markets are highly interdependent, with many intermediaries standing ready to capitalize on rate differences. Furthermore, the jet airplane and telecommunications have put businesses everywhere in touch on a daily basis, if necessary, with banks everywhere. A business executive in Georgia who is dissatisfied with the rates or services offered by his local bank can be in New York or Miami the next day discussing a loan with a large, regional bank. It is little wonder, under these circumstances, that rates tend to center closely on the U.S. average at any point.

If there is any pattern at all, it is for rates in rapidly growing areas to be slightly higher than the U.S. average. The differences are small and probably reflect the balance of supply and demand and the slightly greater risk that may be encountered in a growing, entrepreneurial economy dominated by young and speculative firms. In any event, the pattern is not consistent. Long-term rates in the North Central area in 1977, for example, were higher than corresponding rates in the Southeast.

It would be valuable, although very tedious and expensive, to analyze the true cost of capital to the firm, taking account of equity costs and the rates of return of investment opportunities facing the firm, before concluding that no capital cost differentials exist. Such a step is well beyond the scope of this chapter, or any literature examined to date, however.

Transportation

In an economy dominated by specialization and relatively easy shipment of things long distances, the cost of such shipments is an important factor for many firms. Unfortunately, it is one of the least well measured.

One would think that, with all the regulation of interstate commerce, shipping rates could be easily obtained. The cause of the difficulty in this case is not the lack, but the surplus, of data. The regulations and tariffs have become so complicated and voluminous that it is virtually impossible to obtain a meaningful sample for analytical purposes. As Samuelson put it,

> The use of tariffs is extremely complex. Simple volume of tariffs is one aspect of the problem. A 1962 survey by *Traffic World* magazine found that there were 205,275 active tariffs on file with the Interstate Commerce Commission, stretching 5,366 linear shelf feet in length. There are still more interstate and other non-ICC regulated tariffs which have never been published.[10]

Following the lead of researchers in this field, we will choose to avoid the tangle of published tariffs and focus instead on published estimates of freight costs derived from samples of way bills and other reports from shippers.

To cover a variety of shipping costs, both rail and trucking costs are examined. These two, in combination, amounted to a substantial portion of the total ton-miles shipped in the United States. Furthermore, they tend to represent two slightly different aspects of shipping, with rail tending to handle the bulkier, longer-distance hauls and truckers, although they have been steadily eroding the railroad's long-haul market, tending to have a larger share of local, short hauls in their overall mix.

Rail

An immediate problem in comparing regional differences in transportation costs is deciding what to compare. Costs per vehicle-mile or ton-mile offer a more valid unit cost comparison, but neglect the much greater distances between population concentrations in some parts of the country, and the general remoteness of, say, New England or the West Coast from the center. What we need, instead, particularly for longer-haul rail costs, is a measure of the cost to a business of importing or exporting 100 pounds of freight to or from a typical set of origins and destinations. These costs per hundred weight are presented in table 4-18 for rail shipments.

Fortunately, for rail shipments we can identify specific commodities and thereby separate bulk from high-value items. The table includes, in addition to a U.S. average, subtables for coal (a bulk commodity normally imported by a firm or its local utility as a raw material), chemicals (primarily an interindustry commodity), and electric machinery (which includes appliances and tends to be a high-value, export good). By so doing we span a variety of transportation problems facing businesses.

The table itself, while complicated because of the number of entries, contains a great deal of information. The upper left-hand corner of each subtable is the U.S. average. The first row presents the average cost of shipping the commodity to each region; the first column is the average cost of shipping the commodity from each region.[11] The diagonal elements are the costs of intraregional shipments. Rows and columns may be compared to ascertain the relative advantages of shipping to and from different regions.

The first, and expected, observation is that different commodities cost greatly different amounts to ship regardless of origin and destination. But looking into the bodies of the subtables, we note some substantial regional variations as well. The Southeast has a marked advantage in almost all respects. It tends with great regularity to be the least expensive place to ship things from, to ship things to, and to ship things around in. Most of the coal shipped in 1975, for example, was produced in the Northeast or the South. The South thus has a great advantage relative to other regions in obtaining this increasingly important energy source. Chemicals move to, from, and around the South at least cost, and the South can import electric machinery from its major production center in the Northeast more cheaply than from any other region outside the Northeast. Most major exceptions to this Southern dominance represent minor traffic volumes.

In contrast with the South's advantage is the high cost of getting things to and from the Mountain and Pacific areas. The remoteness of producers and receivers in both areas accounts for a great deal of the higher cost, since we have chosen to compare the total cost of doing business, not just the cost per mile. Unit costs tend to be higher as well, however, and this is reflected

Table 4-18
Cost of Rail Shipments by Origin and Destination of Shipment, 1975
(cents per hundredweight)

	All Commodities to					
From	United States	Northeast	Midwest	Southeast	Southwest	Mountain-Pacific
United States	53	53	48	37	64	86
Northeast	52	39	168	78	213	442
Midwest	50	121	29	133	107	235
Southeast	39	83	132	25	106	310
Southwest	60	145	110	89	39	198
Mountain-Pacific	81	245	94	237	181	45
	Coal to					
	United States	Northeast	Midwest	Southeast	Southwest	Mountain-Pacific
United States	25	24	32	25	34	21
Northeast	23	22	81	36	127	136
Midwest	16	—	16	—	87	—
Southeast	26	38	73	24	58	—
Southwest	28	—	51	—	23	—
Mountain-Pacific	36	89	43	—	106	20
	Chemicals to					
	United States	Northeast	Midwest	Southeast	Southwest	Mountain-Pacific
United States	84	97	81	62	74	118
Northeast	86	72	95	103	195	299
Midwest	70	103	48	129	124	208
Southeast	66	99	110	44	100	297
Southwest	96	142	109	92	59	194
Mountain-Pacific	104	187	117	160	147	69
	Electric Machinery to					
	United States	Northeast	Midwest	Southeast	Southwest	Mountain-Pacific
United States	277	222	307	212	244	493
Northeast	296	217	341	252	293	602
Midwest	330	269	202	352	293	546
Southeast	182	138	329	146	193	617
Southwest	223	307	393	238	125	487
Mountain-Pacific	242	—	422	—	436	86

partly in the higher intraregional costs for most rail commodities. The differences are substantial, in many cases running as much as 50 or 100 percent higher than costs elsewhere in the United States.

The Southwest and the Northeast tend to suffer somewhat from the West's problems but not nearly to the same degree. They tend to be more remote, but are also much closer to the concentrations that account for so much of the shipping volume.

Truck Costs

Trucking costs are not nearly so well measured as rail costs, at least in easily obtainable form. For one thing, more of the truck shipments are intrastate and unregulated. For another, the industry is dominated by many small firms from whom it is more difficult to obtain the necessary data. The best source for regional variations in trucking costs is a report published by the ICC summarizing the cost components of truck operations in different parts of the country.

As indicated above, each mode of transportation has invented its own geography, and we have adopted the trucker's breakdown to facilitate regional comparisons. Since no breakdowns by type of commodity or size of shipment are available, and since trucking tends to be more local and hence less affected by regional remoteness, cost per vehicle-mile offers the best basis for comparison. Table 4-19 presents the cost figures, both for vehicle-miles and for pickup-and-delivery and platform costs. Once again, we note the advantage of the Southeast. It tends to be on the low end of the scale, particularly for the cost of operating a truck for a mile. Also it appears to be controlling the inflation of its costs better than most regions, serving to maintain and even increase its advantage over time.

The Northeast appears to hold a similar advantage both in cost and in the rate of inflation of vehicle-mile costs, belying the notion that differences in labor costs account for variations in trucking costs. The Western states reveal their propensity to be areas with high transportation costs, this time

Table 4-19
Operating Costs of Class 1 and Class 2 Motor Common Carriers of General Commodities[a]

	Cost per Vehicle Mile		Pickup and Delivery Cost (cwt)		Platform Cost (cwt)	
	1975	Recent Annual Percentage Increase	1975	Recent Annual Percentage Increase	1975	Recent Annual Percentage Increase
New England	48[e]	0.7	43	16.5	40	8.6
Mid-Atlantic	66	5.7	35	9.7	51	6.3
Central	69	6.4	28	12.3	39	5.3
Midwest	63	8.6	39	14.0	54	6.0
Southeast	49	4.4	32	10.7	40	8.0
Southwest	54	8.3	35	11.3	38	2.0
Rocky Mountain	69	10.5	42	11.6	53	5.4
Pacific Coast	78	10.8	62	17.4	71	12.9

[a]Rates of increase are based on three- to five-year intervals (varying by region) terminating in 1975. The rates are annualized to facilitate comparison.

Regional Differences in Factor Costs

on a unit-cost basis, and their rate of cost inflation is the highest in the nation.

Summary

From a transportation standpoint, the Southeastern section of the United States holds a distinct advantage in both rail and trucking costs, and the Western states are at a distinct disadvantage, particularly when their remoteness and low density are considered. The rest of the country tends to be near the U.S. average, with the Northeast's remoteness compensated in most cases by its proximity to higher concentrations of business activity.

For selected commodities and destinations, each region may have niches in the rate structure, but those exceptions tend to be in low-volume commodities and/or corridors. For most businesses, the West is a place to avoid if shipping costs are important, and the South is clearly a better choice over most other locations.

Since most of the rates analyzed in this section are established through regulation of one sort or another, the regional differences of today need not extend into the future. Steps toward deregulation or a substantial revision of the rate structure could significantly alter the relative merits of being located in various parts of the country from a transportation standpoint.

Summary

One way to summarize the several threads contained in this chapter is to assume the viewpoint of a firm looking out on the world and rate regions in terms of what they offer in the way of factor costs. Table 4-20 offers such a rating. The first part evaluates regions in terms of the most recent complete data available; the second part assesses the implications of change over time. A " + " in the first part means that the region's costs are significantly lower and hence are a plus to the firm; a " + " in the second part of the table indicates a favorable (lower) rate of growth in a cost factor. A " – ", of course, suggests just the opposite in both cases, and no mark means that the area is not noticeably different from the U.S. average for that dimension.

Any particular firm's reading of this "scorecard" would depend, of course, on the relative importance of the different cost components in its income statement. For the "average" firm, however, the dominant feature is the cost advantages offered by the South in almost all the factor costs examined as well as the relatively high cost of doing business in the West and, to a lesser extent, the Northeast. Trends over time do not significantly alter this conclusion. The South is losing comparative ground somewhat in two

Table 4-20
Comparison of Factor Costs by Region as Viewed by the Firm

	Most Recent Level			
Item	Northeast	North Central	South	West
Retail trade wages			+	−
Wholesale trade wages	−		+	−
Manufacturing wages			+	
Service wages	−		+	−
Residential land			+	−
Agricultural land	−	−	+	+
Capital costs				
Rail shipping costs			+	−
Truck shipping costs	+		+	−

	Recent Changes			
	Northeast	North Central	South	West
Retail trade wages	−			
Wholesale trade wages			−	
Manufacturing wages				
Service wages			−	+
Residential land				−
Agricultural land			−	+
Capital costs				
Rail shipping costs				
Truck shipping costs	+		+	−

of its wage categories, and the West is improving in two of its items, one of which (service labor) put it at a distinct disadvantage. But, by and large, the extent of the differences is much greater than the degree of convergence.

These differences cannot necessarily be considered a direct determinant of growth. For one thing, our list does not include other major items in most firms' income statements, particularly energy and taxes. Nor does it consider opportunities on the revenue side that may or may not draw a firm to an area. The West, for example, continues to thrive despite its distinct cost disadvantages. Such costs must, however, enter into a business executive's thinking as one consideration in selecting a site for a firm. As such, the dominant advantage of the South, and the maintenance of that advantage, along practically all dimensions, will certainly not deter those already flocking to that region, for whatever reason, from continuing to do so.

Notes

1. See U.S. Bureau of the Census, *1967 Enterprize Statistics, Part 3* (Washington: GPO, 1973).

2. This sample represents 82 percent of total U.S. employment.

3. The figure 0.667 is obtained by weighting the average propensity for birth and expansion gains to be made by independent firms by the proportion of all gains from births (40 percent) and expansions (60 percent).

4. The large metropolitan areas are the SMSAs with a population of 1.0 million or more in 1970. The small and medium-sized SMSAs contained less than 1.0 million in 1970.

5. For a detailed description of how the 315 areas are defined, see Allaman and Birch, *Definition of Migration Areas Used in the Inter-Area Migration Project* (Cambridge, Massachusetts: Joint Center for Urban Studies, 1975).

6. In our study of migration flows it became obvious that four areas (New York, Chicago, San Francisco, and Los Angeles) play a special role, and we labeled them supercities. The regional centers include the eighteen largest SMSAs other than the supercities. The break between large and small metropolitan areas is at 350,000 people.

7. These data were not further subdivided by the Census Bureau by size of place because of moderate sample size, and we are unfortunately unable to distinguish land in small towns from land in large metropolitan areas.

8. See, for example, Frieden and Solomon, *The Nation's Housing, 1975-1985* (Cambridge, Massachusetts: Joint Center for Urban Studies, 1977).

9. See, for example, Bierman and Smidt, *The Capital Budgeting Decision* (New York: Macmillan, 1960).

10. See R. Samuelson, *Modeling the Freight Rate Structure* (Cambridge, Mass.: M.I.T. Center for Transportation Studies, 1977).

11. Each mode of transportation has chosen to define its own geographical bases for reporting. The regional breakdown presented in the table corresponds, respectively, to Official, Western, Southern, Southwestern, and Mountain-Pacific breakdown used by the ICC. See Federal Railroad Administration, *1975 Carload Waybill Statistics*, Statement TD-1, 1976.

Appendix 4A: Definition of Place Clusters Defined by Wage Level and Growth Rate[a]

Low Wages Generally, Low Growth

NFK/NPT NS, VA—Rural
Augusta, GA—Rural
Montgomery, AL—Rural
Albany, GA—Rural
Macon, GA—Rural
Columbus, GA/AL—Rural
Memphis, TN/AR—Rural
Bristol, VA/TN—Rural
Cincinnati, OH—Rural
Dubuque, IA—Rural
Minot, ND—Total
Sioux Falls, SD—Rural
Scottsbluff, NB—Total
Lincoln, NB—Rural
Wichita, KS—Rural
Wichita, Fls, TX—Rural

Abilene, TX—Rural
Abilene, TX—Metro
San Angelo, TX—Total
Austin, TX—Rural
Shreveport, LA—Rural
Monroe, LA—Rural
Greenville, MS—Total
Jackson, MS—Rural
Meridan, MS—Total
San Antonio, TX—Rural
Cps Christi, TX—Rural
Brownsville, TX—Total
Pueblo/CoSp, CO—Rural
Boise City, ID—Rural

Low Manufacturing Wages, High Growth

Raleigh, NC—Rural
Wilmington, NC—Metro
Charlotte, NC—Rural
Charlotte, NC—Metro
Columbia, SC—Metro
Charleston, SC—Metro
Augusta, GA—Metro
Gnvl/Jcknvl, FL—Metro
Miami, FL—Total
Tampa/StPet, FL—Rural

Albany, GA—Metro
Chattanooga, TN—Rural
Nashville, TN—Metro
Springfield, MO—Metro
Fort Smith, AR—Metro
Fayetteville, AR—Metro
Austin, TX—Metro
Biloxi/Fpt, AL—Metro
Lafayette, LA—Metro
San Antonio, TX—Metro

---

[a]For a detailed description of the counties contained in each area, see Allaman and Birch, *Definition of Migration Areas Used in the Inter-Area Migration Project* (Cambridge, Mass.: MIT-Harvard Joint Center for Urban Studies, August 1975).

Tampa/StPet, FL—Metro
Ft Myr/Srst, FL—Metro
Tallahassee, FL—Metro

Denver, CO—Rural
Phoenix, AZ—Rural

High Manufacturing Wages, High Growth

Orlando, FL—Metro
Fort Wayne, IN—Metro
Rochester, MN—Metro
Baton Rouge, LA—Metro
Houston, TX—Metro
Denver, CO—Metro

Seattle, WA—Metro
Phoenix, AZ—Metro
San Diego, CA—Total
Los Angeles, CA—Metro
San Francisco, CA—Total
San Jose/SCz, CA—Metro

High Service Wages, High Growth

Washington, DC—Total
Huntsville, AL—Metro
Dallas, TX—Metro
Portland, OR/WA—Metro
Reno, NV—Rural
Reno, NV—Metro
Las Vegas, NV—Total
Snta Barbara, CA—Metro

Appendix 4B: Definition of Sophisticated Industry Categories Used in Analysis of Wage Differences

Sophisticated Manufacturing

Machinery, except electrical/Engines and turbines

Machinery, except electrical/Office and accounting machines

Machinery, except electrical/Electronic computing equipment

Electrical machinery, equipment, and supplies/Radio, TV, and communication equipment

Electrical machinery, equipment, and supplies/Electrical machinery, equipment, and supplies

Transportation equipment/Aircraft and parts

Professional and photographic equipment, and watches/Scientific and controlling instruments

Professional and photographic equipment, and watches/Optical and health services supplies

Professional and photographic equipment, and watches/Photographic equipment and supplies

Professional and photographic equipment, and watches/Watches, clocks, and clockwork-operated devices

Professional and photographic equipment, and watches/Not specified professional equipment

Chemicals and allied products/Drugs and medicines

Sophisticated Services

Air transportation

Radio broadcasting and television

Telephone (wire and radio)

Telegraph and miscellaneous communication services
Banking
Credit agencies
Security, commodity brokerage, and investment companies
Insurance
Real estate, including real estate-insurance-law offices
Finance, insurance, and real estate-allocated
Advertising
Commercial research, development, and testing laboratories
Business management and consulting services
Computer programming services
Offices of physicians
Offices of dentists
Offices of chiropractors
Hospitals
Offices of health practitioners
Health services
Legal services
Colleges and universities
Museums, art galleries
Engineering and architectural services
Accounting, auditing, and bookkeeping services

Part III
Financial Impacts of Regional Change

5
Effects of Regional Shifts in Population and Economic Activity on the Finances of State and Local Governments: Implications for Public Policy

Roy Bahl

Introduction

The shift in economic activity from the Northeastern and Midwestern industrial regions has by now been thoroughly documented.[1] The numerous empirical examinations of this shift have been revealing in describing what has happened and in offering hypotheses about why it happened. Over the same period there has been an outpouring of literature on the financial problems of state and local governments.[2] The relationship between the declining economy and the declining fisc, however, has not been adequately studied,[3] or if it has, public policymakers have not understood the linkage. Perhaps it is because the relationship between the economy and the fisc is so difficult to formulate and because state and local governments have so little control over the performance of the state/local economy that policy analysts have turned in other directions to grapple with fiscal problems. There is probably no more glaring example of this misunderstanding than the proposed solutions to the fiscal problems of the New York City government. Indeed, at least in the early stages much more attention was focused on the financial management issues which surrounded the New York City and State near financial disasters than on the fiscal implications of the economic decline which was taking place. As a result, it would be no great surprise if the remedial management policies implemented do little to deal with the city's long-term fiscal problems.

The objective of this chapter is to analyze the linkage between regional shifts in economic activity and state and local government finances in the growing and declining regions of the United States. An initial assumption of this chapter is that regional shifts in population and employment are not undesirable per se and therefore should not be the object of remedial public policy. Nor is a trend toward interregional income equality or a growing homogeneity in the provision of public services across geographic areas detrimental to the public welfare. What is harmful about regional shifts and what ought to be at the center of concern about public policy to deal with such shifts are the effects on unemployment, poverty, and the fiscal position of state and local governments. In a sense all three of these concerns can be translated into a more general concern for the distribution of income, more specifically to a concern for the share of purchasing power or public services accruing to low-income families. In this context, the problems of decline, those faced by the industrial Northeast and Midwest, would appear more difficult to resolve than the problems of growth, those experienced by the Southern tier states. Certainly there are migration barriers which would cause one to expect a holding of the jobless in central cities in declining regions, and there are institutional barriers which would cause one to expect a worsening fiscal position for jurisdictions in the declining region. This is not to say that there are not severe fiscal and poverty problems in the Southern region, but rather that the adjustment problems associated with regional *shifts* are likely to be more severe in the Northeast.

In any case, this crude statement of the problem suggests that an understanding of the linkages among regional shifts in employment and population, the unemployment problems particularly of large cities and the fiscal problems of state and local governments, is essential to formulating a remedial public policy. This chapter is a modest attempt to deal with one dimension of this linkage, the relationship between regional shifts and state-local finances.

The thesis here is straightforward. The fiscal adjustments to growth in service demands and factor costs in the Southern tier, on the one hand, while painful, as are all fiscal adjustments, could be absorbed in an expanding public sector which had a growing capacity to finance and a governmental structure and tax system more amenable to absorbing such decline. The decline in financial capacity in the Northern tier, on the other hand, did not induce a commensurate contraction in the public sector, partly because service demands did not decline as rapidly as taxable capacity and partly because there are formidable barriers to cutting public service costs (unions and inflation). The result is a narrowing of public service levels on a nationwide basis, but a higher tax burden in the declining region and a more limited ability of governments to deal with public servicing needs in that region. Moreover, there is a real possibility that governments in the declining region will adjust by cutting social service expenditures and service delivery employees. In such a case, the low-income families in the declining sector may suffer disproportionately during the period of decline.

The fiscal implications of a deteriorating economic base for a state which has a highly developed public sector are particularly serious because of the difficulties of downward expenditure adjustment. This is well illustrated by the case of New York State. Public service levels in New York,

Implications for Public Policy

while not adequate by absolute standards in every area, are supported by a high level of expenditure. Public employee compensation, debt service, and certain nonlabor costs (for example, energy-related costs) are not easily controllable, much less reversible; hence, in the face of economic decline, it is not likely that large cutbacks in spending can easily be effected. To the extent that much of New York State's expenditure increase is due to rising compensation rates, the ability to slow down the rate of growth in spending is limited, particularly in a period of inflation. On the other hand, revenues respond dramatically to a slowdown in the rate of economic growth; hence, the resources to finance rising expenditure requirements do not materialize. The result of all this is a cutback in the level of public services and an increase in taxes, which are already thought to be too high. The objectives of this presentation are to support this thesis by quantifying and analyzing the fiscal adjustments actually made in the growing and declining regions and suggesting an appropriate public policy response.

The analysis here is necessarily concerned with regional variations, more specifically with the variation in finances of jurisdictions, state and local, in growing and declining regions. If any regularities are to be ferreted out, some form of aggregation of these jurisdictions must be used. Since the concern here is with how the fisc has been compromised by regional movements in population, jobs, and income, the financing jurisdictions are aggregated here by state and region. In the latter case we follow the general pattern of labeling "Northern tier" the aggregate of the East North Central, Mid-Atlantic, and New England Census regions and "Southern tier" the South Atlantic,[4] East South Central, and West South Central regions.[5]

The danger with such aggregation is that there might remain very wide differences in fiscal structure and performance across states in a region and even across local jurisdictions within a state. For example, in fiscal structure, Texas is more like Ohio than West Virginia, and in terms of economic and population expansion, the city of Atlanta is more like Syracuse than Houston. The reader should remain cognizant of such variations, especially when this analysis is overenthusiastic in identifying "clear" regional variations.

Existing Pattern of Regional Variations

Several characteristics of state fiscal systems are crucial both to an understanding of variations among regions in state/local revenue and expenditure patterns and to an explanation of how these variations have been affected by regional shifts. The most important characteristics would include:

1. The assignment of expenditure and financing responsibility between the state and its local governments
2. The structure of local government and the potential for regionwide service delivery or financing
3. The level and functional composition of expenditures
4. The level of public employee compensation, public employment, and the importance of public employee unions
5. The level of taxation and its composition by major sources
6. The relative use of debt and reliance on federal grants as financing sources
7. Central city/outside central city disparities in local government revenues and expenditures

While these patterns are compared among states and regions below, the existence of substantial intrastate heterogeneity should be kept in mind.

Revenue and Expenditure Assignment

There are two approaches to identifying regional variations in the relative importance of state and local governments. One is to study the characteristics of Southern and Northern states and present whatever pattern emerges. The other is to devise an objective system for classifying all states and examine the results for the two regions. The latter approach was taken in a recent attempt to classify state fiscal systems.[6]

To develop a state fiscal classification scheme, expenditure and financing data were gathered for total state and local expenditures and four specific expenditure functions: education, highways, public welfare, and health and hospitals for 1967 and 1972. From these data, nine specific fiscal characteristics were derived. The first three—*percentage of state and local government expenditures financed by federal, state, and local sectors, respectively*—represent the relative financing responsibilities of the three government levels. The second group of fiscal characteristics—*state and local direct expenditure shares*—describes final spending responsibilities rather than the original source of financing of state and local governments. The sixth characteristic, *per capita expenditures*, is included to capture the scope rather than the division of fiscal responsibilities among the states. The seventh variable is *state grants to local governments as a percentage of total state government expenditure*, and it is geared to separate state governments that dominate financing into two groups: those that retain heavy direct expenditure responsibility and those that pass expenditure responsibility to localities via grant systems. An eighth indicator is revenue effort, defined as *state plus locally financed expenditure expressed as a percentage of state*

Implications for Public Policy

personal income. Finally, the *share of state and local government revenues accounted for by the individual income tax* is designed to roughly approximate the progressivity of state taxation systems.

The fifty state fiscal systems described by these nine characteristics exhibit many varied and distinctive combinations of intergovernmental relationships. Some general patterns, however, also emerge which indicate that although each state may be unique, certain common types of state and local fiscal relationships exist nonetheless.

Based on this analysis, the fifty states were grouped into categories of high, moderate, and low financing responsibilities, expenditure shares, and per capita spending levels. These groupings were used to cross-classify state and local fiscal systems as one of three major types: state-government-dominated in terms of both expenditure responsibility and origin of financing; local-government-dominated; and mixed systems. These results are described in table 5-1.

Although no systematic relationship could be found between Census region and these cross-classifications, it may be noted that nine of the sixteen Southern tier states exhibit a high state financing responsibility and a moderate to a high state expenditure responsibility. Only one Southern state, Texas, is to be found in the locally dominated group. By contrast, only two of the fourteen Northern tier states may be classified as state-dominated—Rhode Island and Vermont—while seven of the fourteen Northern tier states may be classified as locally dominated.

A correlation analysis tends to confirm the argument that Southern states in general tend to have more state-dominated fiscal systems. As may be seen in table 5-2, those states which have a heavier financing and direct expenditure shares tend to be lower income, less urban, and less populous.

Local Government Structure

A second important difference between Northern and Southern tier states is the structure of local government in metropolitan areas. The stereotype difference would be Northern central cities with heavy concentrations of the poor, an antiquated, dilapidated infrastructure surrounded by more affluent suburbs, and with little hope of annexation or consolidation. Many, if not most, Northeastern metropolitan areas would fit this stereotype. The Southern tier cities might be painted as newer, subject to less city and suburb wealth difference, and having been more successful at annexation and consolidation. The examples of Jacksonville, Miami, Nashville, Houston, and Baton Rouge come quickly to mind.

**Table 5-1
Classification of State Fiscal Systems: Nonwelfare Expenditures of State and Local Governments, 1972**

	High State Expenditure Responsibility	Moderate State Expenditure Responsibility	Low State Expenditure Responsibility[a]
High State Financing Responsibility			
High expenditure per capita	Alaska Delaware Hawaii Vermont		
Moderate expenditure per capita	Idaho Utah West Virginia	Louisiana New Mexico	
Low expenditure per capita	Kentucky South Carolina	Arkansas Mississippi North Carolina Oklahoma	
Moderate State Financing Responsibility			
High expenditure per capita	Montana Wyoming	Arizona Maryland Oregon Washington	Minnesota Wisconsin
Moderate expenditure per capita	North Dakota New Hampshire	Connecticut Pennsylvania	Florida
Low expenditure per capita	Maine Rhode Island	Alabama Georgia Tennessee Virgina	Iowa
Low State Financing Responsibility			
High expenditure per capita			California Nevada New York
Moderate expenditure per capita		Colorado Kansas Nebraska South Dakota	Illinois Indiana Massachusetts Michigan Missouri New Jersey
Low expenditure per capita			Ohio Texas

Source: Advisory Commission on Intergovernmental Relations, *Federal Grants: Their Effects on State-Local Expenditures, Employment Levels, Wage Rates* (Washington, February 1977).

[a]High, moderate, and low designations for each category relate to whether the state placed in the top fifteen, middle twenty, or bottom fifteen among states. State expenditure responsibility is the state share of total state and local direct expenditures. State financial responsibility is the share of total state and local expenditures financed by the state. Per capita expenditures is total state and local expenditures per capita.

Table 5-2
Correlations between Fiscal Characteristics of States and Social and Economic Variables, 1972

	Per Capita Income	Percentage Urban	State Population
Federal financing share	−.654*	−.466*	−.382*
State financing share	−.122	−.247	−.327*
Local financing share	.463*	.451*	.461*
State direct-expenditure share	−.340*	−.457*	−.595*
Local direct-expenditure share	.340*	.457*	.595*
Per capita expenditures ($)	.551	.119*	.014
Grants as share of state expenditure	−.189	−.334*	−.583*

Source: Advisory Commission on Intergovernmental Relations, *Federal Grants: Their Effects on State-Local Expenditures, Employment Levels, Wage Rates* (Washington, February 1977).
*Statistically significant at the 5 percent level.

There is more than impressionistic evidence to support this stereotype. Sacks finds striking differences between regions in the percentage of metropolitan area populations residing within the central city. As may be seen in table 5-3, he found an average of 61 percent of metropolitan population residing inside central cities in the South as compared to 34 and 45 percent, respectively, in the East and Midwest.[7] Moreover, he shows that between 1960 and 1973 this percentage increased slightly in the Southern metropolitan areas but declined in all other regions. This in no way allows a conclusion to be drawn that the structure of government in the South is less complicated, but it does show that central cities in the South are a more dominant force in their respective metropolitan areas. In addition to this population advantage, it can be shown that the central cities are both fiscally and economically better off in the Southern tier than in the Northern tier states. Much of this advantaged position of Southern central cities must be ascribed to the greater success of the South in consolidation attempts and/or in using more areawide financing mechanisms. Marando argues that consolidation is essentially a Southern regional phenomenon and that annexation has occurred extensively throughout the United States with the exception of the Northeastern region.[8]

Expenditure Level and Structure

There are wide variations between the Northern and Southern tier states in the level and functional distribution of expenditures. The Northern states spend substantially more—25 percent—on a per capita basis than do the Southern tier states (see table 5-4). This difference holds generally across

Table 5-3
Central-City Population as a Proportion of SMSA Population: 1960 and 1973[a]

	Number of Observations	Mean Value (Percent) of	
		1960	1973
East	18	41	34
Midwest	22	52	45
South	27	59	61
West	18	49	44
Total	85	51	47

Source: Advisory Commission on Intergovernmental Relations, *Trends in Metropolitan America* (Washington, 1977), table 2.
[a]For the eighty-five largest SMSAs.

states within the two regions. Only one Northern tier state (Indiana) spends less than the Southern mean, and only two Southern tier states (Delaware and Maryland) spend above the Northern mean. Only two Southern tier states, again Delaware and Maryland, spend above the national median. This low expenditure level in the South, even in the midst of an increased flow of resources to that region, is important in understanding the possibilities for fiscal adjustment.

In terms of expenditure distribution, the Southern states allocate a slightly greater share of total public resources to education. The same holds true for health and hospitals, but there is much greater variation among states within the two regions. But perhaps the major regional difference in expenditure structure is that the Northern states spend proportionately more for public welfare. No Northern state allocates as little to public welfare as the Southern mean of 11.7 percent.

Public Employment and Wage Levels

On average, there appears to be a greater level of state and local government employment, relative to population, in the South (see table 5-4). Nine of the sixteen states in the Southern tier are above the U.S. median of 476 employees per 10,000 population while only two of the fourteen Northern states are above this median. In general, however, there is much variation among states in both groups, making it difficult to draw a firm conclusion. The variations among the Northern states range from Pennsylvania's 401 state and local government employees per 10,000 population to New York's 563; in the South, the spread is not as great, ranging from Louisiana's 424 to Maryland's 532 employees per 10,000 population.

Implications for Public Policy

There is some evidence that an association exists between the level of local government employment and the rate of population growth. Muller compares twelve growing cities and fourteen declining cities on the basis of common function employment per 1,000 residents.[9] From this relatively small set of observations, he finds declining cities to have 12.1 workers per 1,000 residents as compared to 8.7 in the growing cities.[10] Perhaps even more interesting is his finding that the gap has widened between 1967 and 1972. No such relationship between the level of state and local employment and population growth or decline can be found among the Northern or Southern tier states examined here.

Average public employee wages are higher in the Northern tier (table 5-4). However, there are many problems inherent in a comparison of average wage levels across states. There are not good disaggregated data on the wage levels of public employees at various levels of seniority or in various occupations. The estimates presented in table 5-4 are of average payroll per full-time equivalent employee. Such a measure misses the wide variation in pay levels by class of employees, and since October payrolls are used, mixes nine-month employees (teachers) with twelve-month employees. Moreover, the inclusion of total payroll but only full-time *equivalent* employees introduces distortions created by payments to part-time employees. The variation in this distortion across states is unknown.

Even if payroll per full-time equivalent employee is a reasonable measure of interstate variations in the average wage, there remains the problem of measuring interstate variation in the level of pensions and fringe benefits. Again, there are inadequate data to make these cross-state comparisons, and one must be content to assume that interstate variations in the average wage, as measured above, accurately reflect interstate variations in total compensation. There is good reason to expect that it does not, since most benefits are tied to wage levels, for example, pensions and social security contributions. Hence, it is likely that the regional differences in total compensation are greater than those in average wages.

Finally, even if the payroll per full-time equivalent employee is a reasonable benchmark for comparison, there remains the problem of cost-of-living differentials which may tend to change this pattern of interstate differences. In an attempt to adjust the distribution of average wages for regional cost-of-living differences, we have applied the Department of Housing and Urban Development (HUD) estimated fair-market rent index.[11] When adjusted for living cost differentials in this manner, the advantage of Northern tier average public sector wages over Southern tier *falls* to an almost negligible 2 percent.[12]

If all these caveats are accepted, the greater average wage in the Northern tier suggests that a substantial part of the expenditure difference in the Northern and Southern states is due to public employee compensation dif-

Table 5-4
Expenditure and Employment Characteristics of State and Local Governments by Region, 1975

State and Region	Per Capita Expenditures	Percentage of Current Expenditures			State and Local Government Employees	
		Education	Welfare	Health and Hospitals	Per 10,000 Population[a]	Average Wage[b]
Northern Tier	$1,080[c]	41.1	15.9	7.8	454	$ 780
East North Central	1,000	44.8	14.6	8.0	445	775
Illinois	1,066	43.4	16.1	7.0	446	878
Indiana	827	50.9	10.0	10.2	434	652
Michigan	1,120	42.7	17.7	8.3	470	879
Ohio	894	42.2	13.3	8.1	410	734
Wisconsin	1,091	44.7	16.0	6.5	463	732
Middle Atlantic	1,275	37.7	16.2	8.8	473	901
New Jersey	1,107	40.9	14.5	6.0	456	883
New York	1,611	32.5	17.0	12.8	563	1,004
Pennsylvania	1,007	39.8	17.1	7.6	401	818
New England	1,050	39.6	16.7	7.1	452	723
Connecticut	1,059	40.0	13.2	6.4	417	837
Maine	938	39.4	17.4	4.2	446	610
Massachusetts	1,182	34.3	20.8	11.9	457	829
New Hampshire	924	43.1	14.7	5.4	452	601
Rhode Island	1,044	38.7	19.4	8.3	456	839
Vermont	1,152	42.2	14.9	6.1	486	621

Implications for Public Policy

Southern Tier	863	44.0	11.7	10.2	477	678
South Atlantic						
Delaware	983	44.6	10.5	10.0	489	719
Maryland	1,187	47.7	10.1	6.3	532	764
North Carolina	1,243	42.4	11.6	8.0	496	910
Virginia	826	49.8	10.3	8.9	443	682
South Carolina	974	45.4	11.8	7.4	487	732
Georgia	873	45.2	8.7	13.3	473	602
Florida	925	38.8	13.6	16.6	523	656
West Virginia	944	43.6	7.5	11.4	485	774
	892	44.0	10.7	8.2	478	637
East South Central	839	43.7	12.6	10.8	456	631
Alabama	827	44.4	12.6	12.4	450	660
Kentucky	838	45.2	14.7	7.0	432	645
Mississippi	833	43.0	11.9	11.7	468	565
Tennessee	859	42.3	11.0	12.2	475	658
West South Central	648	43.2	13.1	9.9	474	640
Arkansas	726	43.3	14.1	10.7	424	582
Louisiana	946	40.8	11.4	11.2	517	645
Oklahoma	873	41.0	15.7	8.6	480	610
Texas	838	47.5	11.0	9.2	476	727
U.S. Median	1,008				476	

Source: U.S Bureau of the Census, *Governmental Finances in 1974-75, Series G-F 75, 5* (Washington: Government Printing Office, 1976).

[a]Full-time equivalent employment.
[b]October payroll divided by full-time equivalent employment.
[c]Unless otherwise noted, the regional and tier averages are simple unweighted means over states in the respective group.

ferences. If one accepts a notion that differentials in average wages across regions are not commensurate with productivity differentials in the public sector, then the higher level of per capita spending in the Northern states substantially overstates the difference in the quality of services provided between the two regions. Muller has studied wage variations among local governments using his growth/decline dichotomy and for his sample has determined that average wage levels tend to be higher in older and declining cities. His plausible explanation of this difference is the greater ability of municipal employee associations in older cities to press for more favorable contract terms, coupled with cost-of-living differences and perhaps a necessary premium for what is perceived as a lower quality of life in the older, more congested cities of the Northeast and industrial Midwest.

Sources of Finance

Three aspects of the financing of state and local government expenditures are important in describing regional variations in fiscal systems: reliance on debt, the structure of taxes raised, and the level of revenue effort exerted. With respect to borrowing, the level of general obligation debt in the Northern tier is substantially higher both on a per capita basis and as a percentage of personal income than in the South (see table 5-5). This higher level of per capita debt in the Northern tier suggests a greater fixed commitment for debt service in the annual budget of the states. It is interesting to note that the highest per capita levels of debt, and generally the highest levels of debt as a percentage of personal income, are observed for those states thought to be facing the most serious fiscal crisis, that is, New Jersey, New York, Pennsylvania, and Massachusetts.

To give some rough idea of how the market perceives the quality of this debt, Standard & Poor's ratings of the general obligation bonds of each state are shown in table 5-5. No consistent pattern emerges with respect to variations between regions. From the ratings one might draw the conclusion that the market does not weight the regional shift in economic activity and employment very heavily in gauging the long-term repayment potential of state government. For example, declining New York and growing Florida are both seen as AA credits, while declining New Jersey and growing Texas are both seen as AAA credits.

In terms of revenue structure there are distinct differences between the regions. Southern states are more heavily reliant on sales taxes and Northern states on property taxes (see table 5-6). This difference is largely a reflection of the division of financial responsibility for services between the state and local levels. Where local government involvement in the delivery of ser-

Table 5-5
Debt Levels by Region, 1975

State and Region	Long-Term Debt Outstanding Per Capita	As a Percentage of Personal Income	Bond Rating
Northern Tier	$1,080	17.9	
East North Central	758	12.3	
Illinois	922	12.7	AAA
Indiana	501	8.9	—
Michigan	926	15.0	AA
Ohio	755	13.0	AAA
Wisconsin	684	12.1	AAA
Middle Atlantic	1,547	24.0	
New Jersey	1,208	18.0	AAA
New York	2,194	33.4	AA
Pennsylvania	1,238	20.8	AA
New England	1,117	19.6	
Connecticut	1,566	22.5	AA
Maine	742	15.5	AAA
Massachusetts	1,321	21.7	AA
New Hampshire	743	14.0	—
Rhode Island	994	17.0	AA
Vermont	1,334	26.9	—
Southern Tier	873	16.4	
South Atlantic	897	15.3	
Delaware	1,680	20.7	A
Maryland	1,239	19.1	AAA
North Carolina	427	8.6	AAA
Virginia	737	12.7	AAA
South Carolina	670	14.5	AAA
Georgia	764	15.0	AA
Florida	784	13.9	AA
West Virginia	872	17.7	AA
East South Central	868	18.6	
Alabama	787	17.0	AA
Kentucky	1,135	23.3	AA
Mississippi	647	16.0	A
Tennessee	902	18.4	AA
West South Central	831	16.2	
Arkansas	514	11.1	—
Louisiana	1,054	21.5	AA
Oklahoma	842	16.0	AA
Texas	916	16.3	AAA
U.S. Median	902		

Source: U.S. Bureau of the Census, *Governmental Finances in 1974-75, Series G-F 75,* 5 (Washington: Government Printing Office, 1976); and Standard and Poors Corporation, *Municipal Bond Selector* (New York, 1975).

Table 5-6
Revenue Structure by Region, 1975

State and Region	Percentage of Own-Source Revenues Raised from			Per Capita Federal Aid	Federal Aid as a Percentage of Total General Revenue
	Property Taxes	Sales Taxes	Income Taxes		
Northern Tier	34.8	12.8	14.8	$218	20.5
East North Central	30.6	15.7	16.5	186	18.1
Illinois	31.9	18.5	14.7	196	18.2
Indiana	30.0	20.7	12.2	139	15.3
Michigan	32.5	14.3	14.8	232	20.1
Ohio	28.6	13.0	16.1	164	18.8
Wisconsin	30.1	12.3	24.7	200	18.1
Middle Atlantic	32.2	13.6	16.6	223	18.7
New Jersey	46.3	11.8	3.8	191	17.7
New York	29.2	15.2	23.8	276	18.0
Pennsylvania	21.1	13.9	22.2	201	20.6
New England	39.5	10.1	12.6	242	23.2
Connecticut	42.8	16.8	6.1	190	18.8
Maine	33.4	18.8	8.8	256	27.0
Massachusetts	45.4	4.6	22.7	223	19.0
New Hampshire	47.5	0	6.4	200	23.1
Rhode Island	34.3	14.2	15.9	249	24.0
Vermont	34.1	6.4	15.7	336	27.6
Southern Tier	17.7	18.0	13.9	218	24.1
South Atlantic	19.2	15.8	17.8	217	22.8
Delaware	13.1	0	29.0	225	18.8
Maryland	23.0	10.4	28.4	219	19.1
North Carolina	19.0	16.0	21.2	223	26.4
Virginia	21.6	13.4	18.4	200	21.6
South Carolina	16.1	19.2	16.7	198	24.0
Georgia	23.0	17.7	14.2	232	24.7
Florida	22.6	20.0	3.0	159	18.1
West Virginia	15.3	30.4	11.7	280	29.9
East South Central	14.5	22.4	11.6	225	26.4
Alabama	8.8	21.3	12.2	224	27.0
Kentucky	14.3	16.5	19.8	229	25.6
Mississippi	15.9	27.0	8.8	245	28.7
Tennessee	19.0	25.0	5.7	200	24.6
West South Central	18.2	17.7	8.3	214	24.2
Arkansas	16.6	18.6	15.8	221	29.0
Louisiana	11.1	21.1	6.4	224	22.5
Oklahoma	17.3	13.6	11.2	231	25.5
Texas	27.9	17.8	0	180	20.1

Source: U.S. Bureau of the Census, *Governmental Finances in 1974-75*, Series G-F 75, 5 (Washington: Government Printing Office, 1976).

vices is strong, there tends to be much heavier use of the property tax. But, as shown above, the Southern states tend to be more state-government-dominant; hence there is heavier *reliance* on nonproperty taxation.

The importance of this difference lies with the potential response of the fisc to growth or decline in the economic base. In the South, where there is heavy reliance on sales and income taxes, a combination of real growth and inflation will automatically generate substantial new revenue for expansion of the public sector. In the Northern tier, where reliance is greater on property taxation, even the tax-base growth generated by inflationary increase in income will not be fully or easily captured.[13] Another advantage of such centralization is the controllability of the overall level of state and local government taxation and spending. It is difficult to formulate a long-term state fiscal plan where one-half of all spending and taxing decisions are made by local governments. In terms of the controversial issue of the regional distribution of federal aid, both regions receive about the same per capita amount, but Southern states, because of their lower level of fiscal activity, are more dependent on federal aid as a revenue source.

Local Fiscal Problems

State-to-state variations in fiscal structure and performance mask the differences between regions in the problems facing the largest local governments within the regions. Indeed, the standard stereotype would have central cities in a substantially worse position than their suburbs in terms of income level, public service levels, and concentration of the poor.

Nathan and Dommel have developed a "hardship index" which compares the socioeconomic conditions of fifty-five of the nation's largest central cities with the same conditions both for their surrounding suburban area and with each other.[14] Of the fourteen cities scoring poorest on this hardship index, eleven are in the Northern tier of states while only two, Atlanta and Richmond, are in the South. Of the ten cities found better off, five were in the Southern tier and none were in the North.

Sacks, in his latest compendium of metropolitan fiscal disparities, also supports the stereotype. As may be seen from the data in table 5-7, the Southern cities are more densely populated and wealthier relative to their own suburbs, but are less densely populated and poorer relative to Northern cities. The fiscal disparities which grow out of this socioeconomic disparity are predictable: central cities in the Northeast have greater average tax burdens than their suburbs and apparently provide a lower level of public services.

Table 5-7
City-Suburb Disparities

	Mean Values in 1973		
	Population Density in Central City (Persons per Acre)	Per Capita Income City	Per Capita Income Ratio of City to Suburb
East	16.4	$3,727	0.83
Midwest	8.4	3,756	0.89
South	4.7	3,644	1.06
West	6.3	4,088	1.04
Eighty-five SMSAs	8.5	3,784	0.96

Source: Advisory Commission on Intergovernmental Relations, *Trends in Metropolitan America* (Washington, 1977), tables 4 and 10.

Summary: Regional Variations in State-Local Finances

These data show certain clear differences in fiscal structure and performance between the Northern and Southern tier states. While there certainly are exceptions to this pattern, the general differences observed would appear to hold for most jurisdictions in the two regions. First, the Southern tier states have more state-dominated fiscal systems. This means that they have heavier state government responsibility for both financing and direct expenditures, which in turn means that the growth and distribution of total state and local expenditures are more controllable and that the growth in expenditures is financed with a more elastic revenue source. In the case of the Southern tier states, the sales tax is relied on to a much greater extent than in the North. The Northeastern and Midwestern states, on the other hand, tend to have more local-government-dominated systems. As a result, there is a potential for much greater disparity in public spending levels among jurisdictions within the state, and there is much heavier reliance on the local property tax than in the Southern tier states.

With respect to the level of spending, per capita expenditures are some 25 percent lower in the Southern states than in the Northern states. However, a part of this difference is due to the higher level of welfare expenditures in the Northern tier states. Moreover, since these differences are not adjusted for regional variations in prices, and average public employee wages are much higher in the North, the difference in public service levels may be considerably less than 25 percent. Public employment levels per 10,000 population are greater on average in the Southern states and do not vary systematically with the rate of population growth of a state. Adjustment of average wage differences for differences in the cost of living may all but eliminate the gap in wages between regions.

Implications for Public Policy

There is a major difference between the two regions with respect to the fiscal health of their largest local governments. The Northeast and industrial Midwest regions seem to fit the stereotype of declining and poor central cities surrounded by relatively wealthy and fiscally sound suburbs. The reverse tends to be true in the South, where the per capita income level in the central city is greater than that in the suburbs. This advantaged position of Southern central cities can be attributed in part to the newness of the cities and their resulting local government structure which often tends to encompass growing suburban areas. There would appear to be much less jurisdictional fragmentation in the South, largely because of the greater potential for annexation and consolidation during the rapid growth period of the past two decades. To the contrary, Northern cities which are surrounded by older incorporated jurisdictions find it all but impossible to expand jurisdictional boundaries.

Comparative Fiscal and Economic Growth

An understanding of the fiscal problems resulting from the movement of population and economic activity to the South requires analysis of the structure of the state and local government expenditure and revenue responses to this movement, in both the growing and the declining regions. In the discussion below, we look successively at the growth in the capacity to finance public services and the demand for expansion of public services as measured by the growth in the economic and demographic base of the regions, the expenditure response and the extent to which it was demand- or supply-induced, and the revenue response in terms of its composition by type of tax and changes in the level of tax effort.[15] The results of this analysis suggest that fiscal activity in the South expanded in response to an increase level of population, a demand consideration, and was supported by an increased capacity to finance such activity. In the North, fiscal activity also continued to expand even in the face of a relatively slower-growing, or in some cases a declining, economic base. The expansion of fiscal activity may be attributed to increases in state and local government employment and increases in the average compensation of these employees, demand and supply considerations, respectively.

Growth in the Economic and Poulation Base

The shift in economic activity from the Northern to the Southern states has been well documented in the literature. Jusenius and Ledebur have described this shift in terms of population movement, disaggregating changes into natural increase and inmigration.[16] Greenberg and Valente[17] and Garnick[18] have studied the trends in employment, and the Congressional Budget Office has described the pattern of growth in earnings and personal income.[19] For the purposes of this chapter it is necessary to examine these trends in order to determine their potential effects on the taxable capacity and public servicing requirements of states in each region. Unfortunately, none of these indicators of economic expansion or contraction is an adequate measure of taxable capacity, partly because the tax structures of the fifty states vary so widely. Nevertheless, population movement, employment, and growth in earnings and personal income give some notion of how regional shifts in economic activity enhance or compromise the ability of state and local governments to finance public services.

Per capita income is a composite measure which more than anything else indicates the average level of well-being of citizens in a region. Since per capita income is influenced by changes in population size, it may or may not provide a proxy measure of changes in the capacity to finance. As may be seen in table 5-8, the per capita income growth in the Southern tier was substantially greater than in the North for all three periods considered here. It is interesting to note, however, that the disparity in the rate of growth in per capita personal income was reduced somewhat in the past three years. Between 1962 and 1972, per capita income in the Southern tier was growing about 25 percent faster, but the differential fell to about 14 percent between 1972 and 1975. This narrowing in per capita income growth is due primarily to a relatively heavy loss of population in the Northern tier states and a continued rapid growth of population in the Southern tier states. Hence, as the population shift has continued, there has been a slackening of the rate at which average income levels in the growing and declining region are converging.

The aggregate personal income trends which lie behind these per capita amounts give perhaps a clearer picture of the implications for the capacity to finance. Between 1962 and 1975, there were substantial increases in money income in both regions, but there was relatively little shift in the composition of income. Income originating in manufacturing in Northern states fell from 25 to 21 percent, while income originating in the services rose by about 4 percent. Otherwise, things stayed much the same. Most importantly, the share of income accounted for by all transfer payments—which may provide less taxable capacity than earnings from goods and service production—remained about the same in both regions. These data offer scant evidence that changes in the composition of income have compromised the tax base during the period studied.

However, in the case of local governments, particularly large central-city governments, changes in the composition of personal income may well have had a dampening effect on potential revenue growth. To the extent that local property tax systems include industrial machinery, equipment, and so on, the shift of income composition from manufacturing to services

Table 5-8
Percentage Increase in per Capita Personal Income by Region for Selected Periods

State and Region	1962-1967	1967-1972	1972-1975	1975 Level
Northern Tier	34.1	39.3	28.5	$6,232
East North Central	33.6	39.8	29.1	6,121
Illinois	31.5	38.0	32.3	6,789
Indiana	33.4	38.6	29.4	5,653
Michigan	38.3	43.3	24.7	6,173
Ohio	32.8	40.0	27.2	5,810
Wisconsin	31.8	39.2	32.1	5,669
Middle Atlantic	30.4	39.5	27.7	6,398
New Jersey	29.6	40.4	26.8	6,722
New York	31.0	37.2	25.1	6,564
Pennsylvania	30.5	41.0	31.2	5,943
New England	36.5	38.7	27.8	6,098
Connecticut	31.6	33.5	29.6	6,973
Maine	48.8	43.0	29.6	4,786
Massachusetts	30.2	40.9	26.0	6,114
New Hampshire	32.0	38.1	27.1	5,315
Rhode Island	36.5	36.7	29.5	5,841
Vermont	40.0	40.1	27.7	4,960
Southern Tier	40.3	49.7	30.2	5,292
South Atlantic	39.0	51.0	30.1	5,510
Delaware	27.9	39.8	29.2	6,748
Maryland	30.6	48.6	30.3	6,474
North Carolina	42.8	53.1	28.5	4,952
Virginia	41.0	52.8	31.5	5,785
South Carolina	46.7	52.9	31.7	4,618
Georgia	45.5	52.3	28.2	5,086
Florida	38.5	60.2	25.0	5,638
West Virginia	39.2	48.4	36.5	4,918
East South Central	42.1	52.8	32.0	4,676
Alabama	39.4	53.6	33.7	4,643
Kentucky	38.2	47.4	35.0	4,871
Mississippi	49.9	58.4	27.2	4,052
Tennessee	41.1	51.8	32.0	4,895
West South Central	40.9	44.0	37.4	5,347
Arkansas	42.9	50.3	38.2	4,620
Louisiana	42.7	40.3	37.3	4,904
Oklahoma	39.2	41.4	36.9	5,250
Texas	38.9	44.0	37.3	5,631

Source: Bureau of Economic Analysis, U.S. Department of Commerce, *Survey of Current Business* 56, no. 8 (Washington: Government Printing Office, 1976).

may have depressed the level of property tax revenues. Similarly, the very rapid growth in income generated in the state/local sector in large central cities may not have offset the revenue losses resulting from the outmovement of manufacturing. This is in part due to the exemption of state and local government properties from the real estate tax and the fact that they are not included in the business income tax base.[20]

In terms of changes in the level of employment, the Southern tier states have been growing more rapidly for all three periods considered (see table 5-9). Even though the rate of employment growth has slowed in the Southern states, it still remains considerably higher than that in the North. Perhaps even more important in the context of this analysis is the fact that the relatively low rate of employment growth in the Northern tier between 1967 and 1972 turned to literally no growth and in some cases decline between 1972 and 1975. In the Southern tier, on the other hand, while the growth rate slowed between 1972 and 1975, only one state (Delaware) showed an absolute job loss. Garnick argues that the relative shifts in employment are primarily a Northern central-city phenomenon with central-city counties of the large SMSAs in particular having been subject to absolute declines in employment (especially manufacturing) at least since 1960.[21] When the 1965-1972 pattern of employment growth in metropolitan central cities is examined in the ten largest city counties, declines were registered in New York, Philadelphia, and St. Louis, with only a modest increment in Baltimore. The largest percentage increases in employment were in Denver, Indianapolis, Jacksonville, Nashville, and New Orleans.[22]

Yet a third way to measure the change in economic activity in the two regions is to examine the pattern and trend of population growth. On the revenue side, a declining population may mean a diminishing capacity to finance public services if the population losses are higher-income-earning families. If outmigration is primarily of low-income families, service requirements may be reduced by more than taxable capacity, thereby enhancing the government's fiscal position.

The North-South differentials in population growth rates are predictable. The growth in the Northern tier has slowed steadily since 1962 and was negligible over the 1972-1975 period. Among the Southern states the rate of population growth also slowed but remained well above the Northern rate even during the 1972-1975 period. No state in the Southern tier showed a population decline over the 1972-1975 period while five Northern states (Illinois, New Jersey, New York, Pennsylvania, and Rhode Island) lost population (see table 5-10). With respect to the composition of population change, little data are available by way of the income level and employment characteristics (that is, occupation, industry) of migrants. However, it is known that because of higher fertility rates the Southern tier would have grown faster than the Northern tier even in the absence of migration between the regions.[23]

In terms of population change within metropolitan areas, some evidence is available on the changes by central city/outside central city and by race. These data show that Southern cities tended to increase their share

Implications for Public Policy 179

of metropolitan area population while Northern cities generally tended to decline as a percentage of metropolitan area population. Sacks has shown that the population decline in the major cities of the East between 1960 and 1970 was predominantly an exodus of white population—no major central city in the East showed a gain of white population between 1960 and 1970.

The inference one might draw from these trends is that the declining population in the North likely reduced certain servicing needs, but these reductions may have been offset by increasing concentrations of the poor, particularly in central cities.

Expenditure Growth

Given the relatively slower growth in financial capacity in the Northern states, a slower growth in fiscal activity might have been expected. In fact, expenditure growth in the Northern tier states was not noticeably below that in the Southern states (see table 5-11). Indeed, expenditures grew at a rate approximately 20 to 30 percent higher than personal income in both regions in all three periods considered, except for the 1967-1972 period, when per capita expenditures in the Northern tier grew 87 percent faster than per capita income (see table 5-12). Even in the 1972-1975 period, when total employment increased by about 7 percent in the South and less than 1 percent in the North, per capita expenditures grew by about the same percentage in both regions. From this evidence, one might conclude that there was not a strong relationship between the growth in public expenditures in the two regions and the capacity to finance that growth.

If the growth or decline in taxable capacity does not explain growth of the state and local government sector, then attention might be turned to two other possible explanations: (1) on the demand side, growing requirements for services resulted primarily in increased numbers of public employees and thereby exerted an upward pressure on expenditures; (2) on the supply side, increased public employee compensation resulted from union pressures and inflation and forced up expenditure levels. Either explanation would be consistent with the observed absence of relationship between economic base and public expenditure growth.

There is a wealth of literature on expenditure determinants which attests to the difficulties of separating demand from supply influences to explain expenditure growth and variations.[24] Those difficulties notwithstanding, we proxy the growth in service demand here with three variables: population growth (table 5-10), increase in Aid for Dependent Children (AFDC) recipients (table 5-13), and increase in primary and secondary school enrollments (table 5-14). To the extent that these factors increased over the three periods studied, an increase in state and local government employment levels might

Table 5-9
Growth in Employment by Region

	1962-1967		1967-1972		1972-1975	
State and Region	Change (000s)	Percentage Change	Change (000s)	Percentage Change	Change (000s)	Percentage Change
Northern Tier	4,192.7	15.2	1,835.3	5.8	200.8	0.6
East North Central	2,261.5	19.4	944.3	6.4	294.4	2.0
Illinois	634.9	17.8	117.6	2.8	115.4	2.7
Indiana	315.7	21.6	145.0	8.2	8.4	0.4
Michigan	566.8	24.3	212.9	7.3	10.3	0.3
Ohio	520.6	16.8	318.5	8.8	71.2	1.8
Wisconsin	223.3	18.5	150.3	10.5	89.1	5.6
Middle Atlantic	1,395.8	11.5	632.9	4.7	−204.4	−1.5
New Jersey	324.8	13.4	252.8	10.4	−5.8	0.0
New York	597.0	9.5	171.9	2.5	−239.1	−3.4
Pennsylvania	474.8	12.9	208.2	5.0	40.5	0.9
New England	535.4	14.0	258.1	6.0	110.8	2.4
Connecticut	180.3	19.0	59.5	5.3	30.8	2.6
Maine	37.4	13.4	27.1	8.6	12.1	3.5
Massachusetts	215.8	11.1	98.7	4.6	63.8	2.8
New Hampshire	36.1	17.4	35.7	14.6	13.5	4.8
Rhode Island	40.0	13.4	19.8	5.9	−15.1	−4.2
Vermont	25.8	23.3	17.3	12.7	5.7	3.7

Southern Tier	3,468.0	24.3	3,610.0	20.3	1,498.0	7.0
South Atlantic	1,766.0	25.2	2,016.0	23.0	599.0	5.5
Delaware	41.2	26.9	32.7	16.6	-3.4	-1.5
Maryland	232.9	24.5	175.7	14.9	66.9	4.9
North Carolina	342.4	27.2	323.2	20.2	71.2	3.7
Virginia	248.4	23.0	313.3	23.6	111.5	6.8
South Carolina	144.6	23.7	165.9	22.0	57.5	6.2
Georgia	302.0	27.6	310.3	22.2	19.8	1.2
Florida	428.6	30.9	658.2	36.2	254.9	10.3
West Virginia	56.1	12.5	36.9	7.3	20.6	3.8
East South Central	677.0	23.7	611.0	17.3	206.7	5.0
Alabama	160.0	20.2	120.5	12.7	77.5	7.2
Kentucky	160.9	23.9	152.5	18.3	54.1	5.5
Mississippi	106.2	24.9	106.3	20.0	29.1	4.6
Tennessee	249.4	25.7	196.5	16.1	46.0	3.2
West South Central	1,025.0	23.2	983.0	18.1	693.0	10.8
Arkansas	101.1	25.5	87.5	17.6	34.6	5.9
Louisiana	209.8	26.4	131.5	13.1	62.9	5.5
Oklahoma	104.8	17.4	107.9	15.3	73.2	9.0
Texas	626.9	23.9	638.5	19.6	622.9	16.0

Source: Bureau of Labor Statistics, U.S. Department of Labor, *Employment and Earnings, States and Areas, 1939-75*, Bulletin 1370-12 (Washington: Government Printing Office, 1977).

Table 5-10
Population Level and Growth by Region, 1962, 1967, 1972, and 1975

	Population (000s)				Percentage Increase		
State and Region	1962	1967	1972	1975	1962-1967	1967-1972	1972-1975
Northern Tier	82,785	87,414	90,519	90,362	5.6	3.6	0.2
East North Central	36,874	39,124	40,793	40,901	6.1	4.3	0.3
Illinois	10,260	10,893	11,244	11,160	6.2	3.2	−0.8
Indiana	4,725	5,000	5,287	5,313	5.8	5.7	0.5
Michigan	7,923	8,584	9,014	9,117	8.3	5.0	1.1
Ohio	9,952	10,458	10,722	10,745	5.1	2.5	0.2
Wisconsin	4,014	4,189	4,526	4,566	4.4	8.0	0.9
Middle Atlantic	35,185	36,968	37,621	37,263	5.1	1.8	−1.0
New Jersey	6,385	7,003	7,349	7,316	9.7	4.9	−0.5
New York	17,464	18,336	18,367	18,120	5.0	0.2	−1.4
Pennsylvania	11,336	11,629	11,905	11,827	2.6	2.4	−0.7
New England	10,726	11,322	12,105	12,198	5.6	6.9	0.8
Connecticut	2,640	2,925	3,080	3,095	10.8	5.3	0.5
Maine	990	973	1,026	1,059	−1.7	5.4	3.2
Massachusetts	5,201	5,421	5,796	5,828	4.2	6.9	0.6
New Hampshire	630	686	774	818	8.9	12.8	5.7
Rhode Island	872	900	969	927	3.2	7.7	−4.3
Vermont	393	417	460	471	6.1	10.3	2.4

Southern Tier	56,599	60,634	64,306	67,399	7.1	6.1	4.8
South Atlantic	26,387	28,671	31,168	32,999	8.7	8.7	5.8
Delaware	446	523	571	579	12.2	9.2	1.4
Maryland	3,245	3,682	4,048	4,098	13.5	9.9	1.2
North Carolina	4,736	5,029	5,221	5,451	6.2	3.8	4.4
Virginia	4,187	4,536	4,765	4,967	8.3	5.0	4.2
South Carolina	2,450	2,599	2,688	2,818	6.1	3.4	4.8
Georgia	4,108	4,509	4,733	4,926	9.8	5.0	4.1
Florida	5,392	5,995	7,347	8,357	11.2	22.6	13.7
West Virginia	1,823	1,798	1,795	1,803	−6.5	−0.2	0.4
East South Central	12,407	12,969	13,155	13,544	4.5	1.4	3.0
Alabama	3,342	3,540	3,521	3,614	5.9	−0.5	2.6
Kentucky	3,099	3,189	3,306	3,396	2.9	3.7	2.7
Mississippi	2,276	2,348	2,256	2,346	3.2	−3.9	4.0
Tennessee	3,690	3,892	4,072	4,188	5.5	4.6	2.8
West South Central	17,805	18,994	19,983	20,856	6.7	5.2	4.4
Arkansas	1,875	1,968	2,008	2,116	5.0	2.0	5.4
Louisiana	3,371	3,662	3,738	3,791	8.6	2.1	1.4
Oklahoma	2,435	2,495	2,633	2,712	2.5	5.5	3.0
Texas	10,124	10,869	11,604	12,237	7.4	6.8	5.5

Source: Bureau of the Census, *Current Population Reports, Series P-25*, various issues.

Table 5-11
Indicators of Fiscal Expansion by Region

State and Region	Increases in per Capita General Expenditures (dollars)			Percentage Increases in per Capita General Expenditures		
	1962-1967	1967-1972	1972-1975	1962-1967	1967-1972	1972-1975
Northern Tier	138	342	274	42.8	73.4	34.4
East North Central						
Illinois	132	305	259	40.9	67.8	34.2
Indiana	102	377	271	32.5	90.2	34.2
Michigan	122	242	174	41.9	58.7	26.6
Ohio	162	349	332	46.7	68.5	38.7
Wisconsin	103	244	258	35.3	62.0	40.4
	169	311	260	48.3	59.8	31.3
Middle Atlantic						
New Jersey	157	449	314	47.8	91.4	34.7
New York	115	385	304	38.2	92.5	37.9
Pennsylvania	216	624	372	54.3	101.5	30.1
	139	339	267	51.0	80.2	36.0
New England						
Connecticut	135	320	267	41.8	69.1	34.5
Maine	105	354	233	28.6	74.9	28.2
Massachusetts	122	270	254	41.8	65.3	37.1
New Hampshire	123	426	290	35.8	91.5	32.5
Rhode Island	104	276	241	34.4	68.0	35.3
Vermont	202	228	321	68.9	46.0	44.5
	154	364	262	41.3	69.1	29.4

Southern Tier						
	145	257	248	52.1	64.2	35.9
South Atlantic						
Delaware	162	289	277	56.1	70.3	36.4
Maryland	371	403	177	80.9	66.4	17.6
North Carolina	155	363	408	48.6	76.8	12.9
Virginia	113	201	262	49.5	58.5	46.5
South Carolina	130	258	337	45.5	68.0	52.8
Georgia	103	262	305	64.5	86.0	53.8
Florida	118	302	247	46.1	80.6	36.5
West Virginia	153	222	286	54.0	51.0	43.6
	150	301	190	59.6	75.2	27.1
East South Central						
Alabama	119	240	224	47.0	64.6	36.4
Kentucky	115	240	228	47.0	66.6	38.0
Mississippi	117	215	212	40.0	52.4	34.0
Tennessee	98	284	202	39.3	82.0	32.1
	147	221	252	61.7	57.2	41.5
West South Central						
Arkansas	138	214	214	49.6	51.5	34.4
Louisiana	118	174	215	53.3	51.2	42.0
Oklahoma	152	239	223	45.8	49.0	30.8
Texas	167	205	202	56.2	44.1	30.1
	116	237	216	43.1	61.5	34.7

Source: U.S. Bureau of the Census, *Governmental Finances in 1974-75 (1961-62, 1966-67, 1971-72), Series G-F 75*, 5 (Washington: Government Printing Office, 1976).

Table 5-12
Per Capita Income Elasticity[a] of State and Local Government Expenditures by Region

	Northern Tier	Southern Tier
1962-1967	1.26	1.20
1967-1972	1.87	1.29
1972-1975	1.21	1.19

Source: Computed from tables 5-8 and 5-11.
[a]Percentage increase in per capita expenditure divided by percent income in per capita income.

have been expected. When the states are aggregated by region, it may be seen that the number of AFDC recipients increased at a greater rate in the North than in the South in all years considered, while the reverse was true for total increases in population (table 5-15). Primary and secondary school enrollments increased at a more rapid rate in the North over the 1962-1972 period, but actually declined over the 1972-1975 period. From these aggregates, one might again infer an increasing concentration of high-cost citizens in the North and a considerably greater demand for increased numbers of school personnel, at least during the 1962-1972 period.

In fact, public employment did increase rapidly between 1962 and 1967 in response to relatively high population and school enrollment growth. Between 1967 and 1972, public employment grew at a relatively slow rate, even though the concentration of the poor appeared to increase dramatically in both regions. However, the much greater increase in per capita spending in the 1967-1972 period can be at least partially attributed to the increment in transfer payments necessitated by the growth in AFDC recipients. The 1972-1975 period does not support the demand explanation. While the growth in all three service requirement indicators was relatively low, there was a greater percentage increment in public employment. Although these results do not appear to provide strong support for a demand thesis, it is important to emphasize the very great diversity across states which is disguised in such an aggregate analysis. Particularly in the case of the rate of increase in AFDC recipients, there is great diversity within each region.

These results suggest that the explanations for expenditure increases in the two regions are at least partially to be found on the supply side, that is, in terms of increases in the level of public employee compensation. As may be seen in table 5-16, the percentage increase in payroll per employee was higher in the Northern than in the Southern states over the 1962-1972 period—despite the fact that the capacity to finance such increases in Northern states was declining. By the 1972-1975 period, the rate of increase in average wages in the North had fallen below that in the South.[25] The rates of wage increase observed during these periods tend to support the thesis

Implications for Public Policy   187

that increases in expenditures closely parallel increases in public employee compensation rates.

Revenue Growth

According to the scenario above, the fisc in the Northern states has expanded at about the same rate as that in the Southern states despite very great differences in the growth of their respective economic and demographic bases. As a consequence, revenue effort in the Northern tier states must have increased more rapidly, or the flow of federal aid to the Northern states must have increased. The reality of an increase in revenue effort is borne out by a recent Advisory Commission on Intergovernmental Relations (ACIR) publication which attempts to classify states with reference to both the level and the direction of tax effort.[26] Of the states classified as having high and rising levels of tax effort, nine are in the Northern tier and three are in the South. Similar findings may be found in tables 5-17 and 5-18, where both per capita revenues from own sources and own-source revenues per $1,000 are significantly higher in the Northern tier than in the South.

A comparison of the growth in own-source revenues to the growth in personal income, employment, and population shows a greater revenue-income elasticity in the North in every period (see table 5-18).[27] This means that, on average, the tax on each increment to income was greater in the North, or that the tax reduction of disposable income was largest in the North.

The presentation in table 5-19 disaggregates increases in state and local government revenue by source of increase. The results are helpful in understanding the mechanics of the fiscal response over the period in question. Between 1962 and 1967, Southern states financed expenditure increases through the use of sales and income taxes and through substantial increments in federal aid. In the North, where income and employment growth was slower, the increments were derived relatively more from property taxation and relatively less from federal aid. About the same pattern was observed between 1967 and 1972, when expenditure increases in the North were highest for the period under consideration. Between 1972 and 1975, when the rate of expenditure growth slowed in both regions, essentially the same pattern was observed for the Southern states—heavy reliance on sales and income taxes and relatively little on the property tax as a course of increased revenue. In the North, even though relatively more of the increment was financed from sales and income taxes, the relative use of property taxation remained much greater than in the South. Also noteworthy within the 1972-1975 period was the substantial increase in the reliance on federal grants to finance expenditure increments in the Northern tier. As may be seen in table 5-19, the pattern described above holds true for most states in the two regions.

Table 5-13
AFDC Recipients by Region

State and Region	Level (000s)				Percentage Increase		
	1962	1967	1972	1975	1962-1967	1967-1972	1972-1975
Northern Tier	1,602	2,282	4,986	5,288	42.4	118.5	6.1
East North Central							
Illinois	623	792	2,179	2,314	27.1	175.1	6.2
Indiana	265	275	773	783	3.8	181.1	1.3
Michigan	47	51	168	163	8.5	229.4	-3.0
Ohio	121	183	600	655	51.2	227.9	9.2
Wisconsin	147	222	497	552	51.0	123.9	11.1
	43	61	141	161	41.9	131.1	14.2
Middle Atlantic							
New Jersey	815	1,222	2,216	2,307	49.9	81.3	4.1
New York	83	145	420	444	74.7	189.7	5.7
Pennsylvania	399	786	1,190	1,230	97.0	51.4	1.1
	333	291	606	633	-12.6	108.2	4.5
New England							
Connecticut	164	268	591	667	63.4	120.5	12.9
Maine	43	62	118	126	44.2	90.3	6.8
Massachusetts	22	22	72	83	0.0	227.3	29.2
New Hampshire	70	138	309	356	97.1	123.9	15.2
Rhode Island	4	9	24	27	125.0	166.7	12.5
Vermont	20	29	48	53	45.0	65.5	10.4
	5	8	20	22	60.0	150.0	10.0

Southern Tier						
	1,159	1,418	2,941	2,988	22.3	107.4	1.6
South Atlantic	564	667	1,394	1,440	18.3	109.0	3.3
Delaware	7	17	29	33	142.9	70.6	13.8
Maryland	58	108	220	219	86.2	103.7	-0.5
North Carolina	115	107	151	177	-7.0	41.1	17.2
Virginia	44	58	165	177	31.8	184.5	7.3
South Carolina	34	28	121	138	-17.6	332.1	14.0
Georgia	64	105	341	358	64.1	224.8	5.0
Florida	103	148	300	264	43.7	102.7	-12.0
West Virginia	139	96	67	74	30.9	-30.2	10.4
East South Central	332	377	676	719	13.6	79.3	6.4
Alabama	90	75	150	163	16.7	100.0	8.7
Kentucky	81	106	153	162	30.9	44.3	5.9
Mississippi	79	99	183	187	25.3	84.8	2.2
Tennessee	82	97	190	207	18.3	95.9	8.9
West South Central	263	374	871	829	42.2	132.9	-4.8
Arkansas	25	39	91	103	56.0	133.3	14.3
Louisiana	94	124	250	235	31.9	100.8	-6.0
Oklahoma	71	90	91	100	26.8	1.1	9.9
Texas	73	121	439	391	65.8	262.8	-10.9

Source: U.S. Department of Health, Education, and Welfare, Social Security Administration, August 1975, vol. 38, no. 8: *Social Security Bulletin* (Washington: Government Printing Office, 1975), p. 65; and U.S. Bureau of the Census, *Statistical Abstract of the United States: 1963, 1968, 1973* (Washington, 1963, 1968, 1973), sec. 10

Table 5-14
Primary and Secondary School Enrollment by Region

State and Region	Level (000s)				Percentage Increase		
	1962	1968	1972	1974	1962-1968	1968-1972	1972-1974
Northern Tier	15,393	17,952	19,187	18,676	16.6	6.9	−2.7
East North Central	7,392	8,691	9,212	8,900	17.6	6.0	−3.4
Illinois	1,892	2,188	2,388	2,281	15.6	9.1	−4.5
Indiana	1,038	1,181	1,220	1,191	13.7	3.3	−2.4
Michigan	1,735	2,042	2,193	2,131	17.7	7.4	−2.8
Ohio	1,975	2,359	2,416	2,323	19.4	2.4	−3.8
Wisconsin	752	921	995	974	22.5	8.0	−2.1
Middle Atlantic	6,030	6,942	7,393	7,196	15.1	6.5	−2.7
New Jersey	1,141	1,368	1,514	1,470	19.9	10.7	−2.9
New York	2,856	3,318	3,511	3,426	16.2	5.8	−2.4
Pennsylvania	2,033	2,256	2,368	2,300	11.0	5.0	−2.9
New England	1,971	2,319	2,582	2,580	17.7	11.3	−0.1
Connecticut	516	610	674	660	18.2	10.5	−2.1
Maine	205	229	247	244	11.7	7.9	−1.2
Massachusetts	918	1,084	1,190	1,218	18.1	9.8	2.4
New Hampshire	114	138	168	174	21.1	21.7	3.6
Rhode Island	142	167	190	179	17.6	13.8	−5.8
Vermont	76	91	113	105	19.7	24.2	−7.1

Southern Tier	12,804	14,033	14,388	14,487	9.6	2.5	0.7
South Atlantic	5,961	6,609	6,858	6,960	10.9	3.8	1.5
Delaware	90	118	134	131	31.1	13.6	-2.2
Maryland	650	826	921	896	27.0	10.7	-2.7
North Carolina	1,142	1,193	1,159	1,178	4.5	-2.8	1.6
Virginia	900	1,017	1,069	1,093	13.0	5.1	2.2
South Carolina	631	644	640	606	2.1	-0.6	-5.3
Georgia	996	1,095	1,084	1,081	9.9	-1.1	-0.3
Florida	1,106	1,300	1,437	1,571	17.5	10.5	9.3
West Virginia	446	416	414	404	-6.7	-0.5	-2.4
East South Central	2,891	2,968	2,916	2,860	2.7	-1.8	-1.9
Alabama	807	831	783	764	3.0	6.3	-2.4
Kentucky	651	680	715	705	3.6	5.1	-1.4
Mississippi	585	583	526	513	-0.3	-9.8	-2.5
Tennessee	848	874	892	878	3.1	2.1	-1.6
West South Central	3,952	4,456	4,614	4,667	12.8	3.5	1.1
Arkansas	436	451	459	454	3.4	1.8	-1.1
Louisiana	733	840	847	842	14.6	0.8	-0.6
Oklahoma	557	593	614	591	6.5	3.5	-3.7
Texas	2,226	2,572	2,694	2,780	15.5	4.7	3.2

Source: U.S. Bureau of the Census, *Statistical Abstract of the United States: 1965* (Washington, 1965), sec. 4; and National Education Association, Division of Research, *Estimates of School Statistics, Research Report: 1967, 1972, 1975* (Washington, 1967, 1972, 1975).

Table 5-15
Indicators of Growth in Servicing Requirements

	Percentage Increases					
	1962-1967		1967-1972		1972-1975	
	North	South	North	South	North	South
AFDC	42.4	22.3	118.5	107.4	6.1	1.6
Population	5.6	7.1	3.6	6.1	0.2	4.8
Enrollment	16.6	9.6	6.9	2.5	-2.7	0.7
Public employees	24.4	31.5	20.3	24.7	33.9	30.6
Per capita expenditures	42.8	52.1	73.4	64.2	34.4	35.9

Source: Computed from tables 5-13, 5-9, 5-14, 5-15, and 5-4.

This pattern of revenue increase may reflect the greater automatic responsiveness of tax systems in the South which rely more on sales and less on property taxes. While detailed comparisons are not readily available, it would seem reasonable to assume that relatively more of the revenue increase in the North was the result of discretionary changes in the tax system.

Implications for Public Policy

It is important to separate the general fiscal problems of state and local governments from those which have been exacerbated by the regional shifts that lie at the heart of this discussion. It is particularly important to separate the fiscal problems and public service deficiencies which are primarily attributable to low income—the Southern problem.

The basic dilemma faced by several of the declining states in the Northeast is that their public sector has become overdeveloped relative to financial capacity. As a result, tax burdens are thought to be too high, there is little additional public money to be devoted to what are thought to be serious city fiscal problems, fixed debt and pension commitments are high, union compensation demands will likely parallel cost-of-living increments, and there seems to be no short-term reversal of existing economic trends. To be sure, this pattern does not fit all state and local governments in the Northeastern and Midwestern regions and likely describes some Southern metropolitan-area governments. But the pattern tends to hold for many governments in the Northern tier and tends not to hold for most in the Southern rim.

The strategies for dealing with these fiscal problems would seem to be of four types: reversal of the economic decline, in both the central cities and the region; assistance during the transition period; a strengthening of fiscal position of the poorest local jurisdictions through a grants program and federal welfare assumption; and fiscal planning in the declining region to

Table 5-16
Percentage Increase in State and Local Government Employment and Employee Wages by Region

	Total Employment		
State and Region	1962-1967	1967-1972	1972-1975
Northern Tier	24.4	20.3	33.9
East North Central	25.6	17.8	36.2
Illinois	27.1	20.4	32.6
Indiana	31.9	10.0	35.7
Michigan	26.3	16.3	37.0
Ohio	20.4	16.5	31.3
Wisconsin	22.3	26.0	44.5
Middle Atlantic	26.4	17.9	23.9
New Jersey	26.9	21.7	33.5
New York	25.2	17.1	14.1
Pennsylvania	27.1	14.8	24.1
New England	22.5	23.6	37.0
Connecticut	28.6	19.6	28.0
Maine	18.0	27.1	32.4
Massachusetts	16.9	20.3	26.9
New Hampshire	34.2	21.1	58.9
Rhode Island	23.6	22.1	23.5
Vermont	23.4	31.5	52.2
Southern Tier	31.5	24.7	30.6
South Atlantic	34.7	28.4	30.3
Delaware	36.9	39.8	17.9
Maryland	40.8	26.6	24.0
North Carolina	32.7	24.7	32.3
Virginia	36.8	25.4	39.1
South Carolina	31.2	31.1	40.3
Georgia	36.1	32.2	30.4
Florida	39.0	31.3	34.3
West Virginia	24.3	15.8	24.4
East South Central	29.2	22.6	28.3
Alabama	29.4	21.2	28.7
Kentucky	26.7	18.9	36.1
Mississippi	28.2	23.1	24.8
Tennessee	32.6	27.2	23.7
West South Central	27.5	19.3	33.6
Arkansas	27.5	21.2	37.3
Louisiana	25.0	16.2	27.6
Oklahoma	28.5	16.0	34.9
Texas	28.8	23.7	34.5
	Employment per 10,000 Population		
Northern Tier	18.1	15.3	9.4
East North Central	18.3	16.0	8.3
Illinois	17.8	16.5	8.3
Indiana	23.0	4.0	7.7
Michigan	18.1	29.9	12.8
Ohio	15.5	12.9	6.6
Wisconsin	17.3	16.8	6.3

Table 5-16 *(cont.)*

State and Region	Employment per 10,000 Population		
	1962-1967	1967-1972	1972-1975
Middle Atlantic	19.7	15.7	10.5
New Jersey	15.2	15.7	17.1
New York	19.4	19.4	7.8
Pennsylvania	24.4	11.9	6.5
New England	17.1	14.4	9.7
Connecticut	15.4	13.5	7.9
Maine	18.6	20.1	3.4
Massachusetts	11.9	12.7	6.1
New Hampshire	21.6	7.8	16.6
Rhode Island	20.5	13.6	13.1
Vermont	14.5	18.9	11.3
Southern Tier	22.6	19.4	8.6
South Atlantic	23.8	21.0	8.3
Delaware	22.2	29.4	0.9
Maryland	23.6	14.9	10.0
North Carolina	19.7	24.8	7.8
Virginia	28.0	19.4	14.3
South Carolina	23.5	27.9	10.4
Georgia	23.2	26.3	10.8
Florida	26.0	8.4	4.1
West Virginia	24.2	16.9	8.2
East South Central	23.0	20.6	8.2
Alabama	21.2	22.2	8.9
Kentucky	22.6	14.9	15.6
Mississippi	23.5	27.7	1.6
Tennessee	24.5	17.6	6.5
West South Central	20.1	15.1	9.8
Arkansas	19.3	20.6	8.7
Louisiana	15.1	14.4	10.4
Oklahoma	26.1	9.9	8.0
Texas	19.9	15.4	12.1
	Payroll per Employee		
Northern Tier	29.3	36.7	6.5
East North Central	26.4	38.2	5.7
Illinois	20.5	43.6	3.1
Indiana	25.8	24.9	3.0
Michigan	26.7	49.2	4.4
Ohio	28.0	37.1	8.9
Wisconsin	30.8	36.0	9.0
Middle Atlantic	29.6	39.6	7.7
New Jersey	26.1	37.8	6.3
New York	27.2	45.9	8.4
Pennsylvania	35.6	35.1	8.6
New England	31.6	34.1	6.5
Connecticut	30.5	38.4	3.2
Maine	31.6	30.2	1.7
Massachusetts	31.7	32.2	4.5
New Hampshire	29.3	34.0	9.2

Table 5-16 (cont.)

	Payroll per Employee		
State and Region	1962-1967	1967-1972	1972-1975
Rhode Island	33.1	37.4	9.0
Vermont	33.4	32.4	11.6
Southern Tier	28.1	33.4	10.6
South Atlantic	28.5	35.9	9.5
Delaware	23.9	39.0	9.2
Maryland	33.1	37.6	14.2
North Carolina	25.4	32.0	3.8
Virginia	30.0	37.1	7.3
South Carolina	28.7	38.1	4.1
Georgia	31.6	30.1	16.1
Florida	29.0	45.3	10.0
West Virginia	26.3	27.7	11.1
East South Central	28.3	33.2	11.5
Alabama	33.6	31.2	15.6
Kentucky	23.4	37.1	2.6
Mississippi	25.6	32.8	15.0
Tennessee	30.7	31.8	12.6
West South Central	27.0	28.7	12.0
Arkansas	33.0	24.4	14.8
Louisiana	24.9	31.0	10.4
Oklahoma	23.0	30.3	5.6
Texas	27.1	28.9	17.1

Source: U.S. Bureau of the Census, *Public Employment in 1975, Series GE 75*, 5 (Washington: Government Printing Office, 1976); U.S. Bureau of the Census, *Compendium of Public Employment in 1972 (1962, 1967)* 3, no. 2 (Washington: Government Printing Office, 1974).

bring about a better balance between the size of the public sector and the size of the economic base available to support that public sector.

An alternative strategy would be to take no action to correct the fiscal problems of governments in the declining region. The argument would go that market forces are already underway which are correcting regional disparities in income, employment, and population and that the regional disparities in public service levels also should narrow. Eventually, as the resource base continues to grow slowly, the public sector in the Northeast will also grow slowly. The problem with this line of reasoning is that shrinkage in the public sector in the Northeast will likely mean a cutting of service levels in those areas where expenditures are greatest—health, education, and welfare. This may imply that much of the painful burden of the transition to a lower level of public services will be borne by lower-income residents in the declining regions.

Given these strategies, there would seem to be five policy directions open: cut services, raise taxes, increase productivity, increase federal

Table 5-17
Levels of Revenue Effort: Selected Northern and Southern Tier States, 1975

State and Region	Revenues from Own Sources per $1,000 of Personal Income	Per Capita Revenue from Own Sources
Northern Tier	$153.98	$ 911.51
East North Central	148.95	826.69
Illinois	141.52	881.21
Indiana	148.19	770.98
Michigan	153.49	897.12
Ohio	128.24	706.15
Wisconsin	173.33	901.32
Middle Atlantic	163.14	1,021.48
New Jersey	142.28	890.48
New York	205.18	1,236.01
Pennsylvania	141.96	773.84
New England	153.58	859.33
Connecticut	127.47	821.03
Maine	152.50	692.10
Massachusetts	165.57	948.59
New Hampshire	135.85	663.46
Rhode Island	145.93	788.03
Vermont	194.17	878.51
Southern Tier	144.86	690.66
South Atlantic	146.40	721.41
Delaware	156.30	975.57
Maryland	156.07	926.55
North Carolina	135.26	620.77
Virginia	138.14	728.80
South Carolina	146.87	625.60
Georgia	149.82	705.47
Florida	137.31	719.93
West Virginia	151.46	657.66
East South Central	137.47	623.43
Alabama	99.45	603.53
Kentucky	151.40	664.81
Mississippi	162.19	611.08
Tennessee	136.85	613.97
West South Central	149.17	685.66
Arkansas	131.72	539.08
Louisiana	176.35	768.87
Oklahoma	147.58	675.25
Texas	141.00	687.53
U.S. Median	152.50	794.81

Source: U.S. Bureau of the Census, *Governmental Finances in 1974-75*, Series G-F 75, 5 (Washington: Government Printing Office, 1976).

Table 5-18
Overall Responsiveness of Revenues to Economic Activity, 1963-1975

	Northern Tier			Southern Tier		
	1962-1967	1967-1972	1972-1975	1962-1967	1967-1972	1972-1975
Percentage increase in revenues from own sources	47.0	82.0	29.0	54.0	80.9	38.5
Percentage increase in personal income	40.0	44.2	27.8	49.4	61.2	38.6
Revenue-income elasticity	1.2	1.9	1.0	1.1	1.3	1.0
Percentage increase in total employment	15.2	5.8	0.6	24.3	20.3	7.0
Percentage increase in population	5.6	3.6	0.2	7.1	6.1	4.8

Source: Computed from tables 5-8, 5-9, 5-10, and 5-19.

assistance, or improve the local economy. The first three are options for state and local government action while the last two require federal action.

State and Local Government Options

Increased productivity in the public sector is a favorite policy recommendation in that it solves fiscal problems without requiring governments to either raise taxes or cut services. While there is clearly room for improved management at the local government level, large savings (relative to projected deficits) from increased productivity in the public sector is not a realistic expectation.[28]

Revenues might be increased through further increase in the effective tax rate. The argument against this is the possible retarding effect on economic development. State and local government revenue effort in the Northeastern and Midwestern regions is already high relative to the South, a difference that would reinforce the argument to lower rather than raise taxes for competitive reasons. While this pattern certainly does not hold for all states in the declining region—Connecticut and Ohio have revenue efforts among the lowest in the United States—it fits many of the large industrial states. Service level-reductions are the most likely route. While there will continue to be absolute cutbacks in some areas and reductions in the scope of some services, this will mostly take the form of services not expanding to accommodate increasing needs and increasing unit cost of provision. This does not mean that expenditures will decline. Increasing wages and benefits can drive up expenditures by a significant amount, without raising service levels.

Table 5-19
Increases in General Revenues of State and Local Governments

	1962-1967			1967-1972			1972-1975		
	Percentage of Increase due to			Percentage of Increase due to			Percentage of Increase due to		
	Sales and Income Taxes	Property Taxes	Federal Aid	Sales and Income Taxes	Property Taxes	Federal Aid	Sales and Income Taxes	Property Taxes	Federal Aid
Northern Tier	21.0	23.1	19.5	22.7	29.5	20.0	28.7	16.7	31.0
East North Central									
Illinois	25.0	20.4	17.6	25.7	27.8	19.2	40.1	9.0	26.1
Indiana	21.2	21.4	19.3	28.4	23.1	32.3	40.8	15.6	12.5
Michigan	32.0	22.3	14.9	17.6	35.9	15.2	51.2	-1.4	20.4
Ohio	19.1	17.4	20.8	28.7	23.5	18.5	22.4	23.6	34.8
Wisconsin	13.4	29.3	18.8	24.8	21.6	15.4	40.0	7.4	29.9
	39.4	11.6	14.6	28.8	35.0	14.7	46.1	-0.5	33.1
Middle Atlantic									
New Jersey	24.4	22.9	18.6	24.9	24.3	20.3	34.8	20.0	26.8
New York	20.3	28.6	15.9	13.5	37.5	20.1	20.0	32.5	23.7
Pennsylvania	31.6	20.6	17.6	30.8	21.2	22.6	42.7	19.0	22.5
	21.3	19.5	22.4	30.3	14.1	18.4	41.8	8.7	34.2
New England									
Connecticut	15.9	25.4	21.5	10.0	33.4	20.5	25.8	21.4	37.2
Maine	11.1	33.5	19.8	21.1	35.5	14.7	31.5	22.6	42.9
Massachusetts	23.7	17.5	28.1	19.5	39.2	26.2	29.6	7.8	40.2
New Hampshire	25.0	20.3	20.2	22.1	35.3	21.8	33.6	35.6	22.9
Rhode Island	1.3	40.5	14.8	1.6	39.4	17.0	14.5	23.9	39.3
Vermont	15.5	21.8	29.9	31.5	22.2	21.6	27.1	22.8	34.9
	19.0	18.7	16.3	18.5	29.0	21.5	18.4	15.7	42.9

Southern Tier	21.3	13.8	26.4	25.4	12.1	23.4	31.2	9.7	28.7
South Atlantic									
Delaware	25.7	15.6	23.4	26.1	13.7	22.2	32.7	11.2	29.3
Maryland	15.4	9.6	17.9	18.8	8.9	23.6	36.8	10.2	20.9
North Carolina	25.1	13.3	14.7	36.5	14.9	17.5	36.6	10.0	28.3
Virginia	25.3	14.3	23.2	24.0	14.4	22.6	35.9	9.5	42.8
South Carolina	34.1	14.3	23.1	26.2	17.0	19.2	29.0	13.9	27.2
Georgia	58.4	24.3	21.8	27.8	15.6	24.0	32.2	9.1	29.1
Florida	20.2	17.4	23.1	21.2	15.5	23.7	32.0	17.5	30.4
West Virginia	12.1	22.1	21.1	23.5	16.4	15.5	21.1	12.7	24.1
	15.1	9.9	42.6	30.8	7.1	31.7	38.2	7.9	32.0
East South Central									
Alabama	21.0	9.2	31.1	27.1	8.6	26.0	31.2	7.7	28.5
Kentucky	25.5	6.6	25.5	20.8	4.8	32.2	33.1	3.2	24.9
Mississippi	19.8	8.2	36.9	35.4	7.8	20.2	27.7	7.7	28.3
Tennessee	18.4	10.4	33.1	29.2	7.7	30.1	32.2	9.4	31.8
	20.1	11.5	28.8	22.9	14.0	21.5	32.0	10.4	28.8
West South Central									
Arkansas	12.9	14.6	27.8	22.3	12.4	23.3	28.0	8.6	27.7
Louisiana	15.2	11.2	33.4	20.9	10.4	27.8	38.6	8.7	32.1
Oklahoma	19.8	8.5	24.0	26.9	11.5	20.2	26.2	0.0	27.4
Texas	8.0	16.4	31.5	20.5	9.0	24.3	30.6	7.8	27.7
	8.8	22.4	27.1	20.9	19.0	21.0	16.5	17.9	23.8

Source: U.S. Bureau of the Census, *Government Finances in 1972 (1962, 1967): Compendium of Government Finances* 4, no. 5 (Washington: Government Printing Office, 1974); and U.S. Bureau of the Census, *Governmental Finances in 1974-75, Series G-F 75*, 5 (Washington: Government Printing Office, 1976).

There is another type of reform which is highly desirable but politically difficult. If the tax base in the suburbs could be tapped more fully so as to balance needs for services with capacity to finance, the fiscal situation in central cities could be markedly improved. History has not shown this to be a viable alternative in the Northern industrial states.

Federal Options

The federal government could increase the flow of aid to the state to prop up the public sector during this period of decline. A program of increased aid during a transition period in which the state sought to balance its long-term spending expectations with its likely future economic growth would be a sane program. On the other hand, federal grants to maintain an overdeveloped public sector would only prolong the period of continuing annual fiscal crisis.

A number of federal policies might be undertaken during the fiscal adjustment period, that period when the public sector in the North is moving to a lower level which is commensurate with its capacity to finance. One element of such a program would be an expansion of the countercyclical revenue-sharing program and the temporary public sector job-related programs. But perhaps the most important ingredient of a fiscal reform would be a higher level of federal financing of public welfare. The removal of a substantial share of welfare costs from the declining states in the Northeast would free substantial resources for other uses. The net effect would be to allow governments in the declining states to maintain a higher level of fiscal activity with respect to other social services.

A similar position might be taken with respect to regional development subsidies. They only prolong the period of transition to a lower, but stable level of activity. The longer the period of this transition, the greater the uncertainty with respect to business investment and the greater the chance for a snowballing effect of the decline.

An often discussed approach to dealing with the problems of decline is the creation of a "Regional Energy and Development Corporation" that would finance regional development projects using federally guaranteed taxable bonds. It is hoped that such an activity would accelerate development of Eastern coal and result in substantial job generation. If regional subsidies worked, they could have a strong positive effect on the finances of governments in the declining region. There are two caveats, however, even to the potentially favorable government finance effects. One is that the fiscal problems in the declining region are very much the fiscal problems of the central cities in those regions. Historically, these cities have not always shared in the economic growth of the region, and therefore it is not clear

Implications for Public Policy

how much their fiscal positions would improve in the event the regional shifts slowed. A second, and related, caveat is that the states in the declining region tend to be more heavily dependent on local property taxation, which may make it difficult to fully capture increases in regional income and employment for the public sector. But the most important issue with respect to regional subsidies is whether they induce any *net* improvement in private sector economic activity.

Finally, it should be noted that a successful federal approach will not likely grow out of political compromise. The problems of state and local governments in the regions are sufficiently different that any remedial program which benefits all is not apt to substantially benefit any. Programs such as general revenue sharing, a formula-based program with something for everyone, is an almost classic case of the "compromise effect."

Improved Fiscal Balance

The fiscal problem of many Northern tier states is that their public sectors are overdeveloped. The state's resource bases will no longer support the high level of public services provided in the state, unless tax rates are continuously increased. While shifts in population and economic activity are tending toward equalizing income across the country, the states have retained dominance in their relative national role in state and local fiscal activity. This can no longer be done. A downward transition must be recognized, and policy should center on selecting priorities in the adjustment of public service levels. With appropriate federal aid, this need not mean severe service cutbacks in all areas, but rather a slow growth in services provided while the rest of the nation catches up.

Lessons for the Growing Region

It is likely that the rapid fiscal expansion in the state-local sector in the South has yet to come. Investments in public infrastructure and human capital often lag behind the growth in population and income level. It is noteworthy that this growth has been particularly rapid over the past five years.

If the Southern tier of states is about to enter a fiscal growth period similar to that experienced in the Northern tier in the 1960s, some of the painful fiscal lessons of that period might be well learned. Much of the problem facing the Northern tier states was not of their own making. The very rapid fiscal expansion in the mid- and late 1960s and early 1970s was to a large extent the result of union pressures for higher employee compensation, a demand that was abetted by a high rate of inflation, and a crowding of high-cost, low-income citizens into the central cities. Much of this expenditure increase would have been difficult to avoid. Other aspects of the expansion, however, were more discretionary—the making of substantial long-term fixed debt and pension commitments, the addition of substantial numbers to the public employee roles, and the buying into federal programs to expand the scope of services offered.

The growing states with rapidly developing public sectors could learn much from this experience. But the lesson is not that public employee unionization should be resisted or that public service levels should be kept at modest levels, but rather that the longer-term consequences of fiscal decisions should be continuously monitored. Moreover, there are conditions in the growing region which may make the growth experience much less painful than in the Northern tier. A more favorable local government structure and a more elastic tax mix that is less reliant on the property tax may allow big, newer cities in the growth region to avoid the central-city financial crisis which is so common in the Northern tier.

Notes

1. See, for example, William H. Miernyk, "The Northeast Isn't What It Used to Be," in *Balanced Growth for the Northeast* (New York State Senate, 1975); Christopher Carlaw, *Boston and the Flight to the Sunbelt* (Boston: Boston Development Authority, October 1976); and David Puryear and Roy Bahl, *Economic Problems of a Mature Economy*, Occasional Paper no. 27 (Metropolitan Studies Program, The Maxwell School, Syracuse University, April 1976).

2. See, for example, George E. Peterson, "Finance," in *The Urban Predicament*, eds. William Gorham and Nathan Glazer (Washington: The Urban Institute, 1976); and Roy Bahl, Alan Campbell, David Gretak, Bernard Jump, and David Puryear, "Impact of Economic Base Erosion, Inflation, and Retirement Costs on Local Governments," in *Fiscal Relations in the American Federal System, Hearings before a Subcommittee on the Committee on Governmental Operations* (Washington: Government Printing Office, July 1975).

3. Notable exceptions here are Richard P. Nathan and Paul R. Dommel, who in "Understanding Central City Hardship" [*Political Science Quarterly* 21, no. 1 (Spring 1976)] argue a relationship between regional shifts and urban fiscal problems, and Tom Muller, who argues that population decline is a reasonable proxy for fiscal distress in "The Declining and Growing Metropolis—A Fiscal Comparison" [in *Post-Industrial America: Metropolitan Decline and Regional Job Shifts*, eds. George Sternlieb and

James W. Hughes (New Brunswick, N.J.: Center for Urban Policy Research, State University of New Jersey, 1975), pp. 197-220].

4. Excluding the District of Columbia.

5. The states included in each region are enumerated in the tables which follow. Some authors have followed a procedure of excluding certain states in these regions on grounds that they are qualitatively different in terms of economic base. For example, Jusenius and Ledebur exclude Maine, Vermont, and New Hampshire because the industrial bases of these states differ in kind and degree from the rest of the major region. See C.L. Jusenius and L.C. Ledebur, *A Myth in the Making: The Southern Economic Challenge and the Northeast Economic Decline* (Washington: Economic Development Administration, U.S. Department of Commerce, November 1976), p. 2.

6. See David Puryear, Roy Bahl, and Seymour Sacks, *Federal Grants: Their Effect on State and Local Expenditures, Employment Levels, Wage Rates* (Washington: Advisory Commission on Intergovernmental Relations, February 1977), chapter 2.

7. Sack's East and Midwest regions correspond approximately to our Northern tier, and his Southern region to our Southern tier, with the following exceptions: in the Midwest he includes Des Moines, Wichita, Minneapolis, Kansas City, St. Louis, and Omaha; in the East he includes Washington, D.C. Advisory Commission on Intergovernmental Relations, *Trends in Metropolitan America* (Washington: Government Printing Office, 1977).

8. Vincent Marando, "The Politics of Metropolitan Reform," in *State and Local Government: The Political Economy of Reform*, eds. Alan Campbell and Roy Bahl (New York: The Free Press, 1976), pp. 24-49.

9. Common municipal functions exclude education, hospitals, and other variable functions as defined by the Census.

10. Muller, "The Declining and Growing Metropolis—A Fiscal Comparison," pp. 203-206.

11. The Department of Housing and Urban Development (HUD) has established fair-market rent levels for about 3,100 areas throughout the nation in conjunction with their Section Eight lease housing program. One might argue the use of the data to construct a cost-of-living index because (1) housing costs make up a large proportion of total consumption and (2) much of the variance in living costs might be attributed to housing. Following this procedure, we have taken the indices computed for 501 formula cities under the HUD community development block program, aggregated and averaged the indices by state, and then compared them to the U.S. average to develop an index. For a discussion of the potential use of the HUD index as a cost-of-living measure in another context, see the Controller-general of the United States, "Why the Formula for Allocating

Community Development Block Grant Funds Should Be Improved" (Washington: General Accounting Office, December 1976).

12. There are not adequate deflators for this purpose. The choices here were between the Bureau of Labor Statistics (BLS) levels of living for low-, intermediate-, and high-income families and the HUD index of rent. We chose the latter because the BLS data are available only for forty-one metropolitan areas and this would not seem to provide adequate regional coverage. See Bureau of Labor Statistics, "Autumn 1976 Urban Family Budgets and Comparative Indexes for Selected Urban Areas," *News* (Washington: Department of Labor, April 27, 1977), pp. 77-369.

13. David Greytak and Bernard Jump, "Inflation and Local Government Expenditures and Revenues: Method and Case Studies," *Public Finance Quarterly*, June 1977.

14. Richard P. Nathan and Paul R. Dommel, "The Strong Sunbelt Cities and the Weak Cold Belt Cities," Hearings before the Subcommittee on the City, of the House Committee on Banking, Finance and Urban Affairs, *Toward a National Urban Policy*, 95th Cong. (Washington: Government Printing Office, 1977), pp. 19-26; and "Understanding Central City Hardship," *Political Science Quarterly* 21, no. 1 (Spring 1976):61-62.

15. For a parallel analysis of the New York State economy and fisc, see Roy Bahl, "The Long Term Fiscal Outlook for New York State," in *The Decline of New York in the 1970's*, ed. Benjamin Chinitz (Binghamton, N.Y.: Center for Social Analysis, State University of New York at Binghamton, May 1977), pp. 95-142.

16. Jusenius and Ledebur, *A Myth in the Making*.

17. Michael R. Greenberg and Nicholas J. Valente, "Recent Economic Trends in the Major Northeastern Metropolises," in *Post-Industrial America: Metropolitan Decline and Inter-Regional Job Shifts*, eds. George Sternlieb and James Hughes (New Brunswick, N.J.: The Center for Urban Policy Research, Rutgers University, 1975), pp. 77-100.

18. Daniel Garnick, "The Northeast States in the Context of the Nation," in *The Decline of New York in the 1970's*, ed. Benjamin Chinitz (Binghamton, N.Y.: Center for Social Analysis, State University of New York at Binghamton, 1976).

19. Congressional Budget Office, "Troubled Local Economics and the Distribution of Federal Dollars" (Government Printing Office, August 1977).

20. These possibilities are examined for New York City in Roy Bahl and David Greytak, "The Response of City Government Revenues to Changes in Employment Structure," *Land Economics* 52, no. 4 (November 1976).

21. Garnick, "The Northeast States in the Context of the Nation," p. 188.

22. Puryear and Bahl, *Economic Problems of a Mature Economy*.

23. Jusenius and Ledebur, *A Myth in the Making*, pp. 1-5.

24. R.G. Ehrenberg, "The Demand for State and Local Government Employees," *American Economic Review* 63, no. 3 (June 1973):366-79; and T.E. Borcherding and R.T. Deacon, "The Demand for Services of Non-Federal Governments," *American Economic Review* 62, no. 5 (December 1972):891-901.

25. It is important to reemphasize that the rates of increase of average wages measure do not total compensation, but only direct wage and salary payments. To the extent there are regional differences in the pension and fringe-benefit component of compensation *increases*, these comparisons are distorted. One view would be that this distortion is in the direction of underestimating growth rates in compensation for employees of Northern states.

26. Advisory Commission on Intergovernmental Relations, *Measuring the Fiscal Blood Pressure of the States* (Washington: Government Printing Office, 1977).

27. Revenue-income elasticity is the percentage increase in revenue divided by the percentage increase in personal income. A more rigorous measure of the revenue-income elasticity would require adjusting the revenue data levels for discretionary changes in both the rates and bases of the tax systems within the several states.

28. A review of the issues surrounding productivity measurement and improvement is presented in Jesse Burkhead and John P. Ross, *Productivity in the Local Government Sector* (Lexington, Mass.: D.C. Heath, 1974).

6

Regional Impact of Federal Tax and Spending Policies

George E. Peterson and *Thomas Muller*

Introduction

Few domestic issues now inflame the passions so thoroughly as the regional distribution of federal spending. In view of the present intensity of political debate, it is remarkable that the federal government in the past has not found it necessary to articulate a regional spending policy or—with rare exceptions—a regional development policy of any other description.

The reason for this apparent omission has been the implicit assumption that both final goods and the factors of production are sufficiently mobile to make the initial distribution of federal dollars relatively unimportant to the economic prosperity of citizens living in particular regions of the country. Interstate highways built in one region make it possible to produce goods where it is cheapest to do so and to deliver them, with cost savings, to markets elsewhere. At the same time, the highway network facilitates the migration of households looking for job opportunities. Military procurement contracts, although initially assigned to firms in one state, spawn a series of subcontracts that diffuse the benefits to firms located in other parts of the country. Any long-term regional employment growth stimulated by defense spending is satisfied, in large part, by the migration of job seekers. These manifest linkages between the regions have been taken in the past to imply that federal expenditures for goods and services could be made where they were most efficient, measured in a narrow economic sense, in confidence that the indirect benefits eventually would be spread throughout the country.

For the first century of the nation's history, when the several regions often found it easier to conduct commerce with European states than with one another, and when the benefits of federally financed investments were captured largely within the region of initial spending, the sectionalism debate in this country loomed large in almost all domestic policy decisions. Until recently it had been supplanted by the view that in the modern United States, regions are of little consequence. The federal government's few explicit regional development policies of the last half-century have been directed at areas of the country that seemed to be plagued by a special immobility of resources which excluded them from the mainstream of national commerce, such as the Appalachian Mountain region or the Tennessee Valley. Much of the literature rationalizing these regional development efforts stressed the obligation to bring such areas fully into the twentieth century by establishing links with the rest of the nation.

The current revival of the regionalism debate is distinguished by the assertion that unbalanced growth troubles the country at large, not merely isolated geographical pockets of economic depression. The mobility solution is said to have failed because for the bulk of the population it has operated too vigorously, while for the unemployed, the poor, and racial minorities it has operated insufficiently.

Other chapters in this book document the accelerating regional redistribution of jobs and population and the indirect social and fiscal costs that this rapid migration has inflicted upon those who remain resident in the declining regions. Despite the strength of the overall population flow, migration has done little to remedy unemployment rates among the poorer population of the older manufacturing states.

The charge leveled against federal spending patterns is that by contributing disproportionately to private sector demand and job creation in the expanding regions, they have exacerbated the imbalance of national growth. For the population to move to areas of job availability, very substantial migration is required, with its unsettling impacts and possible economic inefficiencies. Much of the population, of course, finds the migration option closed to it. Thus a federal expenditure pattern skewed toward regions of economic expansion contributes to rising unemployment among that portion of the labor force unable to move away from areas of declining job availability, and for the rest of the nation involves a long-term commitment to massive population movement. Proposed as a counterbalance to this policy is a deliberate federal effort to bring jobs to where the people are, coupled with a simultaneous effort to use federal monies to shield governments from the adverse fiscal and other impacts associated with population redistribution.

A full evaluation of the chain of reasoning that leads to proposals to better synchronize federal demand creation with existing population patterns would require examination of the entire array of regional development issues. The purpose of this chapter is much more modest. The first section considers the principal routes by which federal tax and spending policy can influence regional growth. The second section examines the present pattern of federal spending. Federal domestic outlays as a whole are moderately tilted toward the South and West. Virtually all this regional advantage comes from federal spending for goods and services rather than federal transfer payments or grants-in-aid to state and local governments. In its market purchases, the federal government has responded to the same cost incentives as the private sector. It has acquired goods and services where it is least expensive to do so and has invested in productive facilities where the prospective rate of return was highest. The cost advantages enjoyed by Southern and Western states have caused this policy to carry with it a de facto regional orientation.

In contrast to federal goods-acquisition policy, where federal dollars have flowed to the areas of lowest production costs, federal aid formulas in recent years have been redrafted to specifically favor those parts of the country that labor under competitive cost disadvantages. The shift in the distribution of federal grant monies has reached dramatic proportions. Since 1970 the Sunbelt states' per capita grant receipts have fallen steadily relative to grant receipts in the mature, industrial states. Formula changes adopted in the last two years have accelerated that shift.

The federal government's policy toward regional development, unarticulated though it may be, then, has been to follow and reenforce private market forces in the acquisition of goods and services, while striving to cushion the consequences of imbalanced growth through transfer payments to individuals and grants-in-aid to state and local governments.

Such a policy does little to address the cost and productivity differentials that underlie the present pattern of regional growth. Accordingly, the final section of the chapter examines the biases in federal tax and investment programs that may have placed the older, industrial sections of the country at a competitive disadvantage in attempting to generate future growth. Systematic federal subsidies for new capital investment in preference to replacement and maintenance of existing capital facilities, including housing, have tended to speed up the process of regional and metropolitan adjustment, by encouraging the premature scrappage of older facilities. Although it would be disingenuous to find in such federal programs more than secondary support for private market forces, correction of the biases built into the federal tax code and federal investment policies would have the threefold advantage of enhancing economic efficiency narrowly construed, slowing regional and metropolitan migration rates, and improving the long-run competitive cost position of the older regions of the country. The alternative is to permanently compensate the older regions for their cost disadvantages through personal or governmental transfers, a policy that almost assuredly would exacerbate the regional confrontation.

Regional Demand and the Distribution of Federal Domestic Outlays

Most public debate over the regional pattern of federal spending has concentrated on total cash flows, either by comparing per capita federal outlays in one region with those in another or by comparing federal outlays with regional tax payments to compute the net flow that the region enjoys in its transactions with the federal government.

On occasion, regional tax and spending comparisons of this sort have been put forward as if an even exchange with the federal government were a plausible criterion of fairness in itself. Both the federal tax system and the premise of federal social programs reject this standard of equity. The tax code has been deliberately constructed to impose progressive tax burdens. It follows automatically that geographical regions containing citizens of above average incomes should pay into the Treasury more than they receive in their prorated share of domestic outlays. The legal and philosophical claims of citizens to transfer payments and other social benefits are based on their status as individuals and have nothing to do with the geographical location of their residence. Since many federal programs are designed to assist the poor, it is to be expected that geographical pockets of the poor will fare better in their exchange with the federal Treasury than other areas.[1]

A somewhat more thoughtful basis for tax and spending comparisons is to be found in regional demand models. It is well known that federal outlays play an important role in influencing the level of national economic activity. When the federal government wishes to pursue a stimulative national economic policy, it allows federal government expenditures to exceed tax receipts, generating in the process a net addition to aggregate demand. In principle, at least, macroeconomic management requires a comparable restrictive policy of increased tax receipts and diminished expenditures during periods of approximate full employment.

Many participants in the regional debate have argued that, by analogy, an excess of federal spending over tax collections, measured on a regional basis, will stimulate regional economic development, whereas a regional tax surplus will be deflationary. Perhaps the clearest expression of this analogy has been offered by Senator Moynihan of New York. After pointing out that the gross national product of New York State in 1976 was approximately the same as the gross national product of Canada and that both government units labored under the strain of recession, Moynihan contrasts the $4 billion deficit that the Canadian government sustained with what he calculates to be the $10.6 billion surplus that the federal government extracted from New York:

> Supposing that instead of a $4 billion deficit, Canada had incurred a $10.6 billion surplus? Would not the economy of Canada have been seriously deflated? Would not unemployment have risen, jobs disappeared? Would not the whole of the international finance community have joined in counseling Canada to do what Canada in fact did do, which was to avoid that deflation? Is not the case of New York State comparable?[2]

There are, of course, pitfalls to pressing the analogy to macroeconomic fiscal policy too far. Relatively little is known about regional leakages of

Federal Tax and Spending Policies

demand to other parts of the country. Moreover, household migration between states is a far more viable response to different rates of regional growth than migration across international boundaries. Comparison of the absolute level of federal outlays with the absolute level of tax payments is likely to be especially deceptive in interpreting the regional impacts of federal policy. Any such comparison must decide how to treat the national deficit, which makes it possible for large numbers of states simultaneously to benefit from federal expansionary fiscal policy. Most published estimates of tax and spending comparisons merely assume that the federal deficit is equivalent to another federal spending function, which must be financed by taxes paid. This procedure requires, by definition, that one state's net receipts from the federal government be offset by all other states' net payments to the Treasury.

Perhaps most importantly, it is the shifts in federal spending from one period to the next that most influence regional growth, rather than a regional surplus or deficit defined with respect to an arbitrary breakeven point. It is of far more significance to the Northeast whether defense spending there is rising or falling relative to the national level than whether the actual spending level exceeds or falls short of the region's prorated tax contribution. And even this concern, of course, will surface only when there is substantial unemployment of regional resources, both human and capital. Otherwise, a decline in defense spending will merely free productive resources for use in the provision of other goods and services, perhaps of more immediate value to consumers.

These qualifications are sufficient to deter one from putting too fine a point on any calculation of the regional impact of federal spending patterns. As a rough guideline to the stimulative effect of federal spending on different regions, however, a regional tally of outlays is a reasonable place to begin.

The Regional Distribution of Federal Spending

Federal expenditures do display a discernible regional pattern. Table 6-1 summarizes the regional distribution of all federal domestic outlays in fiscal 1975 and fiscal 1976.[3] Although these figures have been the subject of considerable controversy and undoubtedly are unreliable in some of their programmatic detail, their broad lines are clear enough. The regional distributions for 1976 and 1976 are highly consistent with each other. A separate attempt by the Congressional Budget Office to examine the regional distribution of a subset of "reliably allocated" spending programs found the same geographic pattern as shown in table 6-1, but with somewhat more sharply delineated regional differences.[4]

Table 6-1 also compares federal spending in each region with federal tax collections.[5] Regions with spending/taxes ratios in excess of 1.0 are net

Table 6-1
Per Capital Federal Spending and Tax Collections, by Region, Fiscal Years 1975 and 1976

Region[a]	Federal Spending per Capita[b] (Dollars), 1975	Federal Tax Collections, per Capita[c] (Dollars)	Spending/Taxes Ratio, 1975	Spending/Taxes Ratio, 1976
New England	1,470	1,533	0.96	0.95
Mid-Atlantic	1,325	1,594	0.83	0.85
East North Central	1,064	1,518	0.70	0.70
West North Central	1,287	1,374	0.94	1.03
South Atlantic	1,454	1,303	1.12	1.11
South Central	1,327	1,137	1.17	1.18
Mountain	1,615	1,238	1.30	1.24
Pacific	1,745	1,497	1.17	1.15
Nation	1,412	1,412	1.00	1.00

Source: Joel Havemann and Rochelle L. Stanfield, "Federal Spending: The North's Loss Is the Sunbelt's Gain," in *The National Journal* 8, no. 26 (Washington: Government Research Corporation, 1976), and Joel Havemann and Rochelle Stanfield, "A Year Later the Frostbelt Strikes Back," in *The National Journal* 9, no. 27 (Washington: Government Research Corporation, 1977), pp. 1028-1037.

[a]Regions follow Census definitions, except for South Central which combines Census regions East South Central and West South Central.
[b]Excludes interest payment or federal debt.
[c]Corporate tax payments allocated by headquarters location.

beneficiaries of federal domestic spending policies and should experience net economic stimulation because of federal activity. Regions with spending/taxes ratios of less than 1.0 tend to suffer net demand deflation because of federal tax and spending policies.

As is evident from the table, substantial regional variations exist, both in per capita spending and in spending/taxes ratios. The Mountain and Pacific states enjoy a favorable allocation pattern on both criteria; the North Central states and the Mid-Atlantic states suffer unfavorable allocation patterns. The balance struck in the other regions depends on whether per capita expenditures or spending/taxes ratios are used as the criterion of evaluation.

A reversal of the existing pattern of federal spending on purchases of goods and services would require explicit alteration of the present policy of buying these products competitively on the national market, wherever they can be acquired at least cost.[6] Either a given proportion of federal spending would have to be held aside, to be allocated among suppliers on other grounds than cost competitiveness, or the external impacts of purchasing and location decisions on local labor markets would have to be taken into account and weighed against potential dollar savings. The long-run consequences of requiring the federal government to make any significant portion of its $145 billion annual purchases at higher than market prices are serious enough to argue for adopting noncompetitive purchasing policies only as a last alternative in the effort to stimulate regional development.

Although demand models encourage the treatment of all federal dollar outlays as comparable to one another, and thus subject to addition across spending categories, greater understanding of the potential impact of federal policy decisions can be secured by distinguishing the principal categories of federal domestic spending. Table 6-2 displays the breakdown of fiscal 1977 federal spending as estimated in the U.S. budget. The single largest category of federal domestic outlays is transfer payments. These consist of social security payments (almost half of total federal transfers), Medicare payments to the elderly, unemployment benefits, civil service retirement, and military retirement, as well as a number of smaller retirement and income-related transfer programs.

The legal and philosophical claims of individuals to such benefits, in their capacity as individuals rather than as residents of a particular region, are so strong as to make it virtually unthinkable that federal transfer spending could be reshaped to serve regional development objectives. The only possible inroad that might be made would be to recognize regional cost-of-living differentials in the computation of individual entitlements for such programs as Medicare and Supplemental Security Income. Any attempt to provide further regional targeting for transfer payments would involve the federal government in restricting the freedom of movement of beneficiaries or scrapping uniform national payment standards. Transfer payments therefore can be safely excluded from most of the regionalism debate.

Table 6-3 separates what might be termed direct federal stimulants to private sector regional economic activity. Excluded from federal spending are both transfer payments to individuals and grants-in-aid. The remaining federal expenditures cover primarily direct job creation through permanent

Table 6-2
Federal Domestic Expenditures, Fiscal 1977[a]
(billions of dollars)

Purchases of goods and services	144.8
Defense	(94.8)
Nondefense	(50.0)
Transfer payments to persons	167.7
Grants-in-aid to state and local governments	68.0
Total	380.5

Source: *Budget of the United States Government, Fiscal Year 1978,* Special Analysis A (Washington: GPO, 1978).
[a]Excludes net interest paid and subsidies to government enterprises.

Table 6-3
Federal Purchases of Goods and Services, per Capita, Fiscal 1976

Region	Per Capita Purchase (Dollars)
New England	850
Mid-Atlantic	645
East North Central	498
West North Central	740
South Atlantic	866
South Central	762
Mountain	967
Pacific	1,166

Source: Derived from tables 6-2, 6-6, and Community Services Administration, *Federal Outlays in Summary* (Washington: GPO, 1977).

federal employment, direct federal spending on the acquisition of privately produced goods and services, and direct federal investment in capital formation. These are the expenditure categories that have the maximum multiplier effects—in terms of both the generation of regional aggregate demand and the strength of their inducement to regional migration. They also are the federal expenditures that most closely resemble private sector spending. In the absence of an explicit policy to the contrary, federal outlays for these purposes might be expected to respond to the same cost comparisons that shape regional demand on private markets.

Table 6-3 confirms that federal spending for purchases of goods and services is more strongly skewed toward the rapidly growing regions of the country than are total federal outlays. On a per capita basis, the Pacific states receive more than twice as much federal revenue as the Great Lakes states and 80 percent more than the Mid-Atlantic states.

Although a detailed examination of federal spending would be necessary to establish the point conclusively, table 6-3 strongly suggests that federal employment, goods and service acquisition, and direct capital investment have been shaped by the same cost and profitability considerations that have influenced private sector demand for regional output. Electronic and defense suppliers in the Southwest and West, for example, enjoy a number of competitive advantages, ranging from clear skies for missile testing to economies from the agglomeration of subcontractors, and from nearness to federal military installations to lower private sector wages, all of which enhance their ability to secure competitive contracts from the federal government, just as they would enhance their competitive position in supplying private sector or foreign government demanders. Federal decisions regarding direct employment are likewise sensitive to traditional cost considerations. The Boston Naval Yard was shut down only after an exhaustive comparison of the costs of replacing the aging capital facilities

Federal Tax and Spending Policies

there with the costs of building new facilities elsewhere. The cost comparisons revealed large capital and operating savings from consolidation of facilities outside New England. The parallels with private sector manufacturing decisions about plant location, expansion, and replacement are obvious.

Defense Spending

The most important single lever that the federal government controls in manipulating its spending pattern is defense outlays. Studies conducted in the mid-1960s indicate that domestic defense spending is relatively effective in generating economic growth. Bolton, for example, calculated that each defense dollar, on average, produces two dollars of personal income—that is, that defense spending has a multiplier of 2.0[7] Other studies have reported even higher multipliers.[8]

During most of the 1950s and 1960s defense expenditures were a powerful engine of growth in many parts of the country. Defense purchases accounted for more than 10 percent of gross national product in 1958 and 9.0 percent a decade later. By fiscal 1978, however, defense spending is projected to have declined to 5.2 percent of gross national product. The steep reduction in defense spending, measured in real terms, has converted defense expenditures into a frequent agent of regional decline, as the closing of military installations in many areas has removed a source of demand for regional production.

Table 6-4 demonstrates the uneven pattern of defense employment reductions over the last decade. Although the number of military and other personnel has declined in each of the major regions, the reductions have been sharpest in the Northeast, where there has been a drop of approximately 30 percent in defense employment. Military personnel, of course, are drawn from the entire national workforce pool, so that there is no simple relation between loss of regional defense employment and increases in regional unemployment. But the multiplier effects of this selective job loss have been substantial.

An alternative measure of the incremental effect of adjustments in the defense budget can be derived from military construction plans for fiscal 1978. Table 6-5 shows that military construction spending is concentrated overwhelmingly in the South and West. The sums of money involved are not especially large, but the regional distribution is revealing as a probable forerunner of future employment distributions.

Direct federal civilian and contractor employment not only stimulates regional economic growth but also serves as a particularly powerful magnet in inducing migration. In Southern states average private wages are roughly

Table 6-4
Regional Defense Employment, 1965 and 1976

Region	Employment, 1965	Employment, 1976	Percentage Change, 1965-1976
Northeast	329,241	212,379	−35.5
New England	(105,599)	(58,623)	−44.5
Mid-Atlantic	(223,642)	(153,756)	−31.2
East North Central	199,789	164,504	−17.7
South (excluding District of Columbia)	1,073,395	965,418	−10.1
West	812,984	780,754	−4.0

Source: Derived from CONEG Policy Research Center, Inc. and Northeast-Midwest Research Institute, *A Case of Inequity: Regional Patterns in Defense Expenditures, 1956-1977*, 1977, table 4.

30 percent lower than those paid by federal agencies.[9] After correction for regional cost-of-living differentials, the real wages paid by federal employers in the South considerably exceed the real equivalent of the same money wage paid in the older industrial states. Wages in manufacturing industries dominated by defense production are also higher than wages in other manufacturing sectors.[10] Thus defense contracting is an especially effective stimulant to regional growth.

Measured as a proportion of regional personal income, federal defense outlays in fiscal 1976 ranged from a high of 10.7 percent in the Pacific states and 8.4 percent in the South Atlantic states to a low of 2.4 percent in the East North Central states. Although these figures allocate defense spending according to prime contracts, without tracing the geographical location of subcontracted activities, data for the late 1960s (when detailed subcontracting records were maintained) indicate that, if anything, prime contract awards tend to understate the ultimate geographic concentration of defense-related activity. Polenske has demonstrated that past shifts in federal defense spending patterns have induced significant changes both in the

Table 6-5
Military Construction Budget, Fiscal 1978, by Region

Region	Amount (000s)	Amount per Capita
Northeast	$ 68,729	$1.39
East North Central	20,669	0.43
South (excluding D.C.)	330,442	4.85
West	358,030	7.49

regional distribution of economic activity and in the sectoral composition of output.[11]

The choice of alternative criteria by which to allocate federal spending is also raised in its clearest form with defense expenditures. Since 1952 there has been in effect Defense Manpower Policy number four which calls for defense spending to be targeted to areas of slow economic growth. In practice, however, this policy has been observed only in the breach. In fiscal 1977 it is estimated that less than 0.2 percent of all defense spending will be spent under this program. Virtually all defense contracts are awarded according to strict procurement criteria, designed to elicit the best product for the least cost, a principle which is applied to determine installation closings and expansions as well.

Federal Grants-in-Aid

If federal purchases of goods and services have followed the market, reinforcing private sector responses to regional cost differentials, federal grants-in-aid have increasingly been designed to work against the market—either by allocating federal assistance in direct proportion to local and regional cost factors or by allocating assistance in proportion to measures of private market economic distress.

One measure of the outcome of this process of grant and formula redefinition is illustrated in table 6-6. As shown there, the per capita grant receipts of a sample of Sunbelt states have fallen steadily and significantly since 1970 relative to the receipts of a sample of Northern industrial states. In 1970 the average $127 per capita received by the Sunbelt states was 22 percent greater than the average federal grant payment to the industrial states. Although federal grants to all state and local governments climbed drastically over the next six years, by 1976 the Sunbelt states in this sample were receiving 9 percent *less* per capita than their Northern industrial counterparts.

More recent data on the geographical distribution of all federal grants are not yet available. However, inspection of the projected receipts of individual cities suggests that the regional and metropolitan aid shifts that occurred in fiscal 1977 and fiscal 1978 are still larger than those in the preceding years. For the sample of cities shown in table 6-7, average per capita receipts in Northern cities will jump by almost 150 percent between fiscal 1975 and fiscal 1978, creating a large gap between the average aid received by these cities and cities in the South and West. It would be a mistake, however, to view these shifts as solely, or even primarily, a regional phenomenon. The new aid programs have tended to channel generous federal funding to the older cities wherever they are located, while

Table 6-6
Per Capita Federal Grants-in-Aid to State and Local Governments
(dollars)

States	1970	1971	1972	1973	1974	1975	1976
Mature, Industrial							
Maine	114	164	190	238	268	276	351
New Hampshire	100	124	123	186	190	211	259
Vermont	174	213	234	295	320	327	372
Massachusetts	131	147	190	216	226	250	313
Connecticut	98	132	145	179	217	217	232
Rhode Island	146	145	185	246	262	267	336
New York	129	180	239	261	285	314	355
New Jersey	87	114	142	167	178	205	254
Pennsylvania	114	118	136	197	199	227	263
Ohio	83	95	112	147	162	167	200
Indiana	66	82	103	127	133	152	188
Illinois	86	112	157	191	201	199	249
Michigan	87	117	148	192	198	232	287
Delaware	93	118	170	211	210	208	276
Total Mature	104	130	164	200	212	232	281
Sunbelt States							
California	153	173	201	224	227	233	270
Arizona	136	135	156	194	217	209	234
New Mexico	204	240	272	366	295	349	366
Texas	103	123	143	176	182	180	208
Oklahoma	156	177	191	223	226	263	249
Arkansas	137	167	202	235	234	242	291
Louisiana	140	174	197	252	252	232	296
Mississippi	173	235	257	298	300	273	333
Alabama	147	186	193	224	233	227	271
Georgia	119	151	179	190	237	239	286
Florida	80	95	117	153	157	159	181
Tennessee	120	154	172	198	204	218	257
South Carolina	103	140	154	208	212	204	245
North Carolina	97	126	143	179	188	193	233
Virginia	100	122	132	173	190	202	236
Total Sunbelt	127	155	173	204	207	228	264
Sunbelt as Percentage of Mature	122	119	105	102	98	93	91

Source: 1970-1974: George D. Brown, "The Relative Importance of Federal Grant-in-Aid Formulas in an Overall Agenda for the Northeast" (Paper Prepared for the Federal Domestic Outlays Working Group, 1976); 1975-1976: Department of Treasury, *Federal Aid to States Fiscal Year 1976* (Washington, 1977).

the newer and growing cities have received lesser support, again without reference to regional location. Although the individual cities' grant receipts aggregate to significant regional differentials, regional location is not the reason for the differences. Moreover, the benefits of federal grant assistance to individual government units (unlike the benefits of federal defense spending) are very largely restricted to the jurisdiction receiving the

Table 6-7
Per Capita Grants from Federal Government, Selected Cities, 1975 and 1978

City	Per Capita Grants, Fiscal Year 1975 (Dollars)	Per Capita Grants, Fiscal Year 1978 (Dollars)	Percentage Increase 1975-1978
Northern			
Boston	$104.81	$185.40	76.9
Buffalo	78.24	193.27	147.0
Chicago	53.61	128.54	139.8
Detroit	124.48	233.37	87.5
Indianapolis	50.71	99.76	96.7
Minneapolis	39.39	146.78	472.6
New York	82.02	166.86	103.4
Philadelphia	72.04	181.78	152.3
Pittsburgh	55.20	202.29	466.5
St. Louis	59.97	208.35	447.4
Average	72.05	174.64	142.4
Southern and Western			
Atlanta	88.41	126.36	42.9
Dallas	30.65	49.03	60.0
Denver	102.10	117.45	15.0
Houston	34.57	62.08	79.6
Los Angeles	42.19	105.23	149.4
Nashville	60.86	86.12	41.5
New Orleans	81.55	152.40	86.9
Phoenix	54.97	98.70	79.6
San Francisco	101.77	243.79	139.5
Average	66.34	115.69	74.3

Source: See table 6-8.

aid. For this reason, regional averages are more than ever an artificial construct. They are likely to be especially deceptive in this context, since the large majority of growth in federal aid since 1975 has been directed at local governments rather than state governments.

The aid shifts identified in table 6-7 are the result of the large amounts of assistance disbursed under the administration's temporary stimulus package, coupled with the introduction of new elements into aid allocation formulas. The most important of these formula elements is the local unemployment rate. In calendar 1977 no less than $16 billion of federal assistance was allocated, at least in part, by state and local unemployment rates. The three most important new aid programs for state and local governments (the Public Service Employment provisions of the Comprehensive Employment and Training Act, local public works assistance, and countercyclical revenue sharing) all tie local entitlements to the unemployment rate. The use of this allocating element strongly favors the Northeast and (perhaps surprisingly) the Pacific states, where unemployment rates have run much above average. For example, the most recent quarterly disbursements under the countercyclical revenue sharing program imply annual per capita receipts of $17.08 in the Mid-Atlantic states, compared to receipts of only $6.28 in the south.

One of the most pressing questions facing older metropolitan regions is whether their current, highly favorable aid levels will be rendered permanent by converting the programs contained in the temporary stimulus package into continuing commitments. Although each of these programs was passed under the guise of anticyclical assistance to the state and local sector, each disburses its aid in rough proportion to the seriousness of secular economic decline or stagnation. In fact, the use of unemployment rates, rather than changes in unemployment rates over the economic cycle, in the allocation formula of the Anti-Recession Fiscal Assistance program makes countercyclical revenue-sharing entitlements more closely related to measures of secular decline than to measures of cyclical oscillation.[12] If the temporary stimulus programs are phased out with economic recovery, the older cities and older regions will suffer a severe reduction in federal assistance.

Table 6-8 illustrates the importance of recent federal aid revisions to the total receipts of selected local governments. The programs listed in the first three columns are the elements of the stimulus package. The fourth column shows city entitlements under the version of the revised community development block grant formula approved by the House. The fifth column shows all other federal aid payments to the cities in question, including general revenue sharing. As can be seen, the temporary stimulus programs in combination with the revised community development block grant program account for as much as two-thirds, or even more, of all federal aid payments to these cities.

The debate over the community development block grant formula perhaps constitutes the best example of the power of allocation formulas to redistribute federal funds. The original block grant allocation formula was heavily weighted toward per capita income, an element which has the effect of steering funds to the South and Southwest where average money incomes lag the rest of the nation. The administration recommended creating an alternative allocation formula which would target funds more effectively to the older cities. The bills passed by the House and Senate; both would use, for the first time, age of housing stock and loss of population as positive factors in determining formula entitlements. Inclusion of these elements has been justified on the grounds that both are indirect proxies for the public sector cost burden involved in providing adequate services in aging and declining urban areas. Communities possessing aging housing and public capital infrastructure presumptively face more costly production conditions than communities with new capital. Communities confronted with population loss typically experience stagnating or deteriorating tax bases as well.

Table 6-8
Composition of Federal Grants, Selected Cities, Fiscal Year 1978
(thousands of dollars)

City	Public Service Employment (CETA)	Local Public Works	Antirecession Fiscal Assistance	Community Development Block Grant (CDBG)	Other (Including General Revenue Sharing)	Stimulus Package and CDBG as Percentage of Total
Northern						
Boston	$ 21,532	$ 15,963	$ 7,599	$ 25,235	$ 47,768	59.6
Buffalo	17,378	13,495	5,739	21,928	20,121	74.4
Chicago	69,248	37,483	22,287	116,800	152,538	61.7
Detroit	39,242	26,805	25,158	57,778	162,563	47.8
Indianapolis	17,158	10,575	3,154	10,941	29,503	58.6
Minneapolis	10,800	7,394	1,619	18,625	17,044	69.3
New York	198,612	191,903	143,117	224,775	490,048	60.7
Philadelphia	42,820	54,734	30,834	63,852	137,877	58.2
Pittsburgh	12,900	16,242	7,162	23,815	32,733	64.7
St. Louis	16,074	15,514	9,681	32,983	35,131	67.9
Southern and Western						
Atlanta	18,078	7,205	2,625	14,125	13,060	76.3
Dallas	8,172	0	0	15,223	16,465	58.7
Denver	9,402	8,527	3,135	11,572	30,287	51.9
Houston	16,646	8,650	1,588	23,634	31,866	61.3
Los Angeles	78,792	45,794	15,585	51,010	95,513	66.7
Nashville	4,596	0	0	8,510	23,323	36.0
New Orleans	11,612	9,508	7,172	20,287	36,768	56.9
Phoenix	22,572	15,335	2,390	10,031	15,306	76.7
San Francisco	25,264	30,030	9,529	26,335	70,962	56.2

Source: CETA allocations for Titles II and VI were obtained from Department of Labor, Office of Information. Two-thirds of the $6.6 billion authorized in May 1977 to cover the remainder of fiscal 1977 and fiscal 1978 were allocated to fiscal 1978.

Emergency Local Public Works grant data for fiscal 1978 are from round II planning targets, published in Economic Development Administration, *Planning Targets for Applicants and Areas under the Public Works Employment Act of 1977*, July 14, 1977.

Antirecession Fiscal Assistance data for fiscal 1978 represent the annualized rate of receipts in quarter 5 of the program (the most recent quarter for which data are available and the first quarter under the new funding authorization of $2.25 billion for the five quarters ending with the close of fiscal 1978).

Community Development Block Grant data are from Richard P. Nathan, Paul R. Dommel, and James W. Fossett, "Targeting Development Funds on Urban Hardship," Testimony before the Joint Economic Committee, July 28, 1977.

Table 6-9 shows the large impact that adoption of the dual formula has on the aid entitlements of different types of cities. Similar debates already have been launched regarding the allocation elements to be used to disburse HUD's urban development action grants, assistance under the administration's proposed urban development bank, Law Enforcement Assistance Agency (LEAA) block grants, and numerous other federal assistance programs.

Analytically, many of these formula questions are not regional issues at all. As is true of the community development block grants, many of the federal programs distribute funds only to local governments. They are not meant to cope with peculiarly regional problems or to compensate local governments for regional conditions. An old city with declining population and high unemployment would receive the same entitlement whether located in the South or in New England. The grants-in-aid formulas have become regional questions primarily through the efforts of political coalitions. More often than not, legislators begin with a desired geographical allocation of funding and work backward to find the allocative criteria that will produce such an outcome yet satisfy enough other geographical areas to command a legislative majority. The "region" then consists of neighboring states likely to be benefited by the same formula elements. In this respect, grants analysis appears to have been contaminated by the basic proposition of flow of funds studies—that what matters is the ultimate allocation of federal dollars, not the process that distributes them. The close attention paid to the computer printouts showing dollar flows into each congressional district may obscure the logical rationale, if any, for distributing federal aid according to one criterion rather than another. Although maneuvering of this type can work well in the short run, it portends permanent confronta-

Table 6-9
Community Development Block Grant Receipts under Current Formula and Proposed Dual Formula, Selected Cities
(millions)

City	Actual Allocation, 1977	Projected 1980 under Current Formula	Projected 1980 under Dual Formula	Gains under Dual Formula vs. Current Formula (1980)
New York	$150.65	$79.56	$255.69	$176.13
Chicago	61.59	73.42	132.88	59.46
Los Angeles	48.69	58.03	n.a.	0
Detroit	28.30	30.74	65.73	34.99
Minneapolis	15.09	7.05	21.19	14.14
St. Paul	18.84	4.73	10.42	5.69
Philadelphia	57.16	39.23	72.64	33.41
San Diego	10.28	12.26	n.a.	0

Source: Community Planning and Development, Department of Housing and Urban Development.

tion between vying regional coalitions rather than exploration of the cost factors that properly deserve compensation through federal assistance programs.

It should be pointed out, perhaps, that a similar ambiguity burdens the discussion of grants programs in most European countries. In West Germany, for example, the federal government has scaled its revenue-sharing payments according to the size of receiving jurisdictions; large cities receive more per capita than smaller cities, on the grounds that there are inevitable diseconomies of scale beyond a certain city size. Public debate, there as here, has oscillated between the analytical question of whether diseconomies of scale are inevitable, and, if so, whether it is in the national interest to subsidize large cities rather than encourage outmigration from them, as well as the political question of whether there is a legislative majority to sustain the greater aid payments to big cities. Unfortunately, the answer to the first question does not always seem pertinent to the head counts that will resolve the formula issue legislatively.

Investment Incentives

Over the long run, federal dollars can accomplish more if used to narrow regional cost differentials than if used to sustain federal demand for goods and services in noncompetitive regions or to compensate governments in these regions for their cost burdens. In the framing of any national policy designed to avoid regional confrontation, a good starting point is the identification of those federal expenditures that unwittingly have contributed to regional imbalance.

One good example is the federal subsidization of new capital investment. A variety of federal policies, from the Internal Revenue Code to authorizations under the Water Pollution Control Act, have favored investment in new facilities over the investment in the repair, maintenance, preservation, and upgrading of old facilities.

The effect of subsidizing new investment has been to shorten the useful life of older capital. The process is analogous to that encouraged by the investment tax credit for purchases of new machinery and equipment. When federal policy subsidizes the cost of acquiring new machinery, it speeds up the capital investment cycle, accelerating the scrappage of old machines, which cannot compete effectively with new equipment at the subsidized cost.

Notes

1. This point has been stressed by Carol L. Jusenius and Larry C. Ledebur, *A Myth in the Making: The Southern Economic Challenge and*

Northern Economic Decline (Washington: Office of Economic Research, Economic Development Administration, November 1976).

2. Senator Daniel Patrick Moynihan, "The Federal Government and the Economy of New York State," June 27, 1977.

3. All data on the geographic distribution of federal domestic outlays are derived from estimates of the Community Services Administration, *Federal Outlays in Summary*, December 1976. Table 6-1 reproduces what is perhaps the best known published source, that compiled by the *National Journal*.

4. Congressional Budget Office, *Troubled Local Economics and the Distribution of Federal Dollars* (Washington, August 1977).

5. State allocations of federal tax receipts are estimated by the Tax Foundation.

6. Just such a policy reversal was recommended in the 1976 report of the Joint Economic Committee.

7. Roger E. Bolton, *Defense Purchases and Regional Growth* (Washington: The Brookings Institution, 1966).

8. For example, multipliers as high as 3.0 are estimated for civilian payrolls related to defense in Report of the President's Economic Adjustment Committee, *The Impact of Defense Cutbacks on American Communities* (Washington: July 1972).

9. For example, comparisons for the year 1974 show the following:

State	Annual Federal Wage (Employees Covered by Social Security)	Annual Private Sector Wage	Percentage Difference
Illinois	$11,839	$10,097	17.3
New York	11,770	10,467	12.5
Louisiana	11,634	8,203	41.9
Texas	11,683	8,343	40.0
Arkansas	11,067	6,872	61.0

10. Richard Dempsey and Douglas Schmude, "Occupational Impact of Defense Expenditures," *Monthly Labor Review* 94, no. 12 (December, 1971):12-15.

11. Karen R. Polenske, *Shifts in the Regional and Industrial Impact of Federal Government Spending* (Washington: Department of Commerce, 1969).

12. For example, per capita receipts are more closely correlated with cities' rate of population loss or gain and with long-term tax-base stagnation than with cyclical fluctuations in the unemployment rate or with recession-induced revenue shortfalls.

Part IV
Energy and Regional Distribution of Economic Activity

7

Role of Energy in the Regional Distribution of Economic Activity

Irving Hoch

Overview

This chapter considers the role of energy in the regional distribution of economic activity, using time as a simple organizing principle to present some complicated materials. The next section focuses on the past; the third on the present; and the fourth, and last, on the future. This bald enumeration needs some qualification. The focus, of course, is not unwavering because I assume we want to learn from the past, and find where we are in the present, to improve our decision-making for the future. Again, by "present," I mean "recent and current," inclusive of roughly the past five years. Finally, by "future," I refer primarily to the near term, starting with tomorrow, and applying a fairly high discount rate, so that the long term does not receive much attention.

I have applied the label "Historical Perspective" to the second section, and organized it into three subsections: chronology, regional resource development projects, focusing on the TVA, and energy and the location of economic activity. The chronology relies on some classic studies to trace the discovery and development of U.S. energy resources and then to consider how the location of those resources, and the transportation and industrial technology based on those resources, affected the pace and location of U.S. economic development. The discussion establishes a metropolitan as well as a regional dimension to the location of energy-related economic activity.

There are several massive regional development projects in the United States, with the development and supply of electric power one of their common purposes. However, the Tennessee Valley Authority (TVA) has received the major share of the attention devoted to those projects, well out of proportion to its relative size. This is probably a concomitant of past ideological concerns and conflicts, now muted; current ideological concerns and conflicts—over environmental issues and nuclear power—pit the TVA against today's "progressives." I have focused on the TVA experience as a case study to consider some issues that seem of more general significance, including the policy of "low" (below market equilibrium) prices and the "consumerist" philosophy that supports the policy; the evaluation of the TVA impact on regional development (that evaluation is hardly definitive, but a review of the tools employed in attempting that evaluation seems useful in assessing the current state of the art); and the questionable theses that industrialization implies development and that it is a regional goal to retain people in the region.

In the first part of second section I suggest that energy has had considerable impact on location, but there are dissenting views, and in the concluding part of the section I review the range of opinion on the relationship. That review suggests that differences of opinion are a matter of degree: all observers see energy as only one of a number of factors influencing location, and the issue is whether it has been of "considerable" or only "limited" importance. The defense of my conclusion of considerable importance leads to a consideration of some general issues in the construction of models to estimate regional impact, including the distinction between effects of exports versus those of internal organization and trade and a suggestion, based on some of my past research, that regional input-output models include households as an endogenous sector. An increase in exports or government spending then generates an ultimate impact in increased household income and spending, corresponding to the inmigration of labor (and population) to the region.

The third section reviews the economic geography of recent U.S. energy production and consumption, in turn, focusing on regional and (to some extent) urban issues. Given the general topic, it seemed natural to critically appraise the available information that was drawn on and to note some suggestions on likely data refinements and extensions.

The discussion of production begins with the presentation of graphic information on production by locale and then Bureau of Mines data on fuel extraction and reserves by state and region. The data on reserves are bound to be conservative because they represent resources that can be recovered profitably at current prices, and prices of both natural gas and oil have been kept below market equilibrium price by regulation. The matter of reserves is generalized to that of supply, and the lack of hard evidence becomes clear from a review of the literature. But there is certainly indirect evidence of considerable supply response in the form of wage rates and employment in coal and in oil and gas extraction; all increased markedly in the post-OPEC period, with consequent impact on per capita income. Growth in per capita income by state was related to fuel extractive activity, and a significant relation emerged after accounting for other regional effects.

The metropolitan dimension in fuel production was explored further at this point, using additional data on earnings in fuel extraction. It turns out that roughly one-quarter of coal production and 60 percent of oil and gas production take place within the borders of metroplitan areas (SMSAs). Causality running both ways can explain these results: metropolitan areas induce some extractive activity, either by making marginal deposits economic or by furnishing transportation and marketing facilities, and extractive activity may itself cause an urban area to grow large enough to receive official designation as an SMSA. Although extractive activity earnings are a small percentage of the total (roughly 1 percent), it seems plausible that they have a disproportionate impact on economic activity.

The consumption discussion draws on a detailed study of quantities of energy used and expenditures on energy by state and region for 1972. Good data on quantities were generally available, but price data, particularly on petroleum products, were sparse and scattered. Consumption data by fuel type, sector, and function are presented. A major goal in the work yielding these results was the development of sectoral information, so that concentric levels of consumtion could be distinguished: households, final demand (households plus government), "customer" use (final demand plus business use exclusive of energy used in producing energy), and all energy use. The results show much more variation between states and regions in use in physical terms than in money terms, interpreted as indicative of considerable demand elasticity. Household consumption in the North was about 25 percent above that in the South and West, in physical terms, suggesting (not surprisingly) that the North was more adversely affected by the Organization of Petroleum-Exporting Countries (OPEC) price increase. The question of demand elasticity was investigated by applying the data developed here, and considerable price elasticity emerged, not only in terms of a negative consumption response to an energy source price increase, but also in terms of a positive response to a price increase for an alternative source of energy.

The section concludes by reviewing available data on price changes since the time of the OPEC price increase. It turns out that energy prices at the consumer level have increased only a modest amount more than the general price level, although the latter has shown a marked increase since 1972. (As noted earlier, price controls have been a factor limiting energy price increases.) At the regional level, available evidence indicates that energy prices have increased the most in the North and the least on the Pacific Coast.

The concluding section considers prospects and policies, beginning with forecasts of energy prices for both the short and long run. By defining the former period as the next five to ten years, it seems most probable that energy prices will remain at roughly their present level relative to the general price level; however, a decline in prices, even a precipitous decline, appears more likely than a further price increase, and contingency planning for such an eventuality might be wise. In the longer run, gradual increases in the real price of energy are to be expected, given depletion of the more accessible deposits of fossil fuels and increased worldwide energy consumption, caused by increased population and per capita income.

Given increased energy prices, both production and consumption adjustments will occur. Most energy production is localized to the South and West, and many parts of the South and West have "better" climate than the North, in terms of lower heating plus air-conditioning costs. Hence, both production and consumption effects should speed up the movement of population and economic activity from North to South and West. In drawing on data developed here, it was estimated that household energy expenditures per capita were $90 greater in New England than on the Pacific Coast in 1972, and the regional differential probably increased to roughly $200 per capita, or $600 per family, in 1977. Similar but smaller differences hold between other pairs of regions. It is plausible that some of the differences involve disequilibrium and that some migration has occurred or will occur as a consequence.

Higher energy prices may also affect population distribution at the local level, probably slowing suburban sprawl within urban places and perhaps accelerating the growth of smaller relative to larger metropolitan areas.

The adjustment process has been distorted by current energy policy, which imposes lower than market prices on domestic oil and gas, discouraging U.S. production of those fuels, and encourages coal production, perhaps as an indirect response to the price regulation effects. At the regional level, the development of Rocky Mountain production is accelerated and Southwestern production is reduced, relative to levels that would hold in a free market. Further, coal production seems likely to generate considerable costs in the form of air pollution (in either increased damages or increased costs of meeting standards), and those costs will fall most heavily on the North, particularly the Northeast. Artificially low energy prices may also limit the potential for alternative energy sources, including nuclear power, shale oil, and solar energy, and may thus be a factor in the slower than expected development of those sources. In sum, the reluctance to let prices rise is paid for by a variety of costly consequences.

I conclude by asking whether we can have freer markets and yet attempt to achieve the purported goals of price regulation, recognizing that once regulation is in effect, it creates a species of property rights. Perhaps we can reconcile those objectives. Compensating those disadvantaged by deregulation, developing alternative approaches to the socially accepted goals of regulation (involving income redistribution on equity grounds and antimonopoly policy on efficiency grounds), and moving to freer markets might well yield viable alternatives to confrontation.

Historical Perspective

Chronology

Let us consider the broad sweep of U.S. development, viewed in relation to its energy resources. In this discussion I rely primarily on classic works by Schurr and Netschert [1], Brocher [2], and Perloff and Wingo [3].

Role of Energy

In the span of a century the composition of the U.S. fuel and power base has twice undergone a complete reversal, with a dominant fuel displaced by a newly emergent rival. Fuel wood was displaced by coal between 1850 and 1920, which in turn was displaced by petroleum and natural gas after World War I, although the absolute level of coal production has been relatively stable since 1920. Waterpower and wind power in the early period, hydroelectric power since 1900, and nuclear power at present have been relatively minor energy sources. Figure 7-1 shows the trends in U.S. energy consumption in graphic form, and table 7-1 presents them in tabular form, explicitly listing British thermal unit (Btu) consumption by energy source [4].

In 1850, the energy basis of the U.S. economy was predominantly fuel wood, with wind power and waterpower furnishing most inanimate mechanical power. The early industrial development of the United States was strongly dependent on waterpower, especially from the innumerable small streams of New England, which furnished the power for textile mills and other small-scale manufacturers. Wood was abundant and cheap and, with

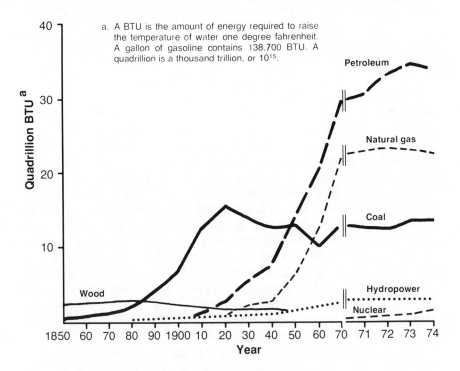

Source: Bureau of the Census, *Historical Statistics of The United States*, presented in U.S. House of Representatives, Subcommittee on Energy and Power of Committee on Interstate and Foreign Commerce, *Energy Information Handbook*, Committee Print 95-18 (Washington, July 1977), figure 15, p. 26.

Figure 7-1. U.S. Energy Consumption by Energy Source, 1850-1974

Table 7-1
U.S. Energy Consumption by Source

a

	Trillion Btu		
Year	Fuel Wood	Windpower and Waterpower	Coal
1850	2,138	500	219
1860	2,641	650	518
1870	2,893	400	1,048

b

			Quadrillion Btu		
Year	Fuel Wood	Coal	Petroleum and Natural Gas	Hydropower	Nuclear Power
1880	2.9	2.0	0.1	—	—
1900	2.0	6.8	0.5	0.3	—
1920	1.6	15.5	3.4	0.8	—
1940	1.4	12.5	10.2	0.9	—
1960	—	10.1	32.8	1.7	—
1970	—	12.7	51.5	2.7	0.3
1976	—	13.7	55.2	3.1	2.7

Source: Table 7-1a: S. Schurr and B. Netschert, *Energy in the American Economy, 1850-1975* (Baltimore, Md.: Johns Hopkins University Press, 1960), pp. 47, 49; table 7-1b: Federal Energy Administration, *Energy in Focus; Basic Facts*, cited in U.S. House of Representatives, Subcommittee on Energy and Power, *Energy Information Handbook*, Committee Print 95-18 (Washington, July 1977), table 9, p. 27.

the clearing of forests for settlement, approached the status of a free good. It has been estimated that a U.S. family around 1850 annually burned wood in fireplaces that was the equivalent of 2.5 tons of coal per person: "All cabin dwellers gloried in the warmth of their fireplaces, exploiting their world of surplus trees where a poor man, even a plantation slave, could burn bigger fires than most noblemen in Europe" [1, p. 49]. This contrasts with modern-day annual domestic heating consumption per capita of around 1 ton of coal equivalent in Btu content, yet this much smaller input is also estimated to yield three times as much useful space heat. Several generalizations are keyed to this example: "profligate" use of energy is a concomitant of low prices and can be viewed as a reasonable response to such relative prices; comparisons in physical units (Btu) generally have limitations and are sometimes misleading; and efficiency in energy technology and utilization has shown massive improvements over time. (See Rosenberg [5] for further examples.)

There are parallels in the careers of coal and of petroleum and natural gas. There was a period of resistance to the introduction of each fuel [1, pp. 58-61, 103, and 125] which was overcome when cost advantages became manifest; commercial production in each case began in Appalachia, followed by movement of production westward; and each fuel has had considerable impact on the location of production and population.

In 1850 the bulk of U.S. coal production occurred in Pennsylvania, with a large portion mined from the great Pittsburgh seam. The early concentration of steel making in western Pennsylvania was a consequence of both the geographic juxtaposition of coal and iron ore and the proximity to U.S. population centers, and hence the major market for iron and steel products. As that market's center of gravity moved westward, and as Minnesota iron ore replaced local ores, the iron and steel industry also shifted westward along the southern rim of the Great Lakes [3, p. 194].

The first commercial oil well was drilled at Titusville, Pennsylvania, in 1859, and Appalachia remained the largest oil-producing region until the start of the twentieth century. Oil was accepted quickly as a source of artificial light and lubricants; its use as a fuel developed more slowly, in part because it first became available near the centers of coal production and hence initially competed with coal where that fuel was cheapest. The discovery of the Lima, Ohio-Indiana fields in the mid-1880s brought lower oil prices, made oil attractive in steam generation and heating, and concomitantly helped establish the Standard Oil Company. By the early 1900s the great centers of oil production were located in California and Oklahoma; by 1930 Texas had become the leading oil-producing state; and in recent years Texas has accounted for roughly 35 percent of U.S. petroleum production, followed by Louisiana with 25 percent of the total.

The first commercial natural gas operation was the shipment of gas from Murrysville, Pennsylvania, to Pittsburgh after a pipeline was built in 1883. Natural gas is often found in oil fields, but most natural gas (80 percent) now comes from fields containing little or no oil [6]. In the early twentieth century, prior to the construction of pipelines from the newly opened oil fields of California and the Southwest, as much as 90 percent of the natural gas discovered was lost or discarded [1, pp. 125-26]. Miernyk notes that for many years it was often more economical to flare off gas at the wellhead than to transport it any distance [7, p. 2]. Technological advances made long-distance interstate pipelines possible and helped generate greatly increased use of natural gas.

The location of large petroleum and gas deposits in the Southwest coincided with rich deposits of salt and sulfur and helped provide the resource base for the chemical industry in that region [3, p. 196].

Historically, pessimism in forecasting oil and gas resources has been common. Just prior to the opening of the California oil fields a petroleum authority discounted that possibility, and in 1925 another expert predicted that our supplies of petroleum would be exhausted by 1945.

There is a metropolitan dimension in the relation between energy resources and locale, and this is developed in some detail by Borchert [2].

He sees a primary economic function of urban areas as collecting, processing, and distributing raw materials and goods produced in their respective hinterlands. Metropolitan growth is seen as dependent both on the size and resource base of the hinterland and on changes in the technology of transportation and industrial energy for the processing of primary resources.

Four technological epochs are identified in U.S. history: sail and wagon, 1770-1830; steamboat and railroad (iron rails), 1830-1870; steel rails and electric power, 1870-1920; and auto-air-amenity-services, 1920 to the present. In the initial period, almost all major urban centers were ports on the Atlantic or seaboard rivers, with high-growth cities located primarily along the inland waterways penetrating the new West—the Erie Canal, lower Great Lakes, and the Ohio River. Manufacturers depended on waterpower, and local waterpower sites influenced industrial location.

The innovations involved in successive transformations of the metropolitan system, and the time path of their impact, are approximated in figure 7-2. Specifically, the figure shows each of ten indicators of technology and energy as a percentage of its peak value in its peak Census year, the indicator taking a value of 100 percent in that year.

The second epoch is identified with the introduction of steam power. The railroad and the steamboat created transportation corridors, which enlarged the hinterlands of ports on inland waterways and the Atlantic. Rail networks were initially tributaries to the water transport system, being built outward from ports to farm, mineral, or timber resources. Steam power made a national transport system possible through the integration of waterways and the rail networks which, in turn, favored the growth of ports with large harbors and proximity to important resource concentrations, at the expense of nearby, smaller ports. In effect, a scale economy was generated, which led to larger population size for the more favored urban places. Steam power was also applied in manufacturing, but the impracticality of long hauls of coal meant that its use in manufacturing was localized, so that the availability of waterpower continued to influence industrial location. It was not until the 1870s, for example, that steam was used universally by cotton mills, coincident with cheaper railroad transportation of coal and other bulk commodities.

This was the start of the third epoch, when steel became relatively low-priced and began to replace iron in railroad tracks. Long-haul rail transport spelled the doom of most passenger and cargo traffic on inland waterways, destroying the economic rationale for many small river ports. Central station electric power became available in the 1880s, and coal and electricity doomed the small, decentralized waterpower sites and helped establish the centralized metropolitan rail centers of much greater scale. Steel and electric power made possible the skyscrapers and rapid transit of the major metropolitan areas. Cities associated with coal grew rapidly; cities at water-

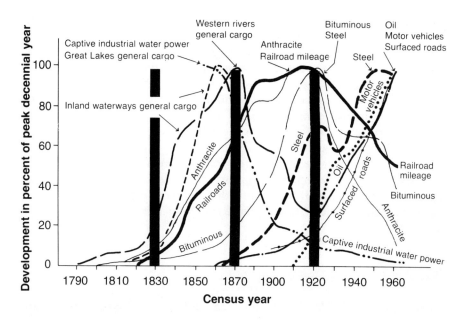

Source: John Borchert, "American Metropolitan Evolution." Reprinted from the *Geographical Review*, vol. 1, 57, 1967, with the permission of The American Geographical Society.

Figure 7-2. Time Path of Indicators of Technology and Energy

power sites in the East and along the great Midwestern rivers grew slowly or declined.

Around the turn of the century, Frank Lloyd Wright predicted that the future would involve a race between the elevator and the automobile, and the wise man would bet on the automobile [8]. The mass use of the automobile heralds the start of the fourth epoch, which Borchert sets at 1920, a reasonable date given the first federal aid to highways in 1916. The internal-combustion engine has had a variety of massive consequences. The tractor in agriculture was a factor in increased farm size and in rural-to-urban migration. Air passenger transport helped centralize the administration of major firms in a few cities. And, of course, the automobile has helped decentralize most metropolitan functions. It has been a factor in the faster growth of newer urban centers in the Southwest and West, which could invest in high-speed roads at lower cost than the more densely settled urban centers in the older parts of the country.

Finally, the internal-combustion engine has had a stimulating impact on the growth of cities near oil fields and a depressing effect on cities near coal fields and at railroad shipping nodes.

Drawing on an argument by Ullman [9], Borchert sees the growth of service employment and concern for amenities as increasingly important factors in location, reducing the role of resources and processing in metropolitan growth. The argument is paralleled by Perloff and Wingo [3, pp. 197-99], who see the location of economic activity as increasingly affected by "amenity resources"—climate, coastline, and land and water features attractive to consumers—and by the increased importance of "footloose" industries not tied to any particular locale and thus able to respond to consumer preferences for the amenity resources.

This brings our chronology to the present. If the role of energy and other resources (nonamenity resources) has been reduced, there is evidence, nonetheless, that they remain of considerable consequence in the location of economic activity, even if we discount the long-term impact of recent energy price increases. Some of that evidence is presented in later sections of this chapter.

Regional Resource Development Projects

Focusing on the TVA. The development and supply of electric power are one of the major purposes common to several massive regional resource development projects in the United States, including the Central Valley Project in California, the Columbia River developments in the Pacific Northwest, and the Tennessee Valley Authority. The electricity supplied by the projects typically is generated initially by hydroelectric power, but there is increasing reliance on other sources. Thus the TVA now generates 80 percent of its power from coal and has considerable nuclear development in prospect [10; 11, pp. 271-74]. Besides power, the multiple purposes of these river basin development projects include navigation, flood control, water supply for irrigation and municipal-industrial use in the Western projects, fertilizer production at TVA, and recreation and wildlife protection.

The TVA electric power system is the nation's largest, supplying electricity to the area shown in figure 7-3, which encompasses most of Tennessee and parts of six adjoining states. The population of the area served is about 7 million, and kilowatthour sales comprise roughly 5 percent of the U.S. total. The Bonneville Power Administration, which is the marketing agent for the Pacific Northwest federal hydroprojects, serves Oregon, Washington, and parts of Idaho and Montana, with a service area population of 6.8 million and sales about 80 percent of those of TVA. In both TVA and Bonneville service areas, electricity prices are considerably below the national average, and consumption is accordingly well above the national average. Residental use per household is roughly twice the national level, and there is a concentration of high-electricity-consuming in-

Source: L. Seeber, "History of the TVA," in *The Tennessee Valley Authority Experience*, ed. H. Knop (Laxenburg, Austria: International Institute for Applied Systems Analysis, 1976), p. 25.

Figure 7-3. The Tennessee Valley Region

dustries—aluminum manufactures, in particular—in both service areas [10, p. 32; 12].

The projects have generated a considerable evaluative and critical literature, much of which deals specifically with the TVA, and I will draw on that literature here, focusing on the TVA experience, but considering some issues that seem of general significance.

Low electricity prices may in part reflect economies of scale, location advantages, and nonmonopolistic behavior by a public monopoly, but they also reflect a conscious policy on the part of Congress in authorizing the projects and a long-term commitment by the agencies in the field in implementing that policy. "'Lowest possible rates' could be identified as TVA's motto. Those words, lifted from the TVA Act, are declared to be primary in every power contract and are used frequently by TVA spokesmen" [13, p. 74].

The low-price policy probably reflected both a belief that low electricity rates could stimulate economic growth and development [14; 15, pp. 74-75] and a belief in an ideology—if it may be graced by the term—of "consumerism," which in its starkest form holds that since low prices to consumers are obviously a good thing, lowering prices is good policy, including price reduction by fiat or regulation. (Critics will ingenuously ask if zero prices are best policy.) In its more sophisticated variant, consumerism involves a countervailing power argument, applying the notion that a "forced" lowering of prices is a counter to monopoly. In a 1945 speech, David Lilienthal expressed this idea: "Workers are organized. Farmers are organized. Finance is organized. Rarely the consumer. . . . It isn't often that anyone has a responsibility to the consumer. . . . that is not so in the case of TVA electricity. . . . [We] owe our first duty to the consumer" [13, p. 74]. The consumerist ideology has much broader currency, of course, than its specific appearance in TVA rate making, since it is a major component of national oil and gas policy, with important consequences that I will discuss in the concluding section of this chapter.

In the case of TVA, low prices seem explainable, in part, as a consequence of subsidy, in effect a subsidy by the rest of the country to the TVA service area. The 1959 TVA self-financing act required payment into the federal treasury of both principal and interest on the appropriated power investment, and such payments have been made since 1961. However, there was the considerable period between 1933 and 1961 before such payments occurred, and the interest rate is that on marketable U.S. debt obligations, which is likely to be below interest rates for utilities in the private sector [13, pp. 175-77].

It seems noteworthy that TVA power rates as a percentage of the average rate charged by seven private power companies in surrounding areas moved from 55 percent in 1960 to 72 percent in 1973, perhaps reflecting what was, in effect, a reduced subsidy [16, p. 70].

Evaluations of TVA performance involve attempts to answer both positive and normative questions: What impacts has TVA had on region and nation? What is the benefit-cost evaluation of those impacts (are they good or bad, on balance)? A review of the literature suggests that these questions are exceedingly difficult and that answers are hardly definitive.

One form of evaluation involves comparing per capita income (from TVA Regional Economics Projects, September 30, 1976), as in this example:

| | *Amount ($)* | | | *Percentage of U.S.* | |
	TVA Region	South-east	United States	TVA Region	South-east
1929	317	359	703	45	51
1974	4,262	4,602	5,448	78	84
1974/1929	13.445	12.819	7.750	1.733	1.647

The TVA region increase is marginally higher than that for the Southeastern United States as a whole, but it is not obvious either that this marginal difference is attributable to TVA operations per se or that it reflects a real rather than a nominal difference. Urban areas typically have a higher cost of living than rural areas, and if the TVA region has had faster rates of urbanization than the Southeast as a whole, real income differences would be correspondingly reduced. Considering such income comparisons, Banner raises the issues of both possible heterogeneity between groups being compared and lack of accounting for changes within the income distribution, for example, how much improvement has there been in the lot of the poor [17, pp. 123, 124]? A more general concern must be that, in the longer run, equilibrating forces should wash out per capita income differences. Thus, if the TVA generated an increment of real income per capita, it would tend to increase employment and population in the region, relative to levels without that increment, until an interregional equilibrium was again attained. Hence, attempts to gauge employment and population impacts may be more meaningful than comparisons of income. Unfortunately, such attempts have generated quite divergent estimates. At one extreme, several observers find little impact. Harry Perry concludes that if there were a regional impact from making low-cost power available, it would have to be small [18]. Netschert draws on a TVA survey that showed a net inmigration to the region of only sixteen manufacturing plants and 1800 employees during the period 1940-1958; he concludes that "it is clear that the role of power in the development of the Tennessee Valley was not dominant" [19, p. 7]. Netschert notes that in this context the TVA was concerned with rebutting charges of "industrial piracy" by "luring" industry from other regions by low prices; further, there appears to be no allusion to the location of the atomic energy complex at Oak Ridge during the period in question.

In contrast to these appraisals, Robock ascribes considerable impact to the TVA, estimating that TVA programs were a significant factor behind one-third of the new industrial jobs and at least one-half of the increased value added by industry, over the period 1929 to 1960, when manufacturing employment increased from 222,000 to 486,000 [20, pp. 113-114].

An early and a late variant of a regional economic simulation model developed by the TVA staff yielded similarly divergent results. The model was an export base model, containing elements similar to an input-output framework but accounting for the impact of price and wage differentials. In a 1974 presentation [21], the effects of changes in relative power costs were traced out, employing the early version of the model. The ratio of TVA industrial power rates relative to that of the seven private companies surrounding the service area was 0.716 in 1973. In the simulation, this ratio was set first at 0.830 and then at 1.093. The estimated impact on manufacturing employment in 1990 was minor, with a decline of 0.3 percent in the first case and 1.0 percent in the second, relative to base-period manufacturing employment. In contrast, the latest version of the model shows much more sensitivity to price changes, ascribed to better econometric modeling, and to greater disaggregation of the manufacturing sector [22]. A 1 percent increase in electricity rates in 1976 would yield a short-run decline in manufacturing employment of 0.8 percent and a long-term (1985) decline of 2.0 percent in manufacturing employment. Since manufacturing comprises about 35 percent of total employment, and since other TVA studies suggest a multiplier effect of one job in services and trade per job in manufactures, it follows that a rough estimate of employment decline in the region would be 1.4 percent per 1.0 percent increase in electricity rates. (It took a 40 percent increase in electricity rates to yield the same employment impact in the earlier vesion of the model.)

In considering the normative evaluation of the TVA, we can begin by again focusing on low electric rates, now from the perspective of national wellbeing. If rates are "artificially" low because of subsidy from outside the region, then from the point of view of income distribution the impacts balance out—the subsidy recipients' gain of a transfer payment is just balanced by the subsidy payers' loss. But from the point of view of the allocation of resources, there is an efficiency loss in the system, because the extra amounts of electricity that we have produced in the TVA region are worth *less* than the things we gave up elsewhere in order to get that extra electricity. Koopmans makes this a central point of his discussion of the TVA [16], arguing that for the rates to be desirable nationally, in terms of their impact on the distribution of industry and population, they must reflect the cost of generation. He stresses the function of prices as a signal system; false signals cause distortion in resource allocation, and net losses in real income, nationally.

However, if TVA rates are relatively low because TVA managers are more efficient than private utility managers, or because private utilities are monopolistic (or more monopolistic), or because there are economies of scale in electricity generation that are captured (only?) by a TVA form of organization, then the country is better off because of TVA: there are net gains in real income nationally.

There is a third possibility, based on observations by Habday that local taxes on electricity are kept low, that TVA's policy and practice is to hold tax equivalents (paid by municipal power systems to their respective municipal governments) at low levels. This policy of "nondiversion" has attained wide acceptance [13, pp. 213, 235, 236]. It suggests that consumers either pay higher "other" taxes or consume lower amounts of public services, having in effect traded some municipal public service for electricity, perhaps by choice.

It is plausible that all three explanations hold to some extent in reality, obviously making normative evaluation extremely difficult. Certainly TVA defenders stress the technical efficiency of TVA management [14, p. 76; 23, p. 250] and infer that it has injected competition into a monopolistic industry: "concentric circles drawn around the area supplied with power from TVA enclose bands of progressively lower rates as it is approached" [23, p. 248]. And Robock argues that TVA is a source of experimentation and innovation that yields external benefits   [20, pp. 116, 119].

Some aspects of the TVA experience are difficult to fit into a widely acceptable normative framework either because tradeoffs are hard to evaluate or because they involve strongly partisan value positions. An example of the former problem is Habday's concern about decision making by technocrats rather than by politicians, that is, by TVA managers and the independent, *appointed* members of electric power boards rather than by *elected* officials responsible and responsive to a constituency   [13]. The latter problem is exemplified by TVA's conflicts with environmentalists and antinuclear power advocates. Thus, "TVA has been in the fore-front of the utilities' long-standing fight with the EPA over whether they should be forced to install scrubbers in the stacks of plants that burn medium sulfur coal" [24; 25 lists some related discussions]. An earlier example is the question of TVA's putative role as a regional planning agency. The authority's founders anticipated that regional planning would be a major function, but that activity became considerably muted over the years, perhaps to the disappointment of both proponents and opponents of the idea   [26, p. 7; 13, pp. 4, 17; 17, pp. 125-28].

In concluding what has been, in effect, a case study of the TVA, it seems worthwhile to critically consider some policy theses set forth by TVA managers or partisans that seem to be widely and strongly held. First, industrialization tends to be identified with economic progress, to the point of viewing the former as cause of the latter   [20, p. 111, for example]. But this can be countered by the argument that rich countries (and perhaps regions) are industrialized because they are rich, and not rich because they are industrialized   [27]. Again, attitudes of "regional chauvinism" seem fairly widespread. In 1972 the TVA board chairman extolled regional growth which "has halted the emigration of the Valley's work force to the crowded industrial centers of the North and East"   [23, p. 254]. Similarly, Koopmans notes that a TVA official "seemed to think it was a shame for the state of Tennessee to lose able-bodied and vigorous young people, if there is no employment for them"   [16, p. 71]. Koopmans again sees proper pricing as the issue here; if migration occurs in response to accurate price signals, so be it. The district form of political organization (for example, congressional districts) may be a factor in regional chauvinism; officials may be concerned about local people *only* if they stay put and remain constituents; such attitudes can be costly, in terms of national well-being.

Energy and the Location of Economic Activity

Following Borchert, I have suggested that energy has had considerable impact on location. But this proposition does not draw universal assent, and even where it is accepted, there is considerable variation in the perceived strength of the relationship. Let me consider some of the points of view and expand on some aspects of the relationship.

Miernyk concludes that, in the past, energy was a significant locational determinant only in a handful of industries; aluminum, chemicals, and glass manufactures are listed among that handful [7, p. 2]. He argues that industrial location was usually determined by minimization of transfer costs, involving the cost of both assembling raw materials and shipping finished products to market. Cheap energy, and hence low transportation costs, meant that energy had little impact on transfer costs, so that "major industrial complexes could develop in the Northeast and Great Lakes, far removed from basic energy sources." In contrast to his perception of past patterns, Miernyk sees a long-term future of expensive energy and considerable impact on location.

Perry makes a similar argument, noting that the price of energy generally was so low that it was not a determining factor in siting industrial plants, save for a relatively small number of energy-intensive industries, including primary metals, chemicals, petroleum refining, food, paper, and stone, clay, and glass. Perry adds that even for energy-intensive industries, other economic factors can be more important than energy; thus in recent years there were some shifts of aluminum plants from the state of Washington to the Ohio River Valley, which had relatively cheap fuel supplies plus good access to aluminum processers [18].

There are some counters to these observations, which at the minimum suggest at least some changes in emphasis. Perry's "relatively small number" of industries is considerably larger than Miernyk's handful, and accounted for 31.5 percent of all wages in manufacturing in 1972 [28]. Aluminum makers in the Pacific Northwest account for about one-third of national output. In 1977, "faced with the choice of leaving the region or paying sharply higher power rates, the aluminum companies have opted to stay. . . . Even at the projected tripled energy costs, the aluminum operations would be paying a 'shade below' the average price paid by non-Pacific Northwest aluminum plants" [28]. Hence, the remaining price differential, though considerably reduced from the original amount, appears strong enough to prevent relocation. Of course, relocation is often an expensive process, accounting for lags in adjustment of considerable duration. In contrast to Miernyk, it could be argued that in their periods of initial development, the major industrial complexes of the Northeast and Great Lakes were *not* far removed from the major energy sources of their time. (See the earlier discussion.) Relocation costs, once fixed capital is in place, as well as the location of the consuming market, could make regional shifting a slow process.

Hence, it seems reasonable to hypothesize that energy is a significant, if not dominant, factor in location. Certainly, other factors of production affect location. As Banner argues, raw materials, skilled labor, and research and development facilities are often more important [17, p. 129]. But current developments may lead to understatement of energy's past role and overstatement of its future role in location.

In the development of theoretical frameworks to explain levels of regional economic activity and growth, a distinction is often made between the impact of exports and that of internal organization and trade. Perloff and Wingo develop the distinction in some detail [3, pp. 199-200], arguing that regional growth typically is promoted by exports to a national market, which expands the regional economy by way of multiplier effects. But the extent of the multiplier effects depends on internal features, such as the quantity and type of labor required by the export industries. Thus, in addition to the export determinants, there are internal determinants of growth, including internal agglomeration economies and economies of scale. Further, Perloff and Wingo hypothesize that the occurrence of rapid self-sustaining growth involves a shift from the dominance of the export market to an increasingly important role for the internal organization of production.

Considering petroleum and natural gas production as a specific case, Perloff and Wingo [3, p. 202] suggest that this extractive industry has relatively weak multiplier effects and linkages, so that changes in the industry will not greatly affect regional economic activity. (They also warn of the negative impact of the depletion of resources.) Their argument is supported in part by the capital-intensive nature of fuel extraction, but this may imply weak multiplier effects only if capital goods employed in fuel extraction are imported from outside the region, rather than being produced locally. Similarly, considerable differences in location impact are to be expected if the profits of an enterprise flow out of the region (to absentee landlords, as in the classic dual economy case) or, rather, are consumed or invested locally.

In two previous studies, I carried out input-output analyses for local economies in which, as a variant, I treated household activity as endogenous. (Labor input appeared as a row entry for all industries and consumer expenditures as a column entry.) Government spending and exports were then the exogenous sectors. Thus, an increase in one dollar of government spending on specific industry A would cause industry A to expand its production and buy more inputs, including labor (assumed to move into the local area from outside the area); the industries furnishing input to industry A would then expand *their* use of inputs (including greater consumption by labor in the role of household consumers), and so on. The studies covered the Chicago area [30] and Napa County, California [31].

The impact of a dollar increase in final demand on households was considerably greater in Chicago than in Napa County, explainable by the much greater reliance on imports in Napa than in Chicago, thus dampening local expansion of production. For Napa county, with a thirteen-sector input-output table, the dollar increase in final demand generated an increase in household activity ranging from $0.39 to $1.55 (over the thirteen sectors). There was less variation in the Chicago results, with the multiplier effect on households ranging from $2.82 to $3.81 (over the forty-six sectors) per dollar increase in final demand. Of some pertinence to the present discussion, the mining sector (in the Chicago table) had a multiplier effect for households of $3.43. Hence, if exports of mining increased by one dollar, the ultimate impact would be $3.43 in new household income and spending, or roughly one additional half-hour of labor brought into the area (at present wage rates). If this result has general applicability, it furnishes another counter to the Perloff-Wingo thesis, for the multiplier effect for mining was near the upper end of the range.

The approach of treating households as endogenous may be useful in attempting to gauge the ultimate impact of expansion to meet an incremental increase in exports or government spending, but it would be an overstatement to argue that all regional employment is "ultimately" to be explained by those variables, neglecting the point that there are internal determinants of economic activity. Certainly, people will locate in an area if there are opportunities for mutually profitable exchange, including situations where all exchange involves commodities produced locally (people, in effect, take in one another's washing, without going outside the area). Perloff and Wingo's discussion suggests that the internal determinants also involve positive externalities, entrepreneurial capacity, and production efficiency, viewed as increasing over time, and development. However, the measurement and operational applications of such variables seem a formidable task. In this connection, I note that in his review of the "early" TVA simulation model, MacKinnon distinguished between export sector effects and (internal) agglomeration economies and argued that there was little accounting for the latter in the model [32]. Unfortunately, he made no specific suggestions on how to get at such determinants.

As a final item here, let me review some detailed evidence to be presented in the next section. I will show that a surprisingly large proportion of U.S. energy extraction (in coal, petroleum, and natural gas) occurs within the boundaries of metropolitan areas. To some extent, such production may have been induced by urban demand, but it seems plausible that the reverse causal relation is also at work: a number of urban places may have grown to metropolitan status because of the economic activity generated by the extractive industry. Although Borchert has argued that the impact of resources on metropolitan growth has been considerably muted, this evidence at least suggests that some of the effect has been remarkably persistent.

Role of Energy

The Economic Geography of Recent U.S. Energy Production and Consumption

In this section I present information on recent U.S. energy production and consumption, focusing on regional and urban issues. The presentation also includes some critical evaluations of available information and some suggestions on data refinements and extensions.

Production

Figure 7-4 through 7-7 present graphic information on U.S. production by locale, with figures reproduced from various sources [33, 34, 35, 36, 37, 38]. Figure 7-4 shows oil and gas fields, and figure 7-5 shows coal-producing areas, including some explicit indication of areas of reserves with little current production. Figure 7-5 displays information on hydroelectric and nuclear power. More than half of developed hydropower and the bulk of potential hydropower is located in the West. Energy-importing regions are more than proportionately represented in the map showing nuclear power plants. There appears to be some clustering near—but not too near—major metropolitan centers. Although there are no formal federal regulations prohibiting location in areas of high population density, the regulatory process nevertheless encourages location in low-density areas, so that in practice nuclear power plants are located at some remove from metropolitan centers [39]. The Indian Point nuclear power facility, located about twenty-five miles from New York City, appears to be the closest nuclear power plant to a major urban area; it has the highest "site population factor" that has been assigned in the regulatory evaluation, indicating the highest population at risk from nuclear accident [39].

Figure 7-6 exhibits trends in oil production by major producing states since 1950; clearly, a few states play a dominant role. Figure 7-7 shows the trend in total U.S. oil production since October 1973. The downward trend is commonly interpreted as exhibiting the depletion of U.S. oil resources, but the effect of price controls in inhibiting production could well be a major—or even *the* major—explanation for the decline. A marked reversal of trend occurs in July 1977 with the advent of north slope Alaskan oil.

Quantitative information on 1974 production and reserves of petroleum, natural gas, and coal is presented in tables 7-2 through 7-5, supplemented by the initial four tables of appendix 7A. Table 7-2 shows state production in trillion Btu, converted from units specific to the three fuels (barrels, cubic feet, and tons, respectively) by applying scale factors from Crump [40]; data in the original units appear as tables 7A-1 and 7A-2. Table 7-3 puts the state data of table 7-2 in percentage form, while table 7-4

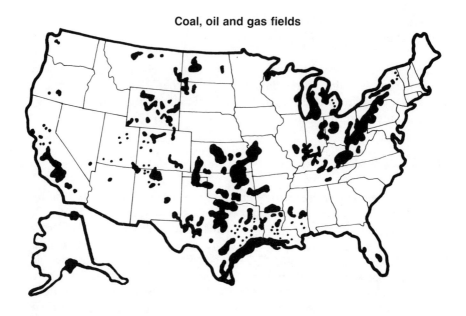

Coal, oil and gas fields

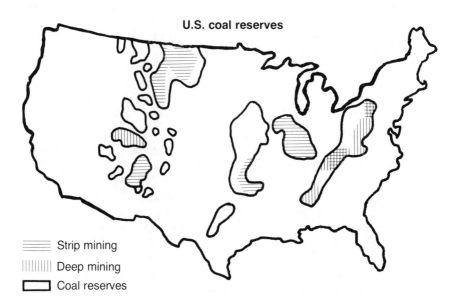

U.S. coal reserves

≡≡≡ Strip mining
||||||| Deep mining
☐ Coal reserves

Source: Coal, oil, and gas fields: Ralph E. Lapp, *America's Energy* (Reproduced in U.S. House of Representatives, *Energy Information Handbook*, figure 26, p. 79 (Washington: USGPO, 1976); U.S. coal reserves: "The Coal Industry's Controversial Move West." Reprinted from the May 11, 1974, issue of *BusinessWeek* by special permission, © 1974 by McGraw-Hill, Inc., Neew York, N.Y. 10020. All rights reserved.

Figure 7-4. Location of U.S. Oil, Gas, and Coal Production

Role of Energy

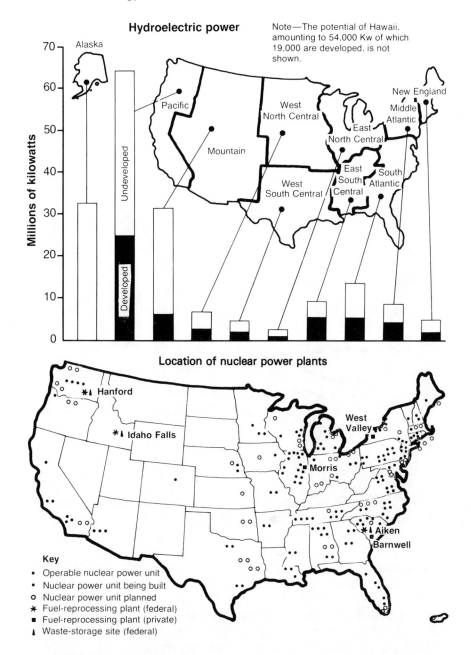

Source: Hydroelectric power: Federal Power Commission, *Hydroelectric Power Resources* of the United States (Washington, 1972), figure 2; nuclear power plants: From "The Reprocessing of Nuclear Fuels," by William Bebbington. Copyright © December 1976 by Scientific American, Inc. All rights reserved.

Figure 7-5. Indicators of Hydroelectric Power and Location of Nuclear Power Plants

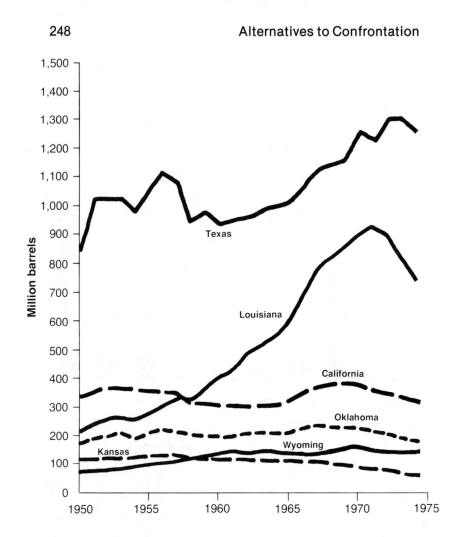

Source: U.S. Bureau of Mines, *Minerals Yearbook 1974*, vol. 1 (Washington, 1976).
Figure 7-6. Long-Term Trends in Oil Production by Major Producing State

aggregates the state data into regional totals, employing the Census classification of regions, and also presents estimated regional fuel reserves as of 1974. Table 7-5 exhibits the corresponding ratios of reserves to production and then shows low-sulfur coal as a percentage of total coal reserves, and coal reserves as a percentage of total reserves. Tables 7A-3 and 7A-4 present corresponding state data on reserves.

In 1974 Texas and Louisiana jointly accounted for roughly 60 percent of oil production and 70 percent of gas production; at a considerable remove, California, Oklahoma, and Wyoming were the next three largest

Role of Energy

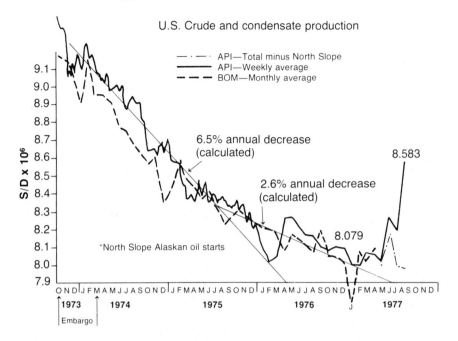

Source: Federal Energy Administration, Office of Oil and Gas, *The Search for Petroleum*, August 22, 1977.

Figure 7-7. Recent Trends in U.S. Oil Production

oil producers, and Oklahoma, New Mexico, and Kansas the next three largest gas producers. Coal production was less concentrated, with Kentucky, West Virginia, and Pennsylvania the leading producing states, followed by Illinois, Ohio, and Virginia. The Rocky Mountain region accounted for roughly 10 percent of coal produced in 1974; the region's percentage of U.S. output has been increasing rapidly, reflecting both the economies of strip mining and the utility of low-sulfur coal in meeting air quality standards. Table 7-5 exhibits the latter factor in dramatic fashion, showing that 80 percent of Western coal is of the low-sulfur variety, in marked contrast to the "dirtier" coal of the East.

Although the West South Central region accounted for more than half the British thermal units produced in 1974, as a consequence of the relative importance of oil and gas in current output, it accounted for only a minor portion of reserves. Here, coal is king, comprising more than 90 percent of total reserves and consequently making the coal-producing regions dominant in the distribution of reserves. The Rocky Mountain region contains 40 percent of listed reserves, with Montana and Wyoming the leading states in that region (see table 7A-3).

Table 7-2
State Production of Fuels, 1974
(trillion Btu)

Region and State	Petroleum	Gas	Coal	Total
New England				
Connecticut	0.000	0.000	0.000	0.000
Maine	0.000	0.000	0.000	0.000
Massachusetts	0.000	0.000	0.000	0.000
New Hampshire	0.000	0.000	0.000	0.000
Rhode Island	0.000	0.000	0.000	0.000
Vermont	0.000	0.000	0.000	0.000
Mid-Atlantic				
New Jersey	0.000	0.000	0.000	0.000
New York	5.197	5.474	0.000	10.671
Pennsylvania	20.172	90.653	2,149.945	2,260.770
East North Central				
Illinois	159.807	1.575	1,326.138	1,487.520
Indiana	28.530	0.193	550.443	579.166
Michigan	104.522	75.839	0.000	180.361
Ohio	52.710	100.984	1,117.061	1,270.756
Wisconsin	0.000	0.000	0.000	0.000
West North Central				
Iowa	0.000	0.000	11.458	11.458
Kansas	357.808	972.800	17.476	1,348.084
Minnesota	0.000	0.000	0.000	0.000
Missouri	0.325	0.036	100.781	101.142
Nebraska	38.344	2.784	0.000	41.128
North Dakota	114.243	34.233	106.572	255.047
South Dakota	2.865	0.000	0.000	2.865
South Atlantic				
Delaware	0.000	0.000	0.000	0.000
District of Columbia	0.000	0.000	0.000	0.000
Florida	210.836	41.836	0.000	252.672
Georgia	0.000	0.000	0.000	0.000
Maryland	0.000	0.146	59.453	59.599
North Carolina	0.000	0.000	0.000	0.000
South Carolina	0.000	0.000	0.000	0.000
Virginia	0.017	7.784	904.833	912.635
West Virginia	15.457	221.930	2,745.982	2,983.368
East South Central				
Alabama	77.273	30.568	510.270	618.111
Kentucky	45.455	78.848	3,493.036	3,617.338
Mississippi	294.518	86.429	0.000	380.948
Tennessee	4.460	0.019	197.574	202.053
West South Central				
Arkansas	95.857	136.001	11.584	243.441
Louisiana	4,276.479	8,505.788	0.000	12,782.267
Oklahoma	1,031.153	1,797.919	59.465	2,888.538
Texas	7,320.331	8,963.365	101.429	16,385.125

Table 7-2 *(cont.)*

Region and State	Petroleum	Gas	Coal	Total
Rocky Mountain				
Arizona	4.292	0.246	140.824	145.362
Colorado	217.546	158.658	153.919	530.123
Idaho	0.000	0.000	0.000	0.000
Montana	200.413	60.196	304.407	565.016
Nevada	0.748	0.000	0.000	0.748
New Mexico	572.431	1,365.523	216.767	2,154.721
Utah	228.305	55.423	149.028	432.756
Wyoming	811.983	358.343	426.896	1,597.221
Far West				
Alaska	409.497	141.442	16.002	566.941
California	1,873.417	400.793	0.000	2,274.211
Hawaii	0.000	0.000	0.000	0.000
Oregon	0.000	0.000	0.000	0.000
Washington	0.000	0.000	75.677	75.677

Source: Developed from data in U.S. Bureau of Mines, *Minerals Yearbook 1974*, vol. 1 (Washington, 1976), pp. 355-6, 845, 864, 978-9, 986; and Lulie H. Crump, *Fuels and Energy Data*, U.S. Bureau of Mines Information Circular 8739 (Washington, 1977), p. 5.

The data on oil and gas reserves ought not be taken too literally. The estimates shown here are "proved" reserves, which represent the most conservative geological estimates and, further, resources that can be profitably recovered at *current* prices. Since prices of both oil and gas have been kept artificially low by regulation, reserves should be understated. Reserve information is reported by the American Petroleum Institute and the American Gas Association on the basis of reports from industry geologists. "Consumerists" and advocates of regulation to keep prices low suggest that gas producers deliberately understate reserve figures as a ploy to obtain increased prices [41]. This "credibility" issue was raised in a 1974 Government Accounting Office report, but in 1976 it was noted that a Federal Energy Administration estimate was in substantial agreement with the industry estimate [42], a finding paralleled by studies by other federal agencies [43].

The Carter administration energy plan calls for an "independent" government information system to develop reliable data on reserves, and the American Petroleum Institute concurs, citing the need for credibility [44]. Whatever the source of estimates, they will be stochastic—past experience suggests there is considerable sampling variability and forecast error in reserve estimation. Thus, forty-five "authoritative" studies made between 1950 and 1965 ranged from zero to 2,959 trillion cubic feet in their estimates of "potential reserves" of gas [45]. (A cubic foot of gas contains 1,097 Btu; hence the high estimate here is ten times the gas reserve listed in table 7-4.)

Table 7-3
State Production of Fuels, as Percentage of U.S. Production, 1974

Region and State	Petroleum	Gas	Coal	Total
New England				
Connecticut	0.000	0.000	0.000	0.000
Maine	0.000	0.000	0.000	0.000
Massachusetts	0.000	0.000	0.000	0.000
New Hampshire	0.000	0.000	0.000	0.000
Rhode Island	0.000	0.000	0.000	0.000
Vermont	0.000	0.000	0.000	0.000
Mid-Atlantic				
New Jersey	0.000	0.000	0.000	0.000
New York	0.028	0.023	0.000	0.019
Pennsylvania	0.109	0.383	14.384	3.951
East North Central				
Illinois	0.860	0.007	8.872	2.600
Indiana	0.154	0.001	3.683	1.012
Michigan	0.563	0.320	0.000	0.315
Ohio	0.284	0.426	7.473	2.221
Wisconsin	0.000	0.000	0.000	0.000
West North Central				
Iowa	0.000	0.000	0.077	0.020
Kansas	1.926	4.105	0.117	2.356
Minnesota	0.000	0.000	0.000	0.000
Missouri	0.002	0.000	0.674	0.177
Nebraska	0.206	0.012	0.000	0.072
North Dakota	0.615	0.144	0.713	0.446
South Dakota	0.015	0.000	0.000	0.005
South Atlantic				
Delaware	0.000	0.000	0.000	0.000
District of Columbia	0.000	0.000	0.000	0.000
Florida	1.135	0.177	0.000	0.442
Georgia	0.000	0.000	0.000	0.000
Maryland	0.000	0.001	0.398	0.104
North Carolina	0.000	0.000	0.000	0.000
South Carolina	0.000	0.000	0.000	0.000
Virginia	0.000	0.033	6.054	1.595
West Virginia	0.083	0.937	18.371	5.214
East South Central				
Alabama	0.416	0.129	3.414	1.080
Kentucky	0.245	0.333	23.369	6.322
Mississippi	1.586	0.365	0.000	0.666
Tennessee	0.024	0.000	1.322	0.353
West South Central				
Arkansas	0.516	0.574	0.078	0.425
Louisiana	23.023	35.896	0.000	22.340
Oklahoma	5.551	7.588	0.398	5.048
Texas	39.410	37.827	0.679	28.636
Rocky Mountain				
Arizona	0.023	0.001	0.942	0.254
Colorado	1.171	0.670	1.030	0.927
Idaho	0.000	0.000	0.000	0.000

Table 7-3 *(cont.)*

Region and State	Petroleum	Gas	Coal	Total
Montana	1.079	0.254	2.037	0.987
Nevada	0.004	0.000	0.000	0.001
New Mexico	3.082	5.763	1.450	3.766
Utah	1.229	0.234	0.997	0.756
Wyoming	4.371	1.512	2.856	2.791
Far West				
Alaska	2.205	0.597	0.107	0.991
California	10.086	1.691	0.000	3.975
Hawaii	0.000	0.000	0.000	0.000
Oregon	0.000	0.000	0.000	0.000
Washington	0.000	0.000	0.506	0.132

Source: From table 7-2 and U.S. totals in table 7-4.

The question of reserves can be generalized to that of supply elasticity. It is my impression that the Carter administration energy policies and proposals are predicated on the belief that gas and oil are highly inelastic in supply. Hence, a price increase is seen as transferring income from consumers to producers, with little or no impact on quantity produced. (The belief could be conducive to the ready acceptance of conservative estimates of reserves.) Some econometric models support this thesis. In an Office of Technology Assessment analysis of the National Energy Plan, a simulation model is employed to reach the conclusions that "oil is highly inelastic above $11.64 per barrel," while for natural gas "if anything, supply is even more inelastic to price changes. However, development of gas in high cost regions will not commence below $1.75 per thousand cubic feet" [46], which just happens to equal the regulated price in the administration plan. Thus, new supply is seen as perfectly elastic at the regulated price and highly inelastic above that price—a very discontinuous result. Rice and Smith found little supply response to oil price decontrol in an application of their econometric model of the oil industry, but they note that substantial improvements in their modeling work are warranted, seeing it as illustrative of the potential value of econometric modeling, rather than a definitive study [47].

In contrast, there is a good deal of anecdotal evidence suggesting considerable supply elasticity on the basis of experience in the unregulated intratstate gas market, where prices reached levels of around $2.00 per thousand cubic feet (mcf), compared to a regulated price of $0.52 for "old" gas and $1.44 for "new" (newly discovered) gas, in the interstate market [48]. Hall and Pindyck present an estimate of 0.2 for supply elasticity; that is, a 5 percent increase in price would yield a 1 percent increase in quantity supplied. They acknowledge that there is some uncertainty over the magnitude and speed of the supply response, but feel their estimate is reasonable [49].

Table 7-4
Production and Reserves of Petroleum, Natural Gas, and Coal by Census Region, 1974

Region	Petroleum	Gas	Coal	Total
Production (Trillion Btu)				
New England	0.00	0.00	0.00	0.00
Mid-Atlantic	25.37	96.13	2,149.95	2,271.44
East North Central	345.57	178.59	2,993.64	3,517.80
West North Central	513.59	1,009.85	236.29	1,759.72
South Atlantic	226.31	271.70	3,710.27	4,208.27
East South Central	421.71	195.86	4,200.88	4,818.45
West South Central	12,723.82	19,403.07	172.48	32,299.37
Rocky Mountain	2,035.72	1,998.39	1,391.84	5,425.95
Far West	2,282.91	542.24	91.68	2,916.83
United States	18,574.99	23,695.83	14,947.02	57,217.84
Production as Percentage of Category Total				
New England	0.00	0.00	0.00	0.00
Mid-Atlantic	0.14	0.41	14.38	3.97
East North Central	1.86	0.75	20.03	6.15
West North Central	2.76	4.26	1.58	3.08
South Atlantic	1.22	1.15	24.82	7.36
East South Central	2.27	0.83	28.11	8.42
West South Central	68.50	81.88	1.16	56.45
Rocky Mountain	10.96	8.43	9.31	9.48
Far West	12.29	2.29	0.61	5.10
United States	100.00	100.00	100.00	100.00

Region	Petroleum	Gas	Coal < 1 Percent Sulfur	All Coal	Total
Reserves (Trillion Btu)					
New England	0.00	0.00	0.00	0.00	0.00
Mid-Atlantic	353.80	1,818.38	195,544.98	828,336.03	830,508.22
East North Central	2,262.00	3,543.32	41,093.34	2,286,357.15	2,292,162.44
West North Central	3,451.00	13,486.09	78,530.55	530,800.30	547,737.37
South Atlantic	1,943.00	2,919.33	437,525.97	1,184,721.26	1,189,583.60
East South Central	2,128.60	2,673.06	188,427.50	752,866.74	757,688.40
West South Central	96,088.60	173,432.19	17,717.71	92,803.41	362,324.20
Rocky Mountain	13,195.00	21,586.89	3,196,039.80	3,982,778.08	4,017,559.92
Far West	79,175.80	40,675.01	273,639.72	304,039.02	423,889.82
United States	198,597.80	260,134.27	4,428,519.57	9,962,702.10	10,421,434.00

Source: From sum of state entries in table 7-2 and table 7A-3.

In a series of editorials, the *Wall Street Journal* has noted the natural gas supply estimate of the MOPPS group at ERDA (the Market Oriented Program Planning Study of the Energy Research and Development. Administration) [50]. At a price of $2.50 per mcf, it was estimated that the

Table 7-5
Reserves Relative to Production, by Region, 1974

	Reserves Relative to 1974 Production				Coal Reserves	
					Coal < 1 Percent Sulfur as Fraction of All Coal	As Fraction of All Reserves
Region	Petroleum	Gas	Coal	Total Fuels		
New England	0.00[a]	0.00	0.00	0.00	0.0000	0.0000
Mid-Atlantic	13.95	18.92	385.28	365.63	0.2361	0.9974
East North Central	6.55	19.84	763.74	651.59	0.0180	0.9975
West North Central	6.72	13.35	2,246.39	311.26	0.1479	0.9691
South Atlantic	8.59	10.74	319.31	282.68	0.3693	0.9959
East South Central	5.05	13.65	179.22	157.24	0.2503	0.9937
West South Central	7.55	8.94	538.05	11.22	0.1909	0.2561
Rocky Mountain	6.48	10.80	2,861.52	740.43	0.8025	0.9913
Far West	34.68	75.01	3,316.31	145.33	0.9000	0.7173
United States	10.69	10.98	666.53	182.14	0.4445	0.9560

Source: From table 7-4.
[a]Zero indicates no reserves and no production.

Table 7-6
U.S. Relative Wages and Employment in Selected Industries and Years, 1929-1975

Industry	1929	1939	1950	1962	1972	1973	1975	1976
Wages and salaries per full-time equivalent employee relative to U.S. value (all industries)								
All industries	100.0	100.0	100.0	100.0	100.0	100.0	100.0	100.0
Coal mining	97.8	96.6	107.1	106.7	128.2	132.5	147.1	146.5
Oil and gas extraction	141.4	131.6	128.1	123.5	124.8	127.5	138.7	141.2
All manufactures	108.1	106.5	109.9	114.7	107.8	107.9	109.8	110.9
Petroleum and coal manufacturing	129.1	144.7	146.9	148.3	141.7	141.6	152.8	154.2
Electric and gas utilities and sanitary services	109.4	135.9	117.4	128.9	130.3	131.4	132.0	135.9
Banking	137.9	153.8	110.5	100.1	91.8	90.4	90.4	90.0
Insurance carriers	158.8	142.7	115.7	110.4	105.2	105.0	105.4	105.3
Legal services	96.5	93.5	79.6	95.3	109.0	112.7	115.4	118.3
Public education	101.2	109.6	92.2	102.5	105.7	105.1	102.6	102.3
Employment in thousands								
All industries	35,338	37,924	48,527	57,743	72,348	75,484	74,290	76,728
Coal mining	622	469	469	150	159	158	210	224
Oil and gas extraction	159	187	259	291	258	263	316	329
All manufactures	10,428	9,967	15,101	16,329	18,548	19,566	17,727	18,480
Petroleum and coal manufacturing	128	135	216	189	183	184	187	194
Electric and gas utilities and sanitary services	493	445	545	597	693	707	706	711
Banking	385	288	418	697	1,073	1,140	1,228	1,257
Insurance carriers	358	414	546	802	1,002	1,028	1,039	1,057
Legal services	90	116	109	147	242	268	310	326
Public education	1,082	1,224	1,536	2,700	4,577	4,753	4,998	5,048

Source: U.S. Bureau of Economic Analysis, *The National Income and Product Accounts* (Washington, 1976), pp. 206-213, and *Survey of Current Business* (Washington, July 1977), tables 6.8 and 6.9, p. 47.

United States would have about forty-five years' worth of natural gas at current levels of consumption, contrasting with present reserve estimates of about ten years' worth (table 7-7). The *Wall Street Journal* reports that ERDA subsequently recalled the MOPPS study and disbanded the seventy-person MOPPS group.

Thus, there seems little hard evidence on supply elasticity, which can be viewed as a concomitant of the uncertainty on reserves [51]. The supply question involves major policy issues, with substantial regional impacts, and I focus on those issues in the concluding section.

It is clear that there has been considerable response to higher prices for coal, and "new" oil and gas, in terms of increased labor wage rates and employment in energy production—the ill wind of OPEC has blown considerable good to domestic production—and those factor shifts seem at least indirect evidence of substantial supply elasticity in energy production. Changes over time in the U.S. labor market for fuel extraction can be viewed in brief in these statistics, derived from table 7-6:

Wage Rates Relative to All U.S. Wage Rates	*1972/1950*	*1976/1972*
Coal	1.197	1.143
Oil and gas	0.974	1.131
Employment		
All U.S. employment	1.491	1.061
Coal	0.339	1.409
Oil and gas	0.996	1.275

Wage rates are measured as wages per full-time equivalent employee; the wage relatives here show the ratio of terminal-year to initial-year wage rates in fuel extraction, in turn divided by the corresponding terminal-year to initial-year wage ratio for all U.S. employment. In the 1950-1972 period, the increase in the relative wage for coal could have been a result of union power and/or an increased recognition of the bodily risks of underground coal mining. Relative wages in oil and gas production slipped a little in that period. Employment in coal dropped precipitously, while that in oil and gas remained essentially unchanged, although total U.S. employment increased by half. During the period there was substantial conversion from coal to oil. Energy prices were stable over most of the period and hence declined in real terms because of considerable increases in the general price level. In the post-OPEC period, wage rates both in coal and in oil and gas extraction increased markedly relative to U.S. wage rates, with roughly the same amount of relative increase in each activity. Employment increases were even more pronounced, with relative increases in coal above those in oil and gas.

Table 7-6 presents more detail on U.S. wage and employment changes over time, exhibiting data for selected industries, as well as for coal and oil and gas, and for a number of years from 1920 through 1976 [52, 53]. Some major trends and structural changes become manifest with such data; for example, note the shifts in public education employment and wages, and the growth in legal services employment.

Some caution must be exercised in comparisons of such data, of course, because of changes in composition within industries. Thus, the banking wage rate trends probably include an effect attributable to greatly increased employment of women, while the wage rate trend in coal probably understates that industry's wage increase because of a substantial decline in anthracite relative to bituminous employment (historically, anthracite wage rates were considerably above bituminous).

Data on labor and proprietor income by industry are available by state and region. In effect, wage rates are multiplied by employment to obtain total wages; then, total wages plus proprietor income plus "other labor income" yield the total income figure that is available in published tables. In 1975 wages accounted for 84 percent, proprietor income for 10 percent, and other labor income for 6 percent of the total. Table 7-7 presents these 1972 and 1975 income figures for all industries, coal, and oil and gas, covering the United States, the Census regions, and major fuel extracting states (actual and potential) [54].

Between 1972 and 1975, labor and proprietor income for all industries increased by 27 percent, while that in coal and oil and gas extraction doubled, moving the share of fuel extraction in income from 0.67 to 1.08 percent of the total. In the case of Texas, the oil and gas share in income moved from 2.96 to 4.43 percent, while for West Virginia the coal share increased from 12.6 to 15.4 percent of total income.

It seemed a plausible hypothesis that the income growth in the fuel industries would yield higher per capita incomes in the major fuel-producing states, by way of both direct and indirect (multiplier) effects. Tables 7-8 and 7-9 show data used in examining this hypothesis, in the form of per capita incomes for 1972 and 1975 by region and state, respectively [55]. Inflation accounted for 27.3 of the 29.7 percent increase in U.S. per capita income in the period; thus for the Mid-Atlantic region and a number of states, per capita real income growth was negative.

The income growth hypothesis was reformulated as a testable association between per capita income growth and fuel extractive activity, and a significant relation emerged. The income ratio of table 7-9 was treated as dependent variable and related to a number of regional dummy variables accounting for major fuel states and for other regional effects. (A dummy variable takes on a value of 1.0 for a state falling in a specific region and a value of 0 for a state not falling in that region.) The regression equation obtained was:

	Coefficient	t Ratio
Constant term	1.322	144.587
Fuel state	.053	3.412
Plains state	.054	3.156
Alaska	.487	12.796
Northeast region	−.046	3.061
North Central region	−.031	2.334

$R^{-2} = .803$ (explained variance)

The fuel states were the seven major fuel-extracting states in table 7-7; the Plains states correspond to the West North Central region (see table 7-9), presumably with high income growth because of a surge in agricultural income; Alaska was treated as a separate variable because of very high income growth, reflecting the construction of the oil pipeline; and the Northeast and North Central regions are major Census regions. (New England plus Mid-Atlantic equals the Northeast, and East North Central plus West North Central equals the North Central region.) The South as a variable was not statistically significant, and the West was omitted as a statistical necessity. (One dummy variable of a set must be omitted.) The regression equation has the following meaning: the ratio of 1975 to 1972 income per capita for states not covered by the dummy variables averaged 1.322; for other states, the ratio averaged 1.322 plus the appropriate coefficient—thus, in the Northeast the ratio would be 1.276 (1.322 − 0.064). Of interest here, a fuel-producing state would have an increase in that ratio of 0.053. All variables listed were statistically significant (t ratio above 1.96), including fuel production. Hence, we can conclude that major fuel-producing states had an estimated 1972 to 1975 per capita income growth of 37.5 percent, in contrast to 32.2 percent for other states in the South and West (a 16 percent differential), and a per capita income growth of 32.9 percent versus 27.6 percent for other states in the Northeast (an 18 percent differential), obtained by adding the fuel coefficient to the previously noted regional base figures.

Of course, association is not necessarily causation, but it is clear from tables 7-6 and 7-7 that much of the estimated increase for fuel-producing states involves the direct effect of high wages in fuel production.

There is an important metropolitan dimension to fuel production, which can be established through the use of additional data on labor and proprietor income. The Bureau of Economic Analysis of the Department of Commerce has issued industry earnings data, consisting of wages plus proprietor income, for metropolitan areas, as well as for states [56].

Table 7-7
Labor and Proprietor Income, by Region and Selected State, 1972 and 1975
(million dollars)

Geographic Unit	All Industries 1972	All Industries 1975	All Industries 1975/1972	Coal 1972	Coal 1975	Coal 1975/1972	Oil and Gas 1972	Oil and Gas 1975	Oil and Gas 1975/1972
Regions									
United States	746,506	950,837	1.274	2,099	4,210	2.006	2,963	6,075	2.050
New England	43,627	53,301	1.222	0	1	—	0	3	—
Mid-Atlantic	147,614	176,562	1.196	408	824	2.020	55	141	2.564
East North Central	156,504	194,345	1.242	363	662	1.824	119	298	2.504
West North Central	56,886	73,596	1.294	27	52	1.926	83	228	2.747
South Atlantic	108,767	139,915	1.286	802	1,387	1.729	54	130	2.407
East South Central	37,221	48,236	1.296	424	1,084	2.557	61	147	2.410
West South Central	60,787	84,700	1.393	9	22	2.444	2,061	3,973	1.928
Rocky Mountain	29,881	40,711	1.362	63	156	2.476	235	579	2.464
Far West	105,223	139,471	1.325	3	18	6.000	295	576	1.953
Major Fuel-Producing States[a]									
Pennsylvania	42,251	52,801	1.250	405	820	2.025	20	69	3.450
West Virginia	5,103	6,719	1.317	641	1,034	1.613	32	86	2.688
Louisiana	10,388	14,198	1.367	0	0	—	553	894	1.617
Texas	37,546	52,859	1.408	1	2	2.000	1,112	2,340	2.104
Oklahoma	7,675	10,453	1.362	6	14	2.333	365	702	1.923
Montana	2,304	3,074	1.334	3	16	5.333	15	36	2.400
Wyoming	1,149	1,785	1.554	9	28	3.111	67	171	2.552

Source: U.S. Bureau of Economic Analysis, *Survey of Current Business*, Washington, August 1974, pp. 34-43, August 1976, pp. 18-27.
[a]Actual and potential.

Table 7-8
Per Capita Income Comparisons, 1972 and 1975, by Region

	Per Capita Income		
	1972	1975	1975/1972
United States	4,549	5,902	1.297
New England	4,774	6,098	1.277
Mid-Atlantic	5,064	6,398	1.263
East North Central	4,766	6,121	1.284
West North Central	4,333	5,785	1.335
South Atlantic	4,265	5,510	1.292
East South Central	3,538	4,676	1.322
West South Central	3,881	5,347	1.378
Rocky Mountain	4,193	5,496	1.311
Far West	4,969	6,520	1.312

Source: 1975 data from U.S. Bureau of Economic Analysis, *Survey of Current Business*, August 1976, table 2, p. 17; 1972 data from *Survey of Current Business*, August 1974, table 2, p. 33.

Table 7-10 shows 1950 and 1970 earnings in fuel extraction that were obtained within metropolitan areas (defined as of 1970), and demonstrates that such earnings comprised a considerable portion of the U.S. total, amounting to about one-third in 1950 and one-quarter in 1970 for coal and roughly 60 percent of the total in both periods for oil and gas. Of course, a metropolitan area is defined as a county or counties encompassing a central city and hence can include a good deal of nonurban territory. Yet the set of SMSAs in 1970 covered only 11 percent of the U.S. land area, in contrast to the much larger share for fuel extraction. The possibility was investigated that office employment in those industries explained much of or all these results, but evidence on mining industry earnings in central administrative offices, warehouses, and research and development laboratories indicated that only a small part of the results could be attributed to that source. So after earnings from that source were eliminated, production worker earnings were estimated to have these percentages of total falling within SMSAs:

	Coal	*Petroleum and Gas*
1950	30.6%	52.0%
1970	22.3	55.0

It is possible that some of this association is accidental or coincidental, but it seems a reasonable hypothesis that causality accounts for much of the association. First, extraction activities near a metropolitan center may be more economic than at remote locations, either because of direct meeting of metropolitan demand or because of advantages conferred by transportation and marketing facilities located in metropolitan areas. Second, the extrac-

Table 7-9
Per Capita Income Comparisons, 1972 and 1975, by State

Region and State	Per Capita Income		
	1972	1975	1975/1972
New England			
Connecticut	5,414.0	6,973.0	1.288
Maine	3,664.0	4,786.0	1.306
Massachusetts	4,825.0	6,114.0	1.267
New Hampshire	4,279.0	5,315.0	1.242
Rhode Island	4,513.0	5,841.0	1.294
Vermont	3,703.0	4,960.0	1.339
Mid-Atlantic			
New Jersey	5,379.0	6,722.0	1.250
New York	5,275.0	6,564.0	1.244
Pennsylvania	4,545.0	5,943.0	1.308
East North Central			
Illinois	5,162.0	6,789.0	1.315
Indiana	4,364.0	5,653.0	1.295
Michigan	4,982.0	6,173.0	1.239
Ohio	4,572.0	5,810.0	1.271
Wisconsin	4,279.0	5,669.0	1.325
West North Central			
Iowa	4,316.0	6,077.0	1.408
Kansas	4,535.0	6,023.0	1.328
Minnesota	4,343.0	5,807.0	1.337
Missouri	4,307.0	5,510.0	1.279
Nebraska	4,451.0	6,087.0	1.368
North Dakota	4,128.0	5,737.0	1.390
South Dakota	3,766.0	4,924.0	1.307
South Atlantic			
Delaware	5,222.0	6,478.0	1.241
District of Columbia	5,827.0	7,742.0	1.329
Florida	4,450.0	5,638.0	1.267
Georgia	3,956.0	5,086.0	1.286
Maryland	5,017.0	6,474.0	1.290
North Carolina	3,868.0	4,952.0	1.280
South Carolina	3,500.0	4,618.0	1.319
Virginia	4,396.0	5,785.0	1.316
West Virginia	3,624.0	4,918.0	1.357
East South Central			
Alabama	3,476.0	4,643.0	1.336
Kentucky	3,634.0	4,871.0	1.340
Mississippi	3,188.0	4,052.0	1.271
Tennessee	3,708.0	4,895.0	1.320
West South Central			
Arkansas	3,345.0	4,620.0	1.381
Louisiana	3,565.0	4,904.0	1.376
Oklahoma	3,837.0	5,250.0	1.368
Texas	4,085.0	5,631.0	1.378

Table 7-9 *(cont.)*

	Per Capita Income		
Region and State	1972	1975	1975/1972
Rocky Mountain			
Arizona	4,273.0	5,355.0	1.253
Colorado	4,600.0	5,985.0	1.301
Idaho	3,711.0	5,159.0	1.390
Montana	4,083.0	5,422.0	1.328
Nevada	5,209.0	6,647.0	1.276
New Mexico	3,512.0	4,775.0	1.360
Utah	3,741.0	4,923.0	1.316
Wyoming	4,269.0	6,131.0	1,436
Far West			
Alaska	5,222.0	9,448.0	1.809
California	5,087.0	6,593.0	1.296
Hawaii	5,153.0	6,658.0	1.292
Oregon	4,339.0	5,769.0	1.330
Washington	4,601.0	6,247.0	1.358

Source: 1975 data from U.S. Bureau of Economic Analysis, *Survey of Current Business*, August 1976, table 2, p. 17; 1972 data from *Survey of Current Business*, August 1974, table 2, p. 33.

tive activity may itself have been an important impetus to growth of an urban area, helping it grow large enough to attain the officially designated status of SMSA. This argument is in large part a restatement of the Borchert thesis, with the added embellishments that the urban hinterland seems fairly concentrated (if roughly coincident with the SMSA boundaries) and that the effect seems remarkably persistent over time.

By drawing on additional data [57], these estimates emerge for the metropolitan share of regional earnings in fuel extraction by major Census region in 1970:

	Fraction of All Earnings that Occur within SMSAs	
	Coal	Oil and Gas
Northeast	0.540	0.708
North Central	0.241	0.354
South	0.147	0.614
West	0.047	0.575
United States	0.243	0.586

The effect is particularly pronounced in the Northeast, perhaps related to the early settlement of that region, but the Southern oil and gas figure is above the U.S. level.

Table 7-10
Fuel Extraction within Metropolitan Areas

Earnings (Millions of 1967 Dollars)

	1950				1970			
	All Industries	Coal Mining	Petroleum and Natural Gas	Total Coal and Petroleum and Natural Gas	All Industries	Coal Mining	Petroleum and Natural Gas	Total Coal and Petroleum and Natural Gas
Total Production								
U.S.	258,747.8	2,284.5	1,734.8	4,019.3	560,822.3	1,439.3	2,592.1	4,031.4
Metropolitan areas	193,246.5	740.5	986.1	1,726.6	441,253.9	349.4	1,519.4	1,868.8
Metroplitan U.S.	0.747	0.324	0.568	0.430	0.787	0.243	0.586	0.464
Location of Metropolitan Production by Region								
Northeast	67,524.2	443.4	14.8	458.2	133,045.0	161.8	26.2	188.0
North Central	58,488.2	111.9	80.6	192.5	120,617.0	61.9	73.0	134.9
South	37,473.5	170.4	745.8	916.2	101,900.7	123.8	1,138.9	1,262.7
West	29,760.6[a]	14.9[a]	144.9[a]	159.8[a]	85,691.2	1.9	281.3	283.2
Total	193,246.5	740.5	986.1	1,726.6	441,253.9	349.4	1,519.4	1,868.8
Major Producing States Metropolitan Areas in:								
Pennsylvania	16,093.7	442.0	9.5	451.5	30,014.2	160.8	9.3	170.1
West Virginia	1,177.9	84.0	8.1	92.1	2,077.2	84.1	11.5	95.6
Alabama	1,888.8	65.4	1.3	66.7	4,339.3	31.0	1.2	32.2
Total	19,160.4	591.4	18.9	610.3	36,430.7	275.9	22.0	297.9
Louisiana	2,263.7	0.0	59.6	59.6	5,222.7	0.0	248.0	248.0
Oklahoma	1,352.3	0.6	114.7	115.3	3,463.0	0.0	188.7	188.7
Texas	8,654.8	0.0	551.8	551.8	22,544.3	0.0	673.3	673.3
Colorado	1,473.1	7.3	6.2	13.5	4,671.1	0.9	56.9	57.8
California	20,198.0	0.0	134.5	134.5	58,874.1	0.0	210.6	210.6
Total	33,941.9	7.9	866.9	874.7	94,775.2	0.9	1,377.5	1,378.4

Source: Developed from data in U.S. Bureau of Economic Analysis, *Population and Economic Activity* (Washington, 1972).
[a]Includes estimated earnings for Alaska and Hawaii.

Appendix 7B presents data on 1970 earnings for individual metropolitan areas. From those data (and corresponding data for 1950), a list of major producing areas was tabulated, arranged in order of 1970 production (table 7-11). Some observations and speculations can be based on these data.

1. Fuel extraction is a major source of earnings within some SMSAs, and the importance of the activity can vary considerably among SMSAs in a region, so that some SMSAs will be much more affected by energy developments than others. Note the variation in oil and gas earnings relative to total earnings for the Texas SMSAs, as shown in appendix 7B.

2. Between 1950 and 1970, earnings in fuel extraction as a percentage of total earnings fell considerably, with the decline somewhat more pronounced for the major producing SMSAs listed in table 7-11 than for the U.S. as a whole. Despite the recent pronounced reversal in this trend, it remains possible that the decline represents a long-term development process, involving both the shift from primary production to secondary (manufactures) and tertiary (services) and the shift from export base to internal organization of production (the Perloff-Wingo thesis). The development process may well be more pronounced within metropolitan than non-metropolitan areas.

3. For the seven major coal-producing SMSAs, fuel extraction earnings in 1970 totaled 0.47 of the 1950 amount; for the nine major oil- and gas-producing SMSAs, the 1970 total was 1.59 that in 1950. For total earnings, the respective 1970-to-1950 ratios were 1.68 for the coal-producing areas and 2.76 for the oil- and gas-producing areas. One can again speculate that there is a causal relation here, so that the decline in coal earnings had considerable effect on total earnings, being a factor in the relatively slow growth of the Appalachian SMSAs (perhaps documenting an inference made by Borchert).

Appendix 7B also lists 1970 earnings in petroleum refining, chemicals, primary metals, and all manufacturing by SMSA. Unfortunately, disclosure rules impose major gaps in these data and limit their use. (As a matter of public policy, the costs imposed by such disclosure rules ought to be weighed against their putative benefits.) For SMSAs with data shown for 1970, the following were the major centers for petroleum refining, ordered by amount of earnings:

SMSA	1970 Earnings (Millions of 1967 Dollars
Philadelphia, Pa.	$ 224.8
Los Angeles, Calif.	209.6
Houston, Tex.	191.3
Beaumont, Tex.	188.0

SMSA	1970 Earnings (Millions of 1967 Dollars)
San Francisco. Calif.	175.0
New York City, N.Y.	154.1
Chicago, Ill.	118.2
Gary, Ind.	81.8
Lake Charles, La.	52.7
Baton Rouge, La.	44.3
Total	$1,439.7

As noted above, it is often hypothesized that the location of chemical and primary metal production will be considerably affected by the availability of energy (that is, by the availability of energy at relatively low prices), and some of the entries in appendix 7B lend some support to that hypothesis. However, data gaps caused by the disclosure rule imply that other sources of data must be tapped to test the hypothesis in rigorous fashion.

Table 7-11
Comparison of Fuel Earnings by Major Metropolitan Producing Areas

	Earnings in (Millions of 1967 Dollars)				Fuel Extraction Earnings Relative To All Earnings	
	All Industries		Major Fuel Extraction			
	1950	1970	1950	1970	1950	1970
Coal						
Pittsburgh, Pa.	4,135.4	6,877.9	191.8	84.8	0.0464	0.0123
Johnstown, Pa.	383.4	542.7	45.0	54.9	0.1174	0.1012
Wheeling, W. Va.	293.7	438.3	38.9	49.5	0.1324	0.1129
Charleston, W. Va.	319.5	653.7	38.6	33.2	0.1208	0.0508
Birmingham, Ala.	926.0	1,865.8	64.9	30.0	0.0701	0.0161
Wilkes-Barre, Pa.	537.2	755.0	156.8	15.2	0.2919	0.0201
Scranton, Pa.	339.6	523.1	40.0	3.3	0.1178	0.0063
Petroleum and Natural Gas						
Houston, Tex.	1,997.1	5,882.6	244.1	318.3	0.1222	0.0541
New Orleans, La.	1,208.1	2,766.5	13.8	151.9	0.0114	0.0549
Tulsa, Okla.	584.9	1,362.2	88.8	122.9	0.1518	0.0902
Los Angeles, Calif.	9,204.2	25,317.9	93.0	96.9	0.0101	0.0038
Dallas, Tex.	1,678.5	5,006.3	111.9	82.5	0.0667	0.0165
Midland, Tex.	73.1	205.1	32.6	71.6	0.4460	0.3491
Oklahoma City, Okla.	676.8	1,815.9	25.8	64.8	0.0381	0.0357
Denver, Colo.	1,229.1	3,797.5	6.2	56.9	0.0050	0.0150
Bakersfield, Calif.	401.5	881.3	23.2	53.7	0.0578	0.0609

Consumption

In this section I draw on a detailed study I made of energy consumption and expenditures by state and region for 1972 [58]. Generally good data were available on quantities consumed by energy source, but data on prices, particularly for petroleum products, were often surprisingly limited. The quantity data posed some specific difficulties, including (1) aggregation of residential and commercial use in some series; (2) differing definitions of industrial, commercial, and household use between series (thus apartment building use of energy was sometimes classified as commercial use); and (3) some discrepancies between series purporting to cover the same form of consumption.

I was concerned with distinguishing between residential and commercial use as part of a broader effort to identify use by sectors: household, federal government, state and local government, and business use. Such a classification is important for applications involving national, regional, and state income accounting, and for input-output analysis; the identification of levels of final demand is crucial in both applications.

Major sources for data on quantities consumed were the U.S. Bureau of Mines [59, 60] and Federal Highway Administration [61] for petroleum products; the American Gas Association [62] and Bureau of Mines [63] for natural gas; the Edison Electric Institute [64] for electricity, hydropower, and nuclear power; and the National Coal Association [65] and Crump and Readling [66] for coal.

An Arthur D. Little study for the Federal Energy Administration [67] was of considerable help in estimating a disaggregated and consistent set of "functional" accounts for energy use within structures (residential, commercial, and industrial use). By applying that source, commercial use was then distributed among office, retail, school, hospital and other use, and those categories, in turn, were distributed among government and business sectors, using information from a number of sources. Other major functional categories included transportation, energy used in energy production, and municipal-institutional use, including street lighting.

I found a number of conflicts between sources. For example, there were some pronounced differences in nongasoline highway fuel use between data series issued by the Bureau of Mines and the Federal Highway Administration. When such conflicts arose, resolution was sometimes obtained by drawing on external evidence and checking for internal consistency. Some sources were treated as more reliable than others, and on occasion I took the simple expedient of averaging the alternative values.

The basic source data on natural gas and electricity included information on expenditures, as well as quantities, so that average prices could be obtained as a matter of course. Because utilities employ a declining rate structure, in which increased levels of use are charged a lower price per unit, the average prices obtained here in a sense overstate the price differentials between locales, which might be better compared at a particular amount (or sets of amounts) used.

In contrast to the availability of data for natural gas and electricity, price data for coal and petroleum products are limited and scattered. The latter posed the major problem, involving, as it does, a great number of products and levels of use, ranging from use at the refinery to use at the retail level. Price information was obtained from a large number of sources, and the attempt was made to integrate those data into a coherent whole, accounting for middlemen margins, state and federal taxes, and geographic differentials. Major sources included industry publications, particularly *Platt's Oilgram, Platt's Oil Price Handbook and Oilmanac*, and *The Oil and Gas Journal*; Foster Associates [68, 69]; surveys of residential costs for various fuels by the Independent Natural Gas Association of America (INGAA) [7]]; and a number of trade associations and federal government agencies. Data on coal prices were obtained from INGAA [70] and the National Coal Association [65]. Finally, in the case of hydroelectric power and nuclear power, fuel-equivalent costs were assigned to the fuel-equivalent energy furnished by those sources, on the basis of the price of the cheapest available alternative fuels. This "shadow price" approach seemed the easiest and most reasonable way of handling those energy sources; it parallels the approach used in Schurr and Netschert [1] in evaluating energy from fuel wood in the 1800s.

I have reviewed my estimating procedures in some detail (in the context of this chapter) both to document the sources of my consumption estimates and to indicate where improved data might best be developed. To reiterate, I see the major problem as limited price data for petroleum products.

Tables 7-12 through 7-18 present some of the key consumption estimates. Table 7-12 shows regional energy consumption and expenditures per capita by energy source, covering all uses of energy (use by households, government, and business). For the United States as a whole, petroleum products account for about half of expenditures (53 percent) followed by electricity (27 percent), natural gas (14 percent), and coal, hydropower, and nuclear power (6 percent). In contrast, the distribution in terms of quantities consumed, measured in British thermal units, is petroleum products (40 percent); natural gas (31 percent); coal, hydropower, and nuclear power (21 percent); and electricity (7 percent). The difference between consumption and expenditure proportions, of course, reflects price differences. This is most pronounced in the case of electricity, primarily reflecting "con-

Table 7-12
Consumption of Energy and Expenditures on Energy, 1972, by Region
(All uses of energy by energy source.)

Region	Petroleum Products	Natural Gas	Electricity	Coal, Hydropower, and Nuclear Power	Total, All Sources
Consumption (Million Btu per Capita)					
New England	215.071	22.311	17.978	14.726	270.086
Mid-Atlantic	154.598	50.664	19.940	65.999	291.201
East North Central	116.010	105.269	25.453	122.825	369.557
West North Central	123.880	131.478	20.672	60.779	336.809
South Atlantic	146.882	49.732	27.024	76.169	299.807
East South Central	106.314	92.576	39.795	147.094	385.780
West South Central	154.195	375.086	31.059	5.040	565.379
Rocky Mountain	137.886	148.914	27.590	89.254	403.644
Far West	139.979	98.462	28.791	55.676	322.909
United States	141.212	110.224	25.852	74.211	351.500
Expenditures (Dollars per Capita)					
New England	334.507	39.826	131.447	8.307	514.088
Mid-Atlantic	256.360	58.980	137.577	31.986	484.903
East North Central	236.136	86.305	138.645	51.494	512.580
West North Central	253.783	79.868	125.877	22.455	481.982
South Atlantic	268.222	39.188	138.666	34.140	480.216
East South Central	242.191	50.692	138.360	51.910	483.153
West South Central	263.840	115.725	138.504	1.276	519.344
Rocky Mountain	286.020	74.902	129.975	23.085	513.982
Far West	247.865	63.229	119.262	12.838	443.195
United States	258.533	68.008	134.089	29.435	490.066

version losses" since roughly two-thirds of Btu content is consumed in the conversion from fuel input to power output. Other important price differences are accounted for by relatively high prices for gasoline, reflecting federal and state taxes, and by the impact of regulation holding down natural gas prices. Clearly, there are different mixes of energy by source between regions, reflecting relative price differences. Total consumption per capita shows considerably more variability than do expenditures per capita; thus, the highest consuming region (the West South Central) consumes twice the amount of the lowest consuming region (New England), but their expenditures are roughly the same, while the extreme difference between regions in expenditures per capita was only 17 percent (West South Central versus Far West). To considerable extent, these patterns reflect relatively low prices for energy in industrial and energy production uses in the high energy-consuming regions.

Table 7-13 exhibits regional household consumption and expenditures

Table 7-13
Consumption of Energy and Expenditures on Energy, 1972, by Region
(household use of energy by source)

Region	Petroleum Products	Natural Gas	Electricity	Coal, Hydropower, and Nuclear Power[a]	Total, All Sources
Consumption (Million Btu per Capita)					
New England	95.626	13.390	6.938	0.140	116.094
Mid-Atlantic	69.452	25.547	5.984	2.231	103.214
East North Central	58.628	42.991	7.506	3.403	112.527
West North Central	59.705	35.545	8.023	1.098	104.370
South Atlantic	58.453	12.727	10.001	1.225	82.406
East South Central	52.146	18.550	12.555	1.855	85.104
West South Central	54.925	23.754	9.895	0.005	88.580
Rocky Mountain	54.941	32.467	7.919	0.180	95.507
Far West	47.618	30.272	8.983	0.011	86.884
Total	60.436	27.387	8.380	1.480	97.683
Expenditures (Dollars per Capita)					
New England	209.314	28.469	59.019	0.199	297.000
Mid-Atlantic	161.866	38.778	53.192	2.881	256.718
East North Central	151.037	46.355	54.806	4.443	256.641
West North Central	147.281	36.512	59.596	1.537	244.926
South Atlantic	156.045	17.752	62.728	1.607	238.133
East South Central	145.375	18.978	57.138	2.230	223.721
West South Central	138.538	21.892	62.430	0.007	222.866
Rocky Mountain	149.939	30.310	51.531	0.222	232.003
Far West	130.787	31.726	48.950	0.014	211.477
Total	152.604	32.107	56.332	1.925	242.967

[a]Coal only in household use.

per capita, by energy source. Note the Frostbelt versus Sunbelt clustering for total consumption, no doubt reflecting the effects of climate. The four Northeastern and North Central regions consume over 100 million Btu per capita; the Rocky Mountain region consumes 95 million Btu per capita; and the Southern and Far West consumption per capita is in the 80 million Btu range. Expenditures per capita show less variation between regions, although New England's expenditures are considerably above those of other regions, running roughly 25 percent above the U.S. average and 40 percent above the amount in the Far West, which was the low expenditure region. That region's level of expenditures per capita was 13 percent below the U.S. average. All other regions had readings that fell within 10 percent of the U.S. average; in particular, per capita expenditures in both the Mid-Atlantic and the East North Central regions were only 6 percent above the average. Hence, it can be concluded that regional variation in energy expenditures is not particularly pronounced.

Table 7-14
Consumption of Energy (Million Btu per Capita), 1972, by State
(household use of energy by energy source.)

Region	Petroleum Products	Natural Gas	Electricity	Coal, Hydropower, and Nuclear Power	Total, All Sources
New England					
Connecticut	82.210	12.711	7.955	0.032	102.908
Maine	110.794	0.963	7.040	0.097	118.895
Massachusetts	97.782	17.312	6.139	0.164	121.397
New Hampshire	109.325	6.750	8.296	0.388	124.759
Rhode Island	87.588	15.464	5.556	0.052	108.660
Vermont	118.344	3.038	10.585	0.435	132.402
Mid-Atlantic					
New Jersey	85.894	23.976	6.285	0.476	116.632
New York	67.003	23.932	5.106	0.727	96.768
Pennsylvania	63.081	29.009	7.152	5.636	104.878
East North Central					
Illinois	53.919	49.608	6.884	3.482	113.893
Indiana	64.988	34.104	8.607	4.493	112.192
Michigan	63.636	42.342	7.213	2.080	115.272
Ohio	53.036	48.316	7.616	3.325	112.292
Wisconsin	66.173	25.606	8.088	4.750	104.618
West North Central					
Iowa	59.832	35.475	8.291	0.728	104.326
Kansas	48.722	45.968	8.061	0.220	102.971
Minnesota	68.629	30.940	7.694	2.180	109.443
Missouri	56.957	36.634	7.885	0.548	102.024
Nebraska	58.335	41.989	8.566	0.654	109.544
North Dakota	57.135	18.769	8.051	2.524	86.479
South Dakota	69.568	20.883	8.344	2.941	101.736
South Atlantic					
Delaware	82.847	15.843	7.744	1.138	107.573
District of Columbia	49.303	24.338	6.692	0.133	80.467
Florida	56.213	3.174	13.365	0.095	72.848
Georgia	55.799	19.365	10.043	0.211	85.418
Maryland	59.683	20.377	6.477	0.482	87.019
North Carolina	65.066	6.829	10.447	1.934	84.277
South Carolina	58.787	8.170	9.981	2.158	79.096
Virginia	62.488	12.993	9.070	2.392	86.944
West Virginia	37.472	34.485	7.379	4.123	83.459
East South Central					
Alabama	51.751	16.016	11.969	0.511	80.247
Kentucky	49.872	26.997	9.337	2.632	88.838
Mississippi	57.700	18.138	9.731	0.709	86.277
Tennessee	51.256	14.110	17.238	3.021	85.623
West South Central					
Arkansas	59.383	24.026	8.414	0.000	91.824
Louisiana	43.649	23.319	9.924	0.000	76.891
Oklahoma	55.768	31.014	9.100	0.000	95.881
Texas	57.595	22.200	10.322	0.009	90.127

Table 7-14 *(cont.)*

Region	Petroleum Products	Natural Gas	Electricity	Coal, Hydropower, and Nuclear Power	Total, All Sources
Rocky Mountain					
Arizona	45.701	19.925	9.511	0.000	75.137
Colorado	53.362	38.913	5.861	0.254	98.390
Idaho	68.592	16.072	12.798	0.000	97.462
Montana	53.800	35.497	8.149	0.559	98.005
Nevada	67.035	20.755	15.082	0.000	102.872
New Mexico	56.666	35.686	4.896	0.000	97.247
Utah	51.277	41.820	5.934	0.444	99.474
Wyoming	78.655	66.661	6.647	0.289	152.251
Far West					
Alaska	71.779	28.710	5.900	0.308	106.696
California	44.605	36.066	6.664	0.000	87.334
Hawaii	34.111	0.000	6.230	0.000	40.341
Oregon	64.254	12.911	17.041	0.046	94.251
Washington	55.904	14.151	18.630	0.029	88.715

Concentration of use by energy source is readily apparent, with relatively high consumption of petroleum products in the Northeast, natural gas in the North Central region, and electricity in the South Atlantic and East South Central regions, primarily reflecting high use in the TVA area and in Florida. Those high use levels can be found in table 7-14, which presents household consumption by state. Table 7-15 presents corresponding expenditure figures. Wyoming consumes the most British thermal units per capita, but Vermont spends the most on energy, per capita; Hawaii has both the lowest consumption and the lowest level of expenditures, per capita.

Besides Vermont and Wyoming, other states with high expenditures per capita include Delaware, New Jersey, Nevada, and most of the New England states. Besides Hawaii, states with low expenditures per capita include Louisiana, West Virginia, North Dakota, and California. It is plausible that climate, income, price, and density of settlement are some of the factors explaining these variations.

In 1972, household energy expenditures as a fraction of personal income averaged 0.0536 for the country as a whole. Regional values for that ratio were:

New England	0.0621
Mid Atlantic	0.0510
East North Central	0.540
West North Central	0.567

South Atlantic	0.0557
East South Central	0.0634
West South Central	0.0573
Rocky Mountain	0.0551
Far West	0.0429

These data suggest, then, that the energy crunch crunched the most in New England and the East South Central region and the least in the Far West.

Table 7-16 exhibits the sectoral distribution of energy consumption and expenditures. Households account for roughly one-quarter of consumption but half of expenditures; business for 70 percent of consumption and 45 percent of expenditures; and government for roughly 5 percent of both consumption and expenditures.

In my system of energy accounting, I distingushed two categories of business use of energy: energy used to produce energy, with electric utility use of fuel a major component, but also including refinery and oil and gas field use of fuel (data were not available to estimate such items as oil company use of electricity) and all other business use, labeled net business use. Table 7-17 shows the distribution of business use among these categories by regions and for the United States, and then exhibits final demand (households and government use only); "customer" use of energy, consisting of final demand plus net business use; and finally, all use of energy, including energy used in energy production. The concentric sets of levels of consumption should "solve" double-counting problems by treating the counting as a matter of convenience, depending on the purposes served. This approach corresponds to the accounting scheme used in input-output tables. Given shipments by an industry to final demand, all other industries, and to itself, output totals can be presented in progressively more inclusive form. United States expenditures in 1972, corresponding to the quantity totals in table 7-17, were final demand, $55.789 billion; customer use, $92.875 billion; and all use, $102.045 billion.

Table 7-18 presents the distribution of energy use per capita by function. Of the uses other than energy production (which accounts for about 30 percent of total consumption and about 10 percent of expenditures) transportation and industrial use each account for about one-quarter of consumption, but the former comprises 40 percent of expenditures in contrast to 15 percent for the latter. Passenger car use by households accounts for half of transportation British thermal units and 60 percent of transportation expenditures. There is considerable variation in regional patterns of use. Thus for New England transportation use and residential use are approximately equal, while for the Far West transportation use is somewhat more than twice residential use in both consumption and expenditures.

Table 7-15
Expenditures on Energy (Dollars per Capita), 1972, by State
(household use of energy by energy source.)

Region	Petroleum Products	Natural Gas	Electricity	Coal, Hydropower, and Nuclear Power[a]	Total, All Sources
New England					
Connecticut	196.681	25.540	65.095	0.053	287.369
Maine	240.313	2.873	56.279	0.151	299.616
Massachusetts	204.721	38.070	55.525	0.227	298.543
New Hampshire	241.199	13.006	64.913	0.511	319.629
Rhode Island	188.652	30.853	50.765	0.073	270.343
Vermont	272.501	5.192	75.930	0.676	354.300
Mid-Atlantic					
New Jersey	190.427	45.635	56.456	0.773	293.292
New York	157.944	36.877	48.742	1.045	244.608
Pennsylvania	150.287	37.477	58.044	7.015	252.822
East North Central					
Illinois	143.702	53.504	57.345	4.665	259.216
Indiana	160.296	38.192	57.508	5.376	261.372
Michigan	154.828	45.375	51.184	2.767	254.154
Ohio	149.444	49.340	53.065	4.223	256.072
Wisconsin	154.665	33.010	56.681	6.662	251.018
West North Central					
Iowa	153.544	38.447	65.044	0.986	258.021
Kansas	123.129	33.562	55.944	0.250	212.885
Minnesota	158.711	37.069	57.422	3.192	256.395
Missouri	146.563	39.186	60.273	0.641	246.663
Nebraska	150.908	40.396	56.284	0.878	248.466
North Dakota	121.522	19.995	61.315	3.612	206.444
South Dakota	156.980	22.964	62.178	4.540	246.662
South Atlantic					
Delaware	204.520	26.243	68.902	1.661	301.325
District of Columbia	127.032	35.755	49.072	0.207	212.065
Florida	156.722	6.426	82.973	0.128	246.249
Georgia	155.895	23.329	56.217	0.285	235.726
Maryland	154.324	32.641	50.080	0.703	237.748
North Carolina	168.454	9.311	62.335	2.585	242.685
South Carolina	160.235	12.679	61.650	2.948	237.513
Virginia	157.004	19.824	55.948	3.096	235.872
West Virginia	109.375	32.230	50.077	4.953	196.636
East South Central					
Alabama	147.545	19.472	58.491	0.638	226.145
Kentucky	140.745	24.469	48.227	3.010	216.452
Mississippi	158.916	19.047	56.045	0.854	234.863
Tennessee	139.755	14.055	63.807	3.737	221.354
West South Central					
Arkansas	156.834	19.707	56.595	0.000	233.136
Louisiana	113.312	20.577	63.401	0.000	197.290
Oklahoma	141.109	26.798	64.059	0.000	231.967
Texas	142.915	21.580	62.757	0.012	227.263

Table 7-15 *(cont.)*

Region	Petroleum Products	Natural Gas	Electricity	Coal, Hydropower, and Nuclear Power[a]	Total, All Sources
Rocky Mountain					
Arizona	132.284	22.730	66.331	0.000	221.345
Colorado	150.141	30.946	44.700	0.296	226.082
Idaho	165.726	22.543	60.037	0.000	248.305
Montana	136.560	32.903	51.508	0.855	221.826
Nevada	197.761	27.988	68.942	0.000	294.691
New Mexico	155.053	31.840	39.112	0.000	226.006
Utah	134.414	39.933	38.992	0.481	213.820
Wyoming	202.966	48.025	48.376	0.355	299.721
Far West					
Alaska	170.507	43.080	57.615	0.598	271.800
California	125.769	36.154	45.683	0.000	207.607
Hawaii	116.587	0.000	51.859	0.000	168.446
Oregon	160.790	20.357	63.822	0.043	245.012
Washington	141.185	19.042	57.433	0.028	217.687

[a]Coal only in household use.

Again, since transportation expenditures per capita increase from east to west, a gasoline tax would be an increasing burden in a westward direction.

Given the data on quantities consumed and prices, it is plausible to consider demand relationships. A number of energy demand studies have been published in recent years, typically finding intuitively plausible values for price and income elasticities. Let me summarize some of the results. Taylor [71] surveys a number of reports on the demand for electricity, noting they show that both price elasticity and income elasticity are much larger in the long run than the short run. Short-run price elasticity estimates range from around -0.15 to -0.80, while all long-run price elasticities are above -1.0; that is, a 1 percent price increase leads to a reduction of more than 1 percent in quantity consumed. Estimated income elasticity is generally positive, but results are more variable. By averaging the results shown, a short-run elasticity of 0.28 is obtained as the average of six cases, and a long-run elasticity of 1.1 as the average of five cases. Taylor notes that given the declining block rate structure in utility pricing (noted earlier), the use of average prices poses econometric problems and may bias these results.

Employing data on states, Nelson [72] and Ferrar and Nelson [73] relate the residential-commercial aggregate use of fuel oil, natural gas, and coal per capita to income, fuel price, electricity price, heating degree-days (which increase as temperature decreases), and urbanization. Fuel price was the weighted average price of the fuel components, while urbanization was measured by the percentage of population

Table 7-16
Consumption of Energy and Expenditures on Energy, 1972, by Region
(use by sector.)

Region	Household Use	Business Use	Federal Government Use	State and Local Government Use	Total Use
Consumption (Million Btu per Capita)					
New England	116.094	138.181	7.682	8.129	270.086
Mid-Atlantic	103.214	176.844	4.349	6.794	291.201
East North Central	112.527	247.590	2.244	7.196	369.557
West North Central	104.370	220.246	2.130	10.064	336.809
South Atlantic	82.406	203.343	6.480	7.578	299.807
East South Central	85.104	290.040	4.771	5.864	385.780
West South Central	88.580	460.896	6.429	9.475	565.379
Rocky Mountain	95.507	290.670	4.937	12.530	403.644
Far West	86.884	217.363	12.251	6.411	322.909
United States	97.683	240.530	5.562	7.725	351.500
Expenditures (Dollars per Capita)					
New England	297.000	189.280	8.070	19.738	514.088
Mid-Atlantic	256.718	202.089	5.742	20.355	484.903
East North Central	256.641	234.349	3.507	18.084	512.580
West North Central	244.926	216.093	3.001	17.962	481.982
South Atlantic	238.133	215.301	7.455	19.328	480.216
East South Central	223.721	240.112	5.596	13.723	483.153
West South Central	222.866	272.592	7.248	16.638	519.344
Rocky Mountain	232.003	253.569	7.334	21.076	513.982
Far West	211.477	201.864	13.177	16.677	443.195
United States	242.967	222.141	6.656	18.301	490.066

residing in urban areas. Best-fitting equations were obtained using logarithms, which yield elasticities directly. With urbanization included, the income elasticity becomes negative. Nelson speculates that "this may be due in part to consumer preferences for electric heating at higher income levels" [72].

The elasticities obtained with urbanization included and then excluded were (the asterisk indicates significance at the 10 percent level):

	Equation 1	Equation 2
Fuel price	−0.1901*	−0.2800*
Electricity price	0.5092*	0.5039*
Income per capita	−0.1584	0.2671
Degree-days	0.5301*	0.4955*
Urbanization	0.2804	—

Table 7-17
Consumption of Energy (Trillion Btu), 1972, by Region
(Use by class of purchasers.)

	Business Use		Use by Final Demand Sectors	Use by Customers (Nonenergy Producers)	Total Use
Region	Net Business Use	Use in Energy Production			
New England	918.34	754.33	1,596.71	2,515.06	3,269.39
Mid-Atlantic	3,858.04	2,795.02	4,302.23	8,160.27	10,955.29
East North Central	6,241.72	3,857.72	4,975.18	11,216.90	15,074.62
West North Central	2,067.52	1,592.53	1,937.05	4,004.57	5,597.10
South Atlantic	3,401.16	3,089.55	3,079.13	6,480.29	9,569.84
East South Central	1,968.24	1,847.24	1,259.46	3,227.69	5,074.93
West South Central	4,600.93	4,609.15	2,087.90	6,688.83	11,297.98
Rocky Mountain	1,368.66	1,212.49	1,003.21	2,371.87	3,584.36
Far West	2,915.10	2,987.40	2,866.10	5,781.20	8,768.60
Total	27,339.71	22,745.41	23,106.97	50,446.68	73,192.09

The estimated elasticity for fuel price is in the -0.2 to -0.3 range; the cross-elasticity for electricity is 0.5; that is, a 1 percent increase in electricity price generates a 0.5 percent increase in fuel use (as a presumed consequence of the substitution of fuel for electricity), and similarly a 1 percent increase in degree-days (increased severity of winter) increases fuel use by 0.5 percent.

The Energy Division of the Oak Ridge National Laboratory presents the estimated long-run electricity demand elasticities for the seven states of the "South Atlantic Division" (the Atlantic Coast from Delaware to Florida) [74], shown in table 7-19. The dependent variable is total consumption, but per capita results should not be much different, given the elasticity of around 1.0 for the "population" measure in each equation (number of customers, population, and value added, respectively).

Employing the data I developed for residential and industrial consumption and (implicitly) prices by state, I have related residential consumption per capita and industrial consumption per worker in manufactures for specific energy sources to a number of explanatory variables using regression analysis. Results for the logarithmic form of the equations appear as tables 7-20 and 7-21. On the whole, these results seem fairly encouraging.

For each energy source, its own price elasticity is negative, while the cross-elasticity for competitive sources is usually positive (and negative cases are not statistically significant). The residential use equations employed "real" disposable income per capita as an explanatory variable; this consisted of disposable income deflated by accounting for price level differences between North and South and between urban areas of different size. The variable performed somewhat better than disposable income, but

Table 7-18
Consumption of Energy and Expenditures on Energy, 1972, by Region
(use by function)

Region	Residential Use	Commercial Use	Industrial Use	Transportation Use	Non-electric Utility	Electric Utility	Municipal Institutional	Total Use
Consumption (Million Btu per Capita)								
New England	75.029	26.301	33.577	71.640	1.899	60.417	1.222	270.086
Mid-Atlantic	67.498	19.397	61.747	66.549	7.130	67.164	1.718	291.201
East North Central	70.670	18.756	111.194	72.491	12.014	82.559	1.874	369.557
West North Central	63.873	19.458	70.610	82.778	19.430	76.401	4.258	336.809
South Atlantic	38.472	16.197	61.853	84.443	3.169	93.621	2.052	299.807
East South Central	44.241	11.608	106.213	81.321	16.962	123.458	1.977	385.780
West South Central	44.959	15.844	168.371	100.113	121.724	108.930	5.438	565.379
Rocky Mountain	53.856	24.981	83.877	99.398	31.114	105.428	4.990	403.644
Far West	45.421	17.349	48.306	100.073	21.670	88.343	1.747	322.909
United States	56.725	18.325	82.752	81.974	22.694	86.539	2.490	351.500
Expenditures (Dollars per Capita)								
New England	173.104	62.070	56.397	177.287	1.628	38.126	5.476	514.088
Mid-Atlantic	145.192	57.572	67.760	164.812	4.385	37.234	7.948	484.903
East North Central	132.965	50.232	97.463	184.778	7.014	34.026	6.102	512.580
West North Central	129.713	48.635	62.200	200.916	7.278	27.911	5.329	481.982
South Atlantic	107.390	50.171	64.892	206.542	1.537	44.283	5.402	480.216
East South Central	103.990	29.779	93.082	204.074	5.772	41.931	4.526	483.153
West South Central	104.409	46.415	82.147	215.224	34.707	30.746	5.695	519.344
Rocky Mountain	110.470	60.760	61.933	234.197	10.011	30.113	6.497	513.982
Far West	92.303	50.968	45.780	208.813	11.154	29.727	4.449	443.195
United States	122.495	50.996	71.900	194.742	8.654	35.385	5.893	490.066

Table 7-19
Long-run Electricity Demand Elasticities for the South Atlantic States

	Electricity Use		
	Residential	Commercial	Industrial
Fuel Price			
Electricity	−1.20	−0.85	−1.75
Natural gas	0.11	0.14	0.15
Gas	—	—	0.22
Consumer income	0.76	1.38	—
Number of customers	0.96	—	—
Population	—	1.11	—
Heating degree-days	0.19	—	—
July mean temperature	0.60	0.19	—
Value added in manufactures	—	—	0.81

the improvement was marginal. Its elasticity was always positive, but statistically significant only in the case of electricity. The climate measures appearing in the individual equations of table 7-20 gave the best statistical results, and all were significant when regional effects were excluded. Degree-days had a positive elasticity for petroleum products and natural gas, while increased summer temperature had a positive effect on electricity use (presumably reflecting increased use of air conditioning) and increased winter temperature had a negative effect (presumably reflecting decreased heating, and perhaps lighting). Urbanization was statistically significant in a number of equations, being measured best by fraction of population in metropolitan areas. (The fraction, rather than the logarithm of the fraction, was used because it performed much better than the alternative measure.) In residential consumption, urbanization had a positive impact on consumption of natural gas and a negative impact on consumption of electricity and petroleum products. The explanation for these differences may be merely the occurrence of pipelines to serve metropolitan areas, or it may reflect lower use of energy as urban size increases. In industrial use, urbanization typically had a negative effect on energy consumption per worker. Perhaps there is less floor area per worker with increased size of metropolitan area, and perhaps industrial processes using energy are correlated with floor area, but this is purely speculative.

    Regional dummy variables were introduced in all the equations and were statistically significant in some cases. They purport to show regional influences not accounted for by the other variables. In the case of natural gas, the New England effect is strongly negative, and the "regional influence" here is likely to be the nonavailability of gas in many areas. In table 7-20 and 7-21, the regional dummy variable are intercept shifters, affecting the constant term in the fitted equation and shifting "base" con-

Table 7-20
Regression Results for Residential Consumption of Energy per Capita

	Log Residential Consumption per Capita		
	Petroleum Products	Natural Gas	Electricity
	Coefficients		
Constant term	−2.948	−2.120*	−3.747*
Log petroleum products price	−0.661	−0.989	−0.008
Log natural gas price	1.707*	−1.886*	0.185*
Log electricity price	0.469	1.111*	−1.244*
Fraction SMSA population	−0.481	0.360*	−0.182*
Log population growth, 1960-1970	−0.809	−0.552	0.883*
Log "real" disposable income	1.289	1.029	0.690*
Log degree-days	0.300*	1.193	—
Log summer temperature	—	—	0.824*
Log winter temperature	—	—	−0.025
Regions			
New England	0.031	−0.442*	0.037
Middle Atlantic	0.119	−0.289*	0.046
East North Central	0.111	−0.187	0.056
West North Central	0.064	−0.230*	0.062
South Atlantic	−0.096	−0.295*	0.013
East South Central	0.164	−0.020	0.017
West South Central	0.219	−0.313*	0.076
Mountain	0.163	−0.060	−0.029
Pacific[a]	0.000	0.000	0.000
	t Ratios		
Constant term	0.905	2.784	2.938
Log petroleum product price	0.918	1.608	0.060
Log natural gas price	4.268	5.516	2.656
Log electricity price	1.479	4.101	16.400
Fraction SMSA population	3.385	2.966	4.909
Log population growth, 1960-1970	0.855	0.682	4.253
Log "real" disposable income	1.418	1.324	3.238
Log degree-days	2.621	1.217	—
Log summer temperature	—	—	1.885
Log winter temperature	—	—	0.280
Regions			
New England	0.195	3.292	0.854
Middle Atlantic	0.706	2.009	0.911
East North Central	0.843	1.622	1.244
West North Central	0.501	2.102	1.081
South Atlantic	0.889	3.189	0.310
East South Central	1.126	0.164	0.262
West South Central	1.514	2.524	1.477
Mountain	1.378	0.596	0.695
Pacific[a]	—	—	—
R^2	0.747	0.851	0.913

[a]Omitted from equation. One dummy variable from a set must be omitted (have coefficient set equal to zero) to avoid exact collinearity.

*Significant at 10 percent level.

—Indicates variable does not enter equation.

Table 7-21
Regression Results for Industrial Consumption of Energy Relative to Manufacturing Employment

	Log Industrial Consumption per Worker in Manufactures			
Variable	Petroleum Products	Natural Gas	Electricity	Coal and Hydropower
	Coefficients			
Constant term	3.290*	2.048*	2.222*	2.274*
Log petroleum products price	−2.924*	1.684	2.991*	−6.295
Log natural gas price	0.963*	−1.977*	0.016	0.824
Log electricity price	0.340	0.409	−0.800*	0.430
Log coal and hydropower price	0.322	0.009	−0.511	−0.434
Fraction SMSA population	−0.619*	−0.052	0.049	−0.998*
Regions				
New England	−0.892*	−0.583*	0.061	−1.478*
Middle Atlantic	−0.819*	−0.115	0.131	0.185
East North Central	−0.845*	0.060	0.012	1.075*
West North Central	−0.585*	0.102	0.062	0.371
South Atlantic	−0.551*	−0.095	0.217	0.350
East South Central	−0.767*	−0.141	0.131	0.931*
West South Central	−0.191	0.198	0.159	−0.449
Mountain	0.232	0.288	0.232*	0.664
Pacific[a]	0.000	0.000	0.000	0.000
	t Ratios			
Constant	8.606	4.430	6.848	1.880
Log petroleum product price	2.248	1.071	2.709	1.530
Log natural gas price	2.677	4.547	0.052	0.725
Log electricity price	1.090	1.084	3.024	0.436
Log coal and hydropower price	0.838	0.019	1.566	0.357
Fraction SMSA population	4.033	0.281	0.375	2.055
Regions				
New England	3.936	2.125	0.317	2.061
Middle Atlantic	3.911	0.455	0.737	0.279
East North Central	5.406	0.316	0.088	2.173
West North Central	3.753	0.542	0.468	0.752
South Atlantic	3.194	0.455	1.484	0.641
East South Central	4.703	0.714	0.943	1.805
West South Central	1.021	0.875	1.001	0.761
Mountain	1.571	1.610	1.849	1.421
Pacific[a]	—	—	—	—
R^2: Measure of explained variance	0.615	0.779	0.618	0.338

[a]Omitted from equation. One dummy variable from a set must be omitted (have coefficient set equal to zero) to avoid exact collinearity.
*Significant at 10 percent level.

sumption. An alternative approach is to utilize the dummy variables as "slope shifters"—in effect varying the elasticities by region. That approach was carried out here employing the four major census regions, with estimates of price elasticity shown in table 7-22.

Table 7-22
Energy Price Elasticities by Type and Region

Region	Petroleum Products	Natural Gas	Electricity
Residential Use			
Northeast	−0.047	−3.137*	−1.198*
North Central	−0.764	−1.496	−1.162*
South	−0.910	−2.752	−1.221*
West	−0.807	−2.052	−1.268
Industrial Use			
Northeast	−2.322	−1.258*	−0.992
North Central	−3.468	−2.552	−0.971
South	−5.006	−1.918*	−0.770
West	−0.809	−2.808	−1.305

*Significant at 10 percent level.

Given the limited sample size, the results must be viewed with some caution. They do show little variation in price elasticity between regions in electricity consumption; marked variability in petroleum and natural gas, with a very low price elasticity for petroleum products in the Northeast; and the expected negative effect of price on consumption in all cases. But a much larger sample size seems necessary before such estimates can be viewed as "reliable"; this case seems illustrative of a general need for energy data in which the unit of observation involves a finer "grain" of analysis, for example, the SMSA or county, rather than the state.

Some data are readily available on price changes post-OPEC (October 1973). In 1972, on a base of 1967 equals 100, the consumer price index was 125.3, the gas and electricity index was 120.5, and the fuel oil and coal index was 118.5. Gasoline prices were about 10 percent above their 1967 levels. In July 1977, the following price indexes held (for the United States at the consumer level) [75]:

Consumer price index, all items	182.6
All fuels and utilities	203.5
Natural gas	240.8
Electricity	192.8
Fuel oil and coal	283.7
Gasoline	190.9

Although OPEC imposed a fourfold increase in crude oil prices, the price increase to households has been much less pronounced, reflecting the substantial middlemen margins in most energy products, the federal government price controls on oil and natural gas, and the great upsurge in the general price level since 1972. Of course, on the latter, it is often argued that much of that increase reflects the multiplier impact of the OPEC price increase. However, the validity of the argument is *not* obvious; evidence presented above showed household spending on energy was about 5 percent of household income, and household use of energy in physical terms was about half of all use by "customers," suggesting that even indirect effects of higher prices would not account for the bulk of the inflationary surge. Russell [76, p. 331] points out that gasoline, electricity, and natural gas prices in real terms are lower than or about the same as they were in 1950, and that energy prices have not risen much since 1973 because of price controls. Hence there has been little incentive for household consumers to reduce consumption.

Drawing on the Bureau of Labor Statistics [75] for fuels and utility prices, and *Oil and Gas Journal* and American Petroleum Institute [77] for gasoline prices, table 7-23 shows estimated regional price indexes for those categories of use. These figures are crude estimates, based on data for major cities in each region. In the case of fuels and utility prices, New England had only one observation (Boston), and the East South Central and Rocky Mountain regions had none. In the other regions, the indices for several cities were averaged. Obviously, the table 7-23 entries are rough indicators, at best. These limited data are again illustrative of the need for better information on prices, particularly at the regional level.

The variations among regions in the gasoline price indices is not particularly pronounced, although the Mid-Atlantic region index is somewhat above the average and the Western values are somewhat below. (Since the

Table 7-23
Estimated Consumer Price Indices for Energy, by Region, July 1977
(1967 = 100)

Region	Fuels and Utilities	Gasoline
New England	207.4	191.3
Mid-Atlantic	216.6	201.4
East North Central	207.8	196.3
West North Central	193.4	192.8
South Atlantic	199.4	192.3
East South Central	—[a]	197.9
West South Central	201.4	191.1
Rocky Mountain	—[a]	188.6
Far West	175.3	187.7

Source: Fuels and Utilities: U.S. Bureau of Labor Statistics, *Retail Prices and Indexes of Fuels and Utilities, Residential Usage*, August 1977; gasoline: *Oil and Gas Journal*, October 8, 1973 and August 22, 1977, and American Petroleum Institute, *Petroleum Facts and Figures* (Washington, 1971).

[a]No data on cities in region listed in source.

gasoline price indexes are based on data on selected cities from [77], it is not surprising they are only roughly consistent with the U.S. price index of 190.9 shown in [75].) Greater variation occurs in the fuel and utility indexes. Again, the Mid-Atlantic region shows the greatest price rise, while the Far West shows the least. The relatively high index for the West South Central region may well reflect the effect of higher unregulated natural gas prices in the intrastate market.

Given the current availability of Alaskan oil on the West Coast, it is likely the price differentials of table 7-23 will persist and perhaps become even more pronounced. Some impacts on the location of economic activity and population can be expected. In conjunction with the relatively low spending on energy in the 1972 base year, these price differentials can be expected to bring about an acceleration of growth in Hawaii, California, Oregon, and Washington.

Prospects and Policies

In this concluding section, I shift my focus to the future, considering what will be (perhaps) and what should be, with emphasis on the regional and urban impacts of energy developments. A forecast of the future trend of energy prices is a precondition for forecasting such impacts. The price forecast, in turn, depends on the preception of such factors as the strength and cohesion of the OPEC cartel; the tendency of non-OPEC countries to act as de facto members of the cartel; the course of future U.S. policies on energy and the environment, which turn on policymakers' evaluations of future prospects, as well as their reading of the beliefs and interests of the electorate; the growth in global population and per capita income affecting worldwide demand for energy; and long-term limits to growth affecting the supply of energy.

Obviously, perceptions vary, most items on the list are stochastic, and a great deal is at stake, so that forecasting becomes an even riskier business than usual. But let me incur some of that risk by making some forecasts in summary fashion and then expanding upon them. Admittedly, I have focused on what I see as key issues, so that a number of topics are only fleetingly touched on. My forecasts run as follows:

1. In labeling energy prices relative to the general price level as "real" prices and defining the short run as the next five to ten years, there is some possibility that real energy prices will decline in the short run, as well as a slim chance that the decline wil be precipitous, in which case the United States could have major adjustment problems. (A reversal of our misfortunes need not be fortunate.) However, it seems most plausible that real energy prices will be fairly stable in the short run, and will exhibit a mild up-

ward trend in the long run, primarily reflecting depletion of lower-cost resources. Note that although much of the OPEC price increase has been dissipated by general price level increases, the forecast of short-run stability involves the expectation that some of the recent real price increase will be maintained.

2. Both consumer and producer responses to real price increases will intensify the shift of population from Frostbelt to Sunbelt. Suburban sprawl should slow down, with some impetus to revival of the central city or at least to the development of high-density subcenters in the suburbs. However, the inverse relation between population growth and size of place will probably continue.

3. The Carter administration energy proposals are likely to be adopted in good part. They involve continued constraint on "old" natural gas and oil production, only limited encouragement of "new" gas and oil, some encouragement of nuclear power, and considerable encouragement of coal production. A number of regional impacts are likely. The impact of expanded production will be accelerated in the West, particularly in the Rocky Mountain states, and to a lesser extent in Appalachia. Contrarily, the expansion of production in the "older" oil and gas areas will continue to be inhibited, with adverse impact (relative to what might have been) on Texas and Louisiana in particular and on other current producing states as well. The policy of regulated pricing of oil and gas below market equilibrium levels obstensibly will benefit the Northeastern and North Central regions, but much of the apparent benefit will be illusory, because of the continued replacement of domestic supplies by higher-priced foreign oil and gas and because of the adverse environmental consequences of the use of coal. The environmental problem should be most serious in the Northeast.

4. Production of nuclear power, Atlantic offshore hydrocarbons, shale oil, and solar energy, although good counterexamples to limits-to-growth extremism, have developed and will probably continue to develop more slowly than enthusiasts hope. All should have some impacts on the location of economic activity when (and if) they come into full flower.

Let me expand upon these themes.

Most economists seem relatively sanguine about the future. For example, William Nordhaus considers two major alternative scenarios: (1) 1974 relative prices are maintained indefinitely, with the OPEC-triggered price increases reflecting "irreversible structural shifts," and (2) there is a return to relative prices, pre-OPEC, by 1985, the energy crisis of 1970-1975+ being viewed as a transient phenomena [78]. He does *not* consider the possibility of still further massive price increases. In contrast, Miernyk argues that "the neo-Victorian optimism so prevalent among economists is notable by its absence in the writings of many geologists and earth scientists" [79, p. 214]. Again, he doubts that present high energy prices will prove to be transitory, suggesting "a more reasonable assumption is that the era of cheap energy is over" [7, p. 3].

Although some of the differences between the Nordhaus and the Miernyk forecasts might be reconciled by the suggestion that they have focused on different lengths of run, it is nonetheless clear that Miernyk is much more pessimistic, whatever the length of run. However, there is a long history of grossly pessimistic geological forecasts [80], and the present temporary (?) "glut" of oil [81] suggests that downward pressures on energy prices can still occur.

Non-OPEC oil producers will likely be as displeased with such downward pressures as the members of the cartel and may well try to avoid generating such pressures. Thus, some observers have noted that Great Britain, Norway, and Mexico have tried to act as if they were de facto OPEC members, and others have commented on U.S. regulatory policies as ultimately encouraging the cartel, perhaps to protect the investment in Alaskan production [50, 82, 83]. More generally, a precipitous fall in energy prices could cause a number of adjustment problems in the United States, and perhaps some contingency planning for that possibility ought to be made. Admittedly, this is not too likely, given Saudi Arabia's share of the market and capacity to restrict production, so that Nordhaus' first scenario seems the most probable for the next decade or so.

Focusing on U.S. production, it seems remarkable that in recent years natural gas supplies have been severely curtailed and oil production has dropped (figure 7-7) at the same time employment in oil and gas extraction has increased substantially (tables 7-6 and 7-7). Of course, this could reflect an extremely rapid rate of depletion so that an ever-larger increment of labor is needed to produce a given increment of oil or gas, but available evidence shows only a limited falling off in the rate of discovery, at worst [84]. It seems much more plausible that low prices for "old" oil and gas have caused many "old" producing wells to be shut down, or curtailed, so that new production accounts for an increasing share of total output. (Consider what would happen if we defined "old" labor as effort expended between 8:00 a.m. and 5:00 p.m., and "new" labor as any additional effort expended between 5:00 p.m. and 8:00 a.m., with the "new" wage rate twice the "old." We might well turn night into day!) It is therefore likely that a great deal of the "new" oil and gas labor and a certain portion of the "old" labor are engaged in exploration and development of "new" oil and gas production. But there is usually a lag of a few years before such production comes onstream, suggesting that there may well be a considerable increase in U.S. production in the next few years. Hence, there may be additional "short-term" pressure on prices from U.S. production increments.

In the longer run, however, most energy economists foresee gradual long-term energy price increases with the depletion of the more accessible, lower-cost deposits of oil and gas, even if the OPEC monopoly does not persist [85, p. 344]. Certainly, with increasing population and per capita income, world demand for energy will increase considerably. In particular, greatly increased consumption of gasoline is to be expected as the rest of the world follows the U.S. lead in automobile ownership and operation. Hence, an outward-moving demand curve should intersect an upwardly sloping supply curve at successively higher prices over time, assuming that technological advance shifting supply outward will lag behind the shifts in demand.

Walt Rostow suggests that long-term (Kondratieff) cycles of forty to fifty years duration may be in play, so that we are in for a long period when energy prices will remain relatively high [86].

Given increased energy prices, both production and consumption adjustments are to be expected, with the obvious prediction that energy-intensive consuming industries, activities, and regions will contract and that energy-producing industries and regions will expand. Since most energy production is localized to the South and West, and since many parts of the South and West have more pleasant climate than most of the North, with lower heating plus air conditioning costs, it follows that increased energy prices will be one of the factors causing the long-term movement of economic activity and population to the South and West. The OPEC price rise should have accelerated the process. Tables 7-13 and 7-15 show a 1972 differential of about $90 per capita in household energy consumption between New England and the Pacific Coast, representing the effect both of higher prices and of greater fuel consumption because of more severe climate. That differential was probably on the order of $200 in 1977, applying the evidence of table 7-22, or around $600 per family, given an average family size of around three persons. Capitalized at even a high rate of interest, the capital value of the difference seems considerable, amounting to $5,000 at an interest rate of 12 percent, perhaps "typical" for consumer loans. Some of the present differential must represent a disequilibrium difference, so that some people will move, or have already moved, from New England to the Pacific Coast in response. My guess is that the differential is large enough to generate substantial migration. Other differentials, though less extreme, also seem likely to have considerable effect on migration. Thus, New England's 1972 per capita consumption was $75 more than that in the South, while both the Mid-Atlantic and East North Central regions' per capita consumption exceeded that in the South and the Pacific Coast by $35 and $45, respectively.

At the metropolitan level, the increased price of gasoline should have some inhibiting effect on automobile commuting, tending to make central location more attractive and thus slowing down, if not reversing, urban sprawl. The effect may be quite mild, given congressionally mandated increases in fuel economy. It can yield some stimulus to central-city revival or, alternatively, help generate some high-density centers in the suburbs [87].

It is not obvious how higher energy prices interact with urban population size. My guess is that the net effect of higher prices is more detrimental to large than small places. However, with less reliance on the automobile, relatively more apartments than single-family dwellings, and smaller housing units, large places tend to consume less energy per family. The urban heat island effect may also be beneficial, on net, since it is primarily a nighttime effect and may lead to more heating savings in winter than additional air-conditioning costs (or discomfort) in summer. On the other hand, there is typically a much longer journey to work in larger places. Transportation costs are a large part of consumer energy costs, and journey-to-work costs are a large part of transportation costs. Further, insofar as increased energy prices contribute to general inflation, there may be an indirect, negative impact on larger places. Large places have higher price levels than small, and therefore pay higher wages for the same work. But the progressive income tax is not geared to price level differences, so taxes will be proportionately higher the larger the place. This effect may be intensified in periods of inflation and could be a factor in the recent inverse relation between metropolitan size and population growth.

Sternlieb and Hughes focus on both the inverse relation between urban size and growth and the "acclerating regional shift," arguing that these new realities can shape planning approaches on all levels for the next decade. They state: "the analogy of the decline of the aging northern industrial crescent with the long-term transformation of the central cities is increasingly—and uncomfortably—apt" [88].

Both Steven McDonald [90] and Miernyk [7] see massive shifts in economic activity between regions as a consequence of changed energy relationships, with stagnation in the North and great spurts of development in the Southwest and West. These scenarios seem to me to involve a certain amount of overstatement, given the earlier conclusions that energy is only one among many location factors and that relative energy price increases have been less pronounced than is commonly perceived.

Further, federal energy policy will distort this development sequence. As noted at several points, artificially low domestic prices for oil and gas now are a central feature of that policy. There are several rationales, including antipathy to "windfall gains" (but *any* change will involve windfall gains and losses) and the view that higher prices help "big oil" and hurt consumers (although much production is by small-scale producers). "Consumerism" has been a factor in energy policy for many years, as indicated in the earlier discussion of TVA and Bonneville price policy. Certainly, many past programs have aimed at helping specific groups of energy producers by keeping prices *above* free-market levels. (But two wrongs do not make a right, nor do two contrary sources of price distortion balance out to yield efficiency or "justice," particularly when separated in time.) Mead documents a number of programs under each policy stance in some detail [85], seeing a common source as responsiveness to special interest groups, including consumers (or those who claim to speak for consumers). No doubt one of the aims of current oil and gas price regulation has been to cushion the impact of higher energy prices in the North. Russell [76, p. 331] and Mead [85, p. 345] cite benefits to the Northeast and New England, but the North Central region, with its dependence on natural gas, is also a putative beneficiary. Unfortunately, the policy leads to a variety of resource malallocations, well documented by Russell and Mead.

Further, the "benefits" are progressively dissipated as high-priced OPEC oil, natural gas from Mexico, and liquefied natural gas from Algeria and Indonesia are increasingly employed, at least in part filling a gap caused by the pricing of domestic fuels below equilibrium levels.

A mandated shift to coal is in progress, and will accelerate if administration policy proposals are fully implemented. This will speed up the development of Western coal, in particular, given its low sulfur content.

But even with greater reliance on low-sulfur Western sources, the shift to coal seems likely to impose considerable environmental impact, in the form of higher costs in attempting to meet environmental quality standards and/or increased morbidity and mortality from air pollution. An advance report on a major Brookhaven and Carnegie-Mellon study [90] concludes that "the implications for the increased use of coal are grim" given the consequences for premature deaths from respiratory disease. Health effects are predicted to be strongest in the Mid-Atlantic region, because discharges from the smokestacks of the Midwest are carried by prevailing winds to the East Coast.

The regulated price limitation on Southwestern production of oil and natural gas has generated considerable concern and bitterness in that region. Thus, Walt Rostow notes, "I cannot convey to you with sufficient force the depth of feeling in the southern energy exporting states about some of the attitudes of the North. At one and the same time, the North appears to be demanding both low energy prices and refusing to develop its own energy resources on environmental grounds. This is seen in the South as a straightforward colonial policy of exploitation . . ." [86].

Accelerated growth in the West has also brought problems and complaints about boomtown conditions and deteriorating quality of life [91]. The governors of Montana and North Dakota have argued that the federal government has an obligation to help communities impacted by federally mandated energy development policies, particularly those calling for increased use of coal from Western states; in a 1977 appearance before a Senate subcommittee, they requested a billion dollars in aid to impacted areas in their states [92]. Markusen has argued against such grants, on the grounds that internalization of unusual infrastructure costs at the site will yield a more efficient allocation of resources [93]. Markusen adds that internalization will reduce economic rent, no doubt affecting the royalties collected by the Western states in the form of coal severance taxes, which have been substantial. Markusen's argument could be generalized to the proposal that new Western production be taxed to cover costs of additional social services as well as new infrastructure (say, streets, sewers, schools) in Western boomtowns. The new social services would respond to putative problems of social disorganization in boomtowns (for example, delinquency, drunkenness, divorce).

Some of the complaints about Western growth seem a species of "crying all the way to the bank." I am willing to entertain the hypothesis that the benefits of growth usually outweigh the associated costs. Hence, it seems to me that the Southwestern complaint is on stronger ground than the Western.

High hopes have been raised for the development of alternative energy sources, but those hopes are likely to be overblown, on the basis of past experience. Nuclear power is a case in point. In 1972 nuclear power accounted for 3.1 percent of electricity generated by electric utilities; electric utility use, in turn, accounted for 24.6 percent of British thermal units consumed in all uses of energy, but for only 7.2 percent of expenditures in all uses. By 1976 nuclear power accounted for 9.4 percent of generated electricity [94]. Despite this considerable increase in generation percentage, there appears general agreement that nuclear power development is slower and costs are higher than expected. Press accounts have stated these conclusions for several years [95], with some parallel testimony by experts. For example, Alvin Weinberg notes: "The euphoria of extremely low-cost nuclear electricity . . . has been superseded by the less euphoric actual turn of events" [96].

Difficulties slowing nuclear development have included a slowing in the growth of electricity use; the constraints imposed by regulation, and the costs of responding to antinuclear power initiatives and legal actions; construction difficulties, including labor problems, design errors, and accidents; and more frequent and longer than anticipated shutdowns of nuclear plants for repairs and maintenance.

Although U.S. electricity use showed a substantial increase between 1975 and 1976, after a period of little growth, some tendency of regulatory commissions to limit rate increases may have tempered industry optimism. It typically takes ten to twelve years to build nuclear plants, although it has been noted that "only" three years are attributable to licensing procedures. The lengthy licensing procedures are part of a complex mosaic of regulations, imposed in large part in response to concern about environmental hazards from nuclear plants. Hazards can be categorized under the headings of (1) radiation, in small amounts on a continuous basis, from normal plant operations, and in large amounts that might be released in an accident; (2) disposal of spent fuel; (3) illegitimate redirection of nuclear fuel for weapons use.

In a review of the sketchy evidence on risks, Ridker and Watson conclude that conventional nuclear reactors (light-water reactors) have about the same degree of risk as coal-fired power plants, which suggests that "nuclear safety should not be an important issue, but public perception of risks is not based only on averages. . . . Marginal impacts, such as those for coal air pollutants, are easier to accept because they are less obvious and less dramatic" [97]. Put another way, the variance for nuclear power risk, by way of nuclear accident, may be perceived as much larger, and hence less acceptable, than risk from coal. William Ramsay has developed some comparisons which suggest that risk (in terms of averages) is considerably lower for nuclear power than it is for coal-fired power. In terms of a range of expected additional death per plant year from power plant operations, his comparisons are [98]:

	Additional Deaths
Light-water nuclear reactors	0.16 – 0.75
Coal fired power plants	
with sulfur scrubbing	0.79 – 13.00
within new source performance standards	0.85 – 29.00
without scrubbing, using 3 percent sulfur coal	2.90 – 00.00

Thus, the upper value of the nuclear range roughly equals the lower value of the range for the "cleanest" form of coal. There is increasing hazard from coal-fired plants with progressively "dirtier" fuel. These data reinforce the earlier inference that the mandated shift to coal will impose higher costs because of environmental impacts.

In the introduction of any new technology, unanticipated problems and delays are common, but such difficulties may have been even more pronounced for nuclear energy than is usually the case. An extreme example is that of General Atomic's "unconventional" high-temperature gas reactor, which may be safer than the conventional light-water reactor but which incurred an estimated billion-dollar loss in its development. And the construction of conventional nuclear reactors apparently is not particularly profitable at present [99].

Nuclear power has a number of consequences involving location, at both the regional and the local levels. The Ridker and Watson study [97] traces some of the regional consequences of nuclear power development as part of a detailed series of simulations ("scenarios") projecting the U.S. economy to the year 2025. By that year, nuclear power is projected as the dominant source of electricity, accounting for 67 percent of generation in a base scenario and 76 percent in an extreme alternative scenario. Power plants are expected to be concentrated near heavily populated areas to minimize transmission losses, but also to have their location responsive to population distribution and meteorological conditions so as to reduce radiation hazard. The heaviest concentrations of nuclear power plants are projected to occur in New England, the Pacific Coast, the Delaware Valley, Florida, and the lower Great Lakes region. Coal-fired plants are expected to be built near coal mining areas. If nuclear power were to be phased out by a nuclear moratorium, additional coal-fired plants would impose a heavy air pollution burden locally on the Ohio, Colorado, and lower Mississippi Valleys, and at considerable remove on the Northeast region, which would suffer from nitrates and sulfates transported from the Ohio and Tennessee Valleys by the prevailing winds.

Weinberg discusses a similar nuclear moratorium simulation [96], projecting that the cost of electricity would rise in most parts of the country, but the increase would usually be no more than 10 to 20 percent, reflecting the relative costs of coal and nuclear power. In California and New England, however, the margin favoring nuclear power is seen as much greater, while in the Rocky Mountain region, coal is projected to be much the cheaper fuel. Hence, the first two regions would be greatly disadvantaged by a moratorium, while the last would be unaffected.

At the local level, nuclear siting constraints have caused a number of small localities to become Eastern boomtowns. A nuclear power plant serving a much larger population typically locates in a small town at some distance from major population centers, and pays a considerable property tax in that small town, rather than in the service area as a whole. This generally results in subtantially lowered property taxes to the local citizenry. There is also expansion of the local employment base, with some hiring of local workers by the power plant. Both factors help generate considerable increases in property values and in population size, as outsiders are attracted by relatively low taxes and by the employment opportunities. The process is described for Plymouth, Massachusetts, and Waterford, Connecticut, in Purdy et al. [100] and for those towns and for Lacey Township, New Jersey, and Morris, Illinois, in several press reports [101]. In those towns, the nuclear plants pay roughly half the property taxes, typically on the order of $200 per capita. That benefit appears to outweigh any discerned increase in risk, from the evidence of the considerable increase in population in all the towns. Plymouth, for example, has doubled in population since 1965, when the nuclear power plant was proposed. Not all host communities benefit, since tax payments reflect state law. A case in point is Peachbottom Township, Pennsylvania, which received little from the two giant reactors located there. And both Connecticut and Massachusetts legislatures are considering redistributing the tax payments on a statewide basis. The case might be made that the host communities bear a disproportionate share of the risk and so ought to receive a corresponding share of the tax benefits. The experience of the Eastern host communities might well provide some lessons for Western boomtowns.

Although forecasts of nuclear reactor requirements have been substantially lowered, given the recent slowdown in nuclear power growth, a considerable increase in nuclear power plants is generally expected over the next fifteen to twenty-five years. (Of course, such squares with the longer-run forecasts of Ridker and Watson [97] and Weinberg [96].) Because of limited availability of suitable sites for nuclear plants, it has been suggested that the number of such sites be increased or that more nuclear plants be placed on individual sites. Cope and Bauman discuss the siting question in some detail, focusing on the possible use of individual sites for clusters of plants forming "nuclear energy centers" [102]. Population density criteria in effect rule out all metropolitan areas in locating such centers [102, p. 8]. Other criteria considered include seismic and geologic conditions, meteorology, electrical demand, and environmental and public acceptance. A review of 110 existing sites yields the conclusion that existing and known potential sites could handle much of any future need for nuclear energy centers. But Cope and Bauman add that large areas of the country do not have any known good sites [102, p. 15]. Further, it can be anticipated that strong opposition to such clustering will be based on environmental grounds [103].

Paralleling the experience with nuclear energy, the production of Atlantic offshore oil and gas, shale oil, and solar energy holds considerable promise for augmentation of energy supplies, but that promise has probably been overstated through overly optimistic appraisals.

Atlantic offshore oil and gas production may come onstream earliest—perhaps within five years. Although subject to environmental lawsuit, some geologists have concluded that "petroleum production off the East Coast would prove ultimately safer to the environment than either drilling in other coastal areas of the United States or importing oil in tankers" [104, p. 791]. Despite that relative safety, drilling leases that were granted in August 1976 were subsequently blocked on environmental grounds in federal court in February 1977, and then the lower court ruling was overturned in August 1977. During the year the case was under adjudication, the oil companies received neither interest on their 1.1-billion-dollar investment for the leases nor permission to carry out preliminary exploration prior to the final court decision. Earliest commercial production is expected in late 1980, although it is not certain that recoverable hydrocarbons exist in the Georges Bank and the Baltimore Canyon basins, which are the major areas of interest.

Estimates of the energy equivalence of shale oil deposits in Utah, Wyoming, and Colorado range from the equal of all U.S. reserves of petroleum and natural gas to several times the world reserves of those fuels [4, pp. 2, 3, 316-18]. Although oil companies have invested some hundreds of millions of dollars in leases, progress toward commercial production has been much slower than anticipated. John M. Blair has argued that the oil companies have been procrastinating on such development in order to obtain public-sector financing and price guarantees [105, p. 62].

When several oil companies dropped out of their shale oil endeavor, they charged that lack of government incentives, air pollution laws, and price controls made oil shale development financially infeasible [106, p. 6]. The first explanation squares with Blair's complaint, but the latter two treat increased risk as a function of national policy, perhaps with some justice.

Darmstadter and Searle note that hopes for an oil shale industry were dashed twice before, in the 1920s and after World War II, because new supplies of crude oil made oil shale development uneconomic [107]. They add that this could happen again.

Although some analysts see solar energy as economically feasible by 1990 [108], the consensus of expert opinion seems to be that the use of solar water heating and space conditioning is hindered by price controls on gas and oil, keeping those fuels relatively low in price, and by a variety of institutional barriers. Solar electric power is technically feasible but will not be economically viable for a considerable span of time [87, pp. 345-46]. There is disagreement among the experts on the regional economics of solar energy. For example, Hans Bethe argues that the Rocky Mountain states comprise the most favorable region for solar energy [109]. In contrast, Schulze, Ben-David, and Balcomb [108] predict that solar residential space heating will first become economically feasible in the Northern tier of states by the late 1970s and then will progress southward, roughly attaining the Mason-Dixon line by 1990; economic feasibility for the Rocky Mountain region arrives fairly late, 1985, in their projection. Economic feasibility for solar hot-water heating is projected for Southern and Eastern states by the early 1980s, but is not seen as occurring even by 1990 for the Rocky Mountain, Midwestern, and Northwestern states.

The consideration of alternative energy sources illuminates the point that considerable amounts of energy are available—at a price. Thus, shale oil production has recently been estimated as economic at twice the world price for oil [106, p. 9]—a considerable escalation over earlier estimates; and large quantities of methane gas and gas from Devonian shale beneath the Eastern and Midwestern part of the country could be extracted at a high enough price [110]. Such information suggests that there are no long-term "limits to growth" *if* we are willing to use the price mechanism and pay a higher price for our energy, as such becomes necessary.

The discussion also reinforces the conclusion that the Rocky Mountain regional economy will be spurred by energy development, given the potential for shale oil production and (perhaps) for solar energy. Further, there is the implication that the New England and Mid-Atlantic regions' energy problems may be somewhat ameliorated by offshore production, perhaps helping to reverse the decline of some Eastern metropolitan areas. (We may someday wonder whether New York City will become another Houston.)

To sum up, the issue of price regulation seems central to energy policy and prospects. The effects of regulated low prices can be summarized as follows:

1. Low prices have curtailed domestic production of oil and gas, leading to some substitution of foreign for domestic production and helping to support OPEC. An extreme case in point is the importation of liquefied petroleum gas from Algeria and Indonesia at considerably higher prices than are set for domestic natural gas.

2. A directed substitution of coal for oil and gas is underway. Such will increase air pollution problems, with costs falling most heavily on the Northeast. But pollution problems should become more widespread generally, so that some areas will shift from negligible to pronounced particulate and sulfate levels. The Southwest seems a prime candidate for such a shift. An increased reliance on coal probably makes the economy more vulnerable to strikes.

3. The acceleration of Western coal production tends to elicit federal funds to aid impacted areas, which seems questionable on equity grounds. The case for internalization of costs seems strong. The financial payments by nuclear power plants to the Eastern towns in which they are located can serve as precedent for similar arrangements in Western boomtowns.

4. Artificial shortages are caused by regulated low prices. Many prospective purchasers of natural gas, our cleanest fuel, are not allowed to buy it. Thus, less urgent uses can receive precedence over more urgent uses.

5. Under the guise of helping the consumer, regulation often evolves into a monopolistic tool for the regulated industry. Perhaps some early signs of that evolution are to be discerned in the energy industry.

6. Low domestic prices for oil and gas may be a factor inhibiting the development of alternative energy sources, including nuclear power, solar energy, and shale oil.

Economists tend to dismiss price regulation as merely a cover for self-interest at the expense of the public interest, usually rationalized by the medieval notion of a "fair" price, which in practice means little more than a traditional or conventional price. But it might be more rational of economists to suggest more efficient alternatives to achieve the purported goals of price regulation, recognizing that once regulation is in force, in effect a set of property rights is created. Hence, a case can be made for compensation if those rights are extinguished. Let me grope toward some suggested answers to the double-barreled question of developing other means to achieve regulatory goals and of compensating vested interests created by regulation. If current consumers of natural gas are viewed as "entitled" to protection from price increases, why not issue them marketable five-year "entitlements" to the difference between annual costs before and after deregulation that begin with the full difference and are reduced by 20 percent annually? Or why not give welfare recipients stock in oil companies? Or use the half billion dollars per year that is now spent in regulation of oil and gas prices [111] in promoting oil and gas competition, paralleling the original TVA yardstock idea? Admittedly, these specifics may well seem outrageous, but I believe the general approach deserves consideration by persons of good will who desire better ways to settle disputes than by confrontation. Perhaps that approach can be applied to regional conflicts, allowing for the development of alternatives to confrontation on regional energy issues.

References

[1] Schurr, Sam H., and Netschert, Bruce C., with Eliasberg, V.F., Lerner, J., and Landsberg, Hans H. *Energy in the American Economy 1850-1975*. Baltimore, Md.: Johns Hopkins University Press, for Resources for the Future, 1960.

[2] Borchert, John R. "American Metropolitan Evolution." *Geographical Review*, July 1967, pp. 301-32.

[3] Perloff, Harvey S., and Wingo, Lowdon. "Natural Resource Endowment and Regional Economic Growth." In J.J. Spengler, ed., *Natural Resources and Economic Growth*. Washington: Resources for the Future, 1960, pp. 191-212.

[4] U.S. House of Representatives, Subcommittee on Energy and Power of Committee on Interstate and Foreign Commerce. *Energy Information Handbook*. Committee Print 95-18. Washington, July 1977.

[5] Rosenberg, Nathan. "Thinking about Technology Policy for the Coming Decade." In U.S. Congress, Joint Economic Committee, *U.S. Economic Growth from 1976 to 1986: Prospects, Problems and Patterns*. Vol. 9: *Technological Change*. Washington, 1977, pp. 25, 26 in particular.

[6] Stanwood, Roger D., Vice President of Transcontinental Gas Pipe Line Corporation. Cited in *New York Times*, February 22, 1977, p. 14.

[7] Miernyk, William H. "Rising Energy Prices and Regional Economic Development." *Growth and Change*, July 1977, pp. 2-7.

[8] Duncan, Hugh Dalziel. "The Chicago School: Principles." In Arthur Siegel, ed., *Chicago's Famous Buildings*. Chicago: University of Chicago Press, 1965.

[9] Ullman, Edward L. "Amenities as a Factor in Regional Growth." *Geographical Review*, 1954, pp. 119-32.

[10] Seeber, L. "History of the TVA," pp. 3-26; "What Has Been Accomplished in the First Forty Years," pp. 27-38; "Electrical Energy for the Tennessee Valley Region," pp. 268-77. In Knop, Hans, ed., *The Tennesee Valley Authority Experience*. Laxenberg, Austria: International Institute for Applied Systems Analysis, 1976.

[11] Knop, Hans, ed. *The Tennessee Valley Authority Experience, Proceedings of the First Conference on Case Studies of Large Scale Planning Projects, 1974*, vols. I and II. Laxenberg, Austria: International Institute for Applied Systems Analysis, 1976.

[12] U.S. Department of Interior, Bonneville Power Administration. *About BPA*. Washington, circa 1976, pp. 5, 15, 17. And *1976 Annual Report*.

[13] Habday, Victor C. *Sparks at the Grassroots: Municipal Distribution of TVA Electricity in Tennessee*. Knoxville: University of Tennessee Press, 1969.

[14] Wengert, Norman. "The Politics of Water Resource Development as Exemplified by TVA." In Moore, John, ed., *The Economic Impact of TVA*. Knoxville: University of Tennessee Press, 1967, pp. 57-80.

[15] Moore, John R., ed., *The Economic Impact of TVA*. Knoxville: University of Tennessee Press, 1967.

[16] Koopmans, T. "Discussion Paper." In Knop, Hans, *The Tennessee Valley Authority Experience*, pp. 69-72.

[17] Banner, Gilbert. "Toward More Realistic Assumptions in Regional Economic Development." In Moore, John, ed. *The Economic Impact of TVA*, pp. 57-80.

[18] Perry, Harry. "The Energy Problem in a Regional-Urban Setting." Appendix B in Irving Hoch et al., "The Relationship between Natural Resources and Patterns of Urban and Regional Development." Report to HUD from Resources for the Future, 1973, unpublished.

[19] Netschert, Bruce C. "Electric Power and Economic Development." In Moore, John, ed., *The Economic Impact of TVA*, pp. 1-24.

[20] Robock, Stefan H. "An Unfinished Task: A Socio-Economic Evaluation of the TVA Experiment." In Moore, John, ed., *The Economic Impact of TVA*, pp. 105-20.

[21] Hinote, Hubert. "TVA Economic Simulation Model." In Knop, Hans, ed., *The Tennessee Valley Authority Experience*, pp. 79-131.

[22] Hinote, Hubert. Telephone conversations with the author, August 23 and 24, 1977.

[23] Owen, Marguerite. *The Tennessee Valley Authority*. New York: Praeger, 1973.
[24] Shapley, Deborah. "TVA Today: Former Reformers in an Era of Expensive Electricity." *Science*, November 19, 1976, pp. 814-18.
[25] Cowan, Edward. "Today's TVA Is Assailed as Threat to Environment." *New York Times*, August 5, 1973, pp. 1, 43. And "Lawsuits Seek TVA Air Compliance." *Coal Age*, August 1977, p. 37.
[26] Clapp, Gordon R. "The Meaning of TVA." In Martin, Roscoe C., ed., *TVA: The First Twenty Years*. University, Ala., and Knoxville, Tenn.: University of Alabama Press and University of Tennessee Press, 1956.
[27] Rottenberg, Simon. "Economic Policy in the Poor Countries." *The Journal of Law and Economics*, October 1959.
[28] *Survey of Current Business*, July 1976, table 6.6, p. 51.
[29] Wong, William. "Aluminum Makers in Pacific Northwest Face End of Industry's Cheapest Power." *Wall Street Journal*, August 15, 1977, p. 18.
[30] Hoch, Irving. "Interindustry Forecasts for the Chicago Region." *Papers and Proceedings of the Regional Science Association*. Vol. 5. Philadelphia: Regional Science Association, 1959, p. 230 in particular.
[31] Hoch, Irving, and Tryphonopoulos, N. *A Study of the Economy of Napa County California*. Giannini Foundation Report no. 303, August 1969, p. 145 in particular.
[32] MacKinnon, R. Discussion of "TVA Economic Simulation Model." In Knop, Hans, ed., *The Tennessee Valley Authority Experience*, p. 133.
[33] Lapp, Ralph E. "America's Energy." Reproduced in U.S. House of Representatives, *Energy Information Handbook*, figure 26, p. 79.
[34] "The Coal Industry's Controversial Move West." *Business Week*, May 11, 1974, p. 134.
[35] Federal Power Commission. *Hydroelectric Power Resources of the United States*. Washington, 1972, figure 2, p. X.
[36] Bebbington, William P. "The Reprocessing of Nuclear Fuels." *Scientific American*, December 1976, p. 40.
[37] U.S. Bureau of Mines. *Minerals Yearbook 1974*, vol. I. Washington, 1976.
[38] Federal Energy Administration, Office of Oil and Gas. "The Search for Petroleum." August 22, 1977.
[39] Kenneke, A.P., Nuclear Regulatory Commission, personal communication, August 30, 1977.
[40] Crump, Lulie H. *Fuels and Energy Data: United States by States and Census Divisions, 1974*. U.S. Bureau of Mines Information Circular 8739. Washington, 1977.

[41] Miller, James Nathan. "Natural Gas Reserves." Reprinted in *Washington Post*, February 13, 1977, p. C1.
[42] General Accounting Office. Report to the Congress by the Comptroller General. *Improvements Still Needed in Federal Energy Data Collection, Analysis and Reporting.* Washington, June 1976, pp. 32-33 in particular.
[43] Cowan, Edward. "The Natural Gas Shortage Was Only Partially Man-Made." *New York Times*, January 30, 1977, sec. 4, p. 1. Statement by John O'Leary, Federal Energy Administrator. CBS television program, *Energy: The Facts, the Fears, the Future*, August 31, 1977.
[44] Rattner, Steven. "Oil and Gas Reserves: Figures Open to Question." *New York Times*, May 8, 1977, sec. 4, p. 3.
[45] Sterba, James P. "Gas Shortage a Fundamental Long Term Economic Threat to U.S., Experts Say." *New York Times*, February 22, 1977, p. 14.
[46] Office of Technology Assessment. *Analysis of the Proposed National Energy Plan*, June 1977, prepublication draft, appendixes B, B-4, and B-6.
[47] Rice, Patricia, and Smith, V. Kerry. "An Econometric Model of the Petroleum Industry." *Journal of Econometrics*, 1977, forthcoming.
[48] *Washington Post*, February 15, 1977, p. D9. *Wall Street Journal*, February 3, 1977, pp. 1, 12.
[49] Hall, Robert E., and Pindyck, Robert S. "The Conflicting Goals of National Energy Policy." *The Public Interest*, Spring 1977, p. 4 in particular.
[50] *Wall Street Journal*, editorials, April 27, 1977, p. 26; May 20, 1977, p. 14; May 27, 1977, p. 10.
[51] This evaluation is supported by the expert opinion of the energy economists at Resources for the Future. Personal discussion with John Schanz, September 2, 1977.
[52] U.S. Bureau of Economic Analysis. *The National Income and Product Accounts.* Washington, 1976, pp. 206-13.
[53] *Survey of Current Business*, July 1977, tables 6.8 and 6.9, p. 47.
[54] *Survey of Current Business*, August 1974, pp. 34-43, and August 1976, pp. 18-27.
[55] *Survey of Current Business*, August 1974, p. 33, and August 1976, p. 17.
[56] U.S. Bureau of Economic Analysis. *Population and Economic Activity in the United States and Standard Metropolitan Statistical Areas.* Washington, 1972.
[57] U.S. Bureau of Economic Analysis. *State Projections of Economic Activity, 1980-2000*, March 1977, tables 1 and 2, "Earnings by Industry."

[58] Hoch, Irving. *Energy Use in the United States by State and Region: A Statistical Compendium of 1972 Consumption, Prices and Expenditures*. Baltimore, Md.: Resources for the Future, 1978.

[59] U.S. Bureau of Mines. *Mineral Industry Surveys, Fuel Oil Sales Annual, Sales of Fuel Oil and Kerosine in 1972*. Washington, 1973.

[60] U.S. Bureau of Mines. *Mineral Industry Surveys, Liquefied Petroleum Gas Sales, Annual Sales of Liquefied Petroleum Gases and Ethane in 1973*. Washington, 1974.

[61] U.S. Federal Highway Administration. *1972 Highway Statistics*. Washington, 1974.

[62] American Gas Association. *1972 Gas Facts*. Arlington, Va., 1973.

[63] U.S. Bureau of Mines. *Minerals Yearbook*. Washington, 1972, vol. I.

[64] Edison Electric Institute. *Statistical Yearbook of the Electric Utility Industry for 1972*. New York, 1973.

[65] National Coal Association. *Steam-Electric Plant Factors*, 1973 ed. Washington, 1974.

[66] Crump, Lulie H., and Readling, Charles L. *Fuel and Energy Data, United States by States and Regions*. U.S. Bureau of Mines Information Circular 8647, 1974.

[67] Arthur D. Little, Inc. for Federal Energy Administration. *Project Independence, Energy Conservation*. Vol. I: *Residential and Commercial Energy Use Patterns 1970-1990*. Washington, D.C.: Arthur D. Little, Incorporated, 1974.

[68] Foster Associates. *Energy Prices, 1960-73*. Cambridge, Mass.: Ballinger Publishing, 1974.

[69] Foster Associates. *Prospective Regional Markets for Coal Conversion Plant Products Projected to 1980 and 1985*. Vol. 3: *Current and Projected Demand, Supply and Price of Energy in the United States, Schedules*. Prepared for Office of Coal Research, 1974. NTIS PB-236 633.

[70] Independent Natural Gas Association of America. *Comparison of Seasonal Househeating Costs for Gas, Fuel Oil, Coal and Electricity*. Washington, 1971 and 1972.

[71] Taylor, Lester D. "The Demand for Electricity: A Survey." *The Bell Journal of Economics*, Spring 1975, pp. 74-110.

[72] Nelson, Jon P. "Climate and Energy Demand: Fossil Fuels." In Terry A. Ferrar, ed., *The Urban Costs of Climate Modification*. New York: Wiley, 1976, pp. 123-38.

[73] Ferrar, Terry A., and Nelson, Jon P. "Energy Conservation Policies of the Federal Energy Office: Economic Demand Analysis." *Science*, February 21, 1975, pp. 644-46.

[74] Oak Ridge National Laboratory. *Energy Division Annual Progress Report*. Oak Ridge, Tennessee: 1976, p. 178.

[75] U.S. Bureau of Labor Statistics. *Retail Prices and Indexes of Fuels and Utilities, Residential Usage*. Monthly; August 1977 release cited.

[76] Russell, Milton. "Energy." In J. Pechman, ed., *The 1978 Budget, National Priorities*. Washington: Brookings Institution, 1977.

[77] *Oil and Gas Journal*, October 8, 1973 and August 22, 1977. American Petroleum Institute. *Petroleum Facts and Figures*, 1971 ed. Washington, 1971.

[78] Nordhuas, William. "The Demand for Energy: An International Perspective." *Proceedings of the Workshop on Energy Demand*. Laxenburg, Austria: International Institute for Applied Systems Analysis, 1976, pp. 573-74.

[79] Miernyk, William H. "Some Regional Impacts of the Rising Costs of Energy." *Papers of the Regional Science Association* 37 (1976):213-27.

[80] "The Doomsayers Debunked." *The Petroleum Independent*, May-June 1976, pp. 21-26.

[81] "The Oil Glut Slows OPEC's Production." *Business Week*, August 22, 1977, p. 23.

[82] "For some reason, we are asked to take our chances on handing OPEC more money than on giving Texas any incentive to find more oil and gas." Hyman, Leonard S., letter to editor. *Science*, July 22, 1977, p. 326.

[83] Davidson, James Dale. "Plain Speech on Energy." *Public Utilities Fortnightly*, August 4, 1977, pp. 4-6.

[84] Rostow, W.W., Fisher, W.L., and Woodson, H.H. *National Energy Policy: An Interim Overview*. The Council on Energy Resources, the University of Texas at Austin, September 12, 1977. Figure 8, p. 29, of that report shows the exploration finding rate for oil and gas from 1947 through 1975, exclusive of north slope Alaska. (The finding rate consists of barrels equivalent discovered per foot drilled.) From 1951 through 1975 the finding rate seems essentially stable, although a small local turndown occurs from 1970 through 1975. The authors predict "a stable fnding rate should persist" (p. 2).

[85] Mead, Walter J. "An Economic Appraisal of President Carter's Energy Program." *Science*, July 22, 1977.

[86] Rostow, W.W. "A National Policy towards Regional Change." Speech presented at Western New England College, Springfield, Mass., March 31, 1977.

[87] The last possibility was noted by Benjamin Chinitz when the conference draft of this chapter was presented.

[88] Sternlieb, George, and Hughes, James W. "New Regional and Metropolitan Realities." *Journal of the American Institute of Planners*, July 1977, p. 228, and editor comment, p. 226.

[89] McDonald, Steven. Cited in "Energy Shortage Is Said to Pose Lasting Economic Threat to North." *New York Times*, February 1, 1977, p. 18.

[90] *New York Times*, July 4, 1977, p. A1.

[91] For example, see "The Mountain States: Cooling the Boom." *Business Week*, January 27, 1975, pp. 108-13. Federation of Rocky Mountain States, Inc. *Energy Development in the Rocky Mountain Region*. New York, N.Y.: 1975. Gilmore, John S. "Boomtowns May Hinder Energy Resource Development." *Science*, February 13, 1976, pp. 535-40. Monaco, Lynne A. *State Responses to the Adverse Impacts of Energy Development in Colorado*. Energy Impacts Project, Laboratory of Architecture and Planning, Massachusetts Institute of Technology, 1977. Cummings, Ronald G., and Mehr, Arthur F. "Investment for Urban Infrastructure in Boomtowns." *Natural Resource Journal*, April 1977.

[92] *Land Use Planning Report*, August 15, 1977, p. 258.

[93] Markusen, Ann R. "What Alternative Boomtown Policies Can Do for Regional Growth Policy, National Energy Policy and Fiscal Federalism." In *The Fiscal Crisis of American Boomtown: An Analysis of State and Federal Policy*. Proposed staff study, Washington: General Accounting Office, 1977.

[94] Data from Edison Electric Institute *Yearbook* for 1972 and 1976, and from table 7-18 of this book.

[95] Some examples of reduced hopes for nuclear power appear in "Nuclear Power Dilemma," editorial. *New York Times*, December 3, 1973, p. 38. Burnham, David. "Hope for Cheap Power from Atom Is Fading." *New York Times*, November 16, 1975, p. 1. "A-Plant Growth Declines Sharply." *Washington Post*, January 24, 1976, p. A20. "An Electricity Crisis Looms . . ." *Wall Street Journal*, February 24, 1977, p. 1; Burnham, David. "Atom Plants Produce Much Less than Expected, Researchers Say." *New York Times*, May 24, 1977, p. 61. "A Slapdash Approach to Nuclear Plants," editorial. *New York Times*, August 29, 1977, p. 26.

[96] Weinberg, Alvin M. "Can We Do without Uranium?" In Oak Ridge Associated Universities, *Future Strategies for Energy Development*. Oak Ridge, Tennessee: 1977, p. 260. Weinberg was director of the Oak Ridge National Laboratory for more than twenty-five years.

[97] Ridker, Ronald, and Watson, William. "To Choose a Future." Resources for the Future, processed 1978.

[98] Ramsay, William. "Unpaid Costs of Electrical Energy: Health and Environmental Impacts from Coal and Nuclear Power." Resources for the Future, processed 1978.

[99] "General Atomic's Gas Plant Ends in Fiscal Debacle; Total Loss Is $1 Billion." *Wall Street Journal*, February 25, 1976, p. 1, and *Wall Street Journal*, February 24, 1977, p. 1.

[100] Purdy, Bruce J.; Peele, Elizabeth; Bronfman, Benson H.; Bjornstad, David J. *A Post Licensing Study of Community Effects at Two Operating Nuclear Power Plants, Final Report*. Oak Ridge, Tennessee: Oak Ridge National Laboratory, 1977.

[101] "Connecticut Town Discovers Its Nuclear Plant Is a Mixed Blessing." *New York Times*, May 22, 1973, p. 37. "A-Waste Doesn't Faze Town." *Washington Post*, January 4, 1976, p. A4. "It May Be a Hazard but Lacy Township Loves Nuclear Plant." *Wall Street Journal*, August 11, 1976, p. 1. "A-Plant Brings Growth, Change to Plymouth, Mass." *Newsday*, reprinted in *Washington Post*, November 25, 1976, p. B17.

[102] Cope, D.F., and Bauman, H.F. *Expansion Potential for Existing Nuclear Power Station Sites*. Oak Ridge, Tennessee: Oak Ridge National Laboratory, 1977.

[103] An early example of such opposition appears in a study by the Center for Environmental Studies at Princeton University, which concluded that there was no good site for a cluster of nuclear reactors in New Jersey. *New York Times*, February 20, 1976, p. 1.

[104] Travers, William B., and Luney, Percy R. "Drilling, Tankers and Oil Spills on the Atlantic Outer Continental Shelf." *Science*, November 19, 1976, pp. 791-96. Also reply to letters to editor, *Science*, January 13, 1978, pp. 131-32.

[105] Blair, John M. "Decentralizing Oil Cartels." *Society*, March/April 1977, pp. 57-62 (an abridgement of *The Control of Oil*, Pantheon Books, 1976).

[106] Monaco, Lynne A. State Responses to the Adverse Impacts of Energy Development in Colorado. Energy Impacts Project Laboratory of Architecture and Planning, Massachusetts Institute of Technology, 1977.

[107] Darmstadter, Joel, and Searle, Milton F. "The Outlook for U.S. Energy Demand and Supply." Statement before the Subcommittee on International Economics of the Joint Economic Committee, U.S. Congress, November 8, 1973.

[108] Schulze, W.D., Ben-David, S. and Balcomb, J.D. *The Economics of Solar Home Heating*. Prepared for the Joint Economic Committee, 95th Congress, March 13, 1977.

[109] Bethe, Hans. Cited in "Energy Debate: Nuclear vs. Solar." *New York Times*, February 9, 1977, p. D1.

[110] Maugh, T.H. "National Gas: United States Has It if the Price Is Right." *Science*, February 13, 1976, pp. 549-50.

[111] The half-billion-dollar estimate appears in both Mead, "An Economic Appraisal of President Carter's Energy Program," and Russell, "Energy." In a personal discussion, Russell noted that this figure was an estimate of the cost of regulation per se based on available information.

Appendix 7A: Supplementary Information on State and Regional Fuel Production and Reserves

Table 7A-1
State Fuel Production and Reserves, 1974

	1974 Production			Reserves as of 1974			
	Petroleum (Million bbl)	Gas (Billion ft³)	Coal (Million Tons)	Petroleum (Million bbl)	Gas (Billion ft²)	Coal (Million Tons)	
Region and State						<1 Percent Sulfur	All
New England							
Connecticut	0.000	0.000	0.000	0.0	0.0	0.0	0.0
Maine	0.000	0.000	0.000	0.0	0.0	0.0	0.0
Massachusetts	0.000	0.000	0.000	0.0	0.0	0.0	0.0
New Hampshire	0.000	0.000	0.000	0.0	0.0	0.0	0.0
Rhode Island	0.000	0.000	0.000	0.0	0.0	0.0	0.0
Vermont	0.000	0.000	0.000	0.0	0.0	0.0	0.0
Mid-Atlantic							
New Jersey	0.000	0.000	0.000	0.0	0.0	0.0	0.0
New York	0.896	4.990	0.000	11.0	165.5	0.0	0.0
Pennsylvania	3.478	82.637	80.462	50.0	1,492.1	7,318.3	31,000.6
East North Central							
Illinois	27.553	1.436	58.215	160.0	399.4	1,095.1	66,664.8
Indiana	4.919	0.176	23.726	24.0	64.1	548.8	10,622.6
Michigan	18.021	69.133	0.000	82.0	1,458.3	4.6	118.2
Ohio	9.088	92.055	45.409	124.0	1,308.2	134.4	21,077.2
Wisconsin	0.000	0.000	0.000	0.0	0.0	0.0	0.0
West North Central							
Iowa	0.000	0.000	0.590	0.0	80.5	1.5	2,844.9
Kansas	61.691	886.782	0.718	395.0	11,704.7	0.0	1,388.1
Minnesota	0.000	0.000	0.000	0.0	3.2	0.0	0.0
Missouri	0.056	0.033	4.623	0.0	17.7	0.0	9,487.3
Nebraska	6.611	2.538	0.000	27.0	54.6	0.0	0.0
North Dakota	19.697	31.206	7.463	173.0	432.7	5,389.0	16,003.0
South Dakota	0.494	0.000	0.000	0.0	0.2	103.1	428.0

Appendix 7A

South Atlantic						
Delaware	0.000	0.000	0.000	0.0	0.0	0.0
District of Columbia	0.000	0.000	0.000	0.0	0.0	0.0
Florida	36.351	38.137	0.000	303.0	308.9	0.0
Georgia	0.000	0.000	0.000	0.0	0.0	0.5
Maryland	0.000	0.133	2.337	0.0	42.0	135.1	1,048.2
North Carolina	0.000	0.000	0.000	0.0	0.0	0.0	31.7
South Carolina	0.000	0.000	0.000	0.0	0.0	0.0	0.0
Virginia	0.003	7.096	34.326	0.0	44.7	2,140.1	3,649.9
West Virginia	2.665	202.306	102.462	32.0	2,265.6	14,092.1	39,589.8
East South Central							
Alabama	13.323	27.865	19.824	69.0	507.4	624.7	2,981.8
Kentucky	7.837	71.876	137.197	37.0	844.0	6,558.6	25,540.6
Mississippi	50.779	78.787	0.000	261.0	1,079.4	0.0	0.0
Tennessee	0.769	0.017	7.541	0.0	5.9	204.8	986.7
West South Central							
Arkansas	16.527	123.975	0.455	106.0	2,113.4	81.2	665.7
Louisiana	737.324	7,753.681	0.000	4,227.0	64,052.4	0.0	0.0
Oklahoma	177.785	1,638.942	2.356	1,232.0	13,390.3	275.0	1,294.2
Texas	1,262.126	8,170.798	7.684	11,002.0	78,540.7	659.8	3,271.9
Rocky Mountain							
Arizona	0.740	0.224	6.448	0.0	1.4	173.3	350.0
Colorado	37.508	144.629	6.896	289.0	1,881.7	7,475.5	14,869.2
Idaho	0.000	0.000	0.000	0.0	0.0	0.0	0.0
Montana	34.554	54.873	14.106	207.0	901.3	101,646.6	108,396.2
Nevada	0.129	0.000	0.000	0.0	0.0	0.0	0.0
New Mexico	98.695	1,244.779	9.392	625.0	11,944.9	3,575.3	4,394.8
Utah	39.363	50.522	5.858	251.0	1,031.4	1,968.5	4,042.5
Wyoming	139.997	326.657	20.703	903.0	3,917.4	33,912.3	53,336.1
Far West							
Alaska	70.603	128.935	0.700	10,094.0	31,866.6	11,458.4	11,645.4
California	323.003	365.354	0.000	3,557.0	5,194.6	0.0	0.0
Hawaii	0.000	0.000	0.000	0.0	0.0	0.0	0.0
Oregon	0.000	0.000	0.000	0.0	0.0	1.5	1.8
Washington	0.000	0.000	3.913	0.0	17.2	603.5	1,954.0

Table 7A-2
Regional Fuel Production and Reserves, 1974

	1974 Production			Reserves as of 1974			
							Coal (Million Tons)
Region	Petroleum (Million bbl)	Gas (Billion ft³)	Coal (Million Tons)	Petroleum (Million bbl)	Gas (Billion ft³)	1 Percent Sulfur	All
New England	0.00	0.00	0.00	0.00	0.00	0.00	0.00
Mid-Atlantic	4.37	87.63	80.46	61.00	1,657.60	7,318.30	31,000.60
East North Central	59.58	162.80	127.35	390.00	3,230.00	1,782.90	98,482.80
West North Central	88.55	920.56	13.39	595.00	12,293.60	5,493.60	30,151.30
South Atlantic	39.02	247.67	139.13	335.00	2,661.20	16,367.60	44,320.10
East South Central	72.71	178.55	164.56	367.00	2,436.70	7,388.10	29,509.10
West South Central	2,193.76	17,687.40	10.49	16,567.00	158,096.80	1,016.00	5,231.80
Rocky Mountain	350.99	1,821.68	63.40	2,275.00	19,678.10	148,751.50	185,388.80
Far West	393.61	494.29	4.61	13,651.00	37,078.40	12,063.40	13,601.20
Total	3,202.58	21,600.57	603.40	34,241.00	237,132.40	200,181.40	437,685.70

Table 7A-3
State Reserves of Fuels, 1974
(trillion Btu)

			Coal		
Region and State	*Petroleum*	*Gas*	*< 1 Percent Sulfur*	*All Coal*	*Total Fuels*
New England					
Connecticut	0.00	0.00	0.00	0.00	0.00
Maine	0.00	0.00	0.00	0.00	0.00
Massachusetts	0.00	0.00	0.00	0.00	0.00
New Hampshire	0.00	0.00	0.00	0.00	0.00
Rhode Island	0.00	0.00	0.00	0.00	0.00
Vermont	0.00	0.00	0.00	0.00	0.00
Mid-Atlantic					
New Jersey	0.00	0.00	0.00	0.00	0.00
New York	63.80	181.55	0.00	0.00	245.35
Pennsylvania	290.00	1,636.83	195,544.98	828,336.03	830,262.87
East North Central					
Illinois	928.00	438.14	24,946.38	1,518,624.18	1,519,990.30
Indiana	139.20	70.32	12,732.16	246,444.32	246,653.84
Michigan	475.60	1,599.76	108.56	2,789.52	4,864.88
Ohio	719.20	1,435.10	3,306.24	518,499.11	520,653.41
Wisconsin	0.00	0.00	0.00	0.00	0.00
West North Central					
Iowa	0.00	88.31	29.13	55,247.96	55,336.27
Kansas	2,291.00	12,840.06	0.00	33,786.35	48,917.41
Minnesota	0.00	3.51	0.00	0.00	3.51
Missouri	0.00	19.42	0.00	206,823.14	206,842.55
Nebraska	156.60	59.60	0.00	0.00	216.50
North Dakota	1,003.40	474.67	76,954.92	228,552.84	230,000.91
South Dakota	0.00	0.22	1,546.50	6,420.00	6,420.22
South Atlantic					
Delaware	0.00	0.00	0.00	0.00	0.00
District of Columbia	0.00	0.00	0.00	0.00	0.00
Florida	1,757.40	338.86	0.00	0.00	2,096.26
Georgia	0.00	0.00	7.71	12.85	12.85
Maryland	0.00	46.07	3,436.94	26,666.21	26,712.28
North Carolina	0.00	0.00	0.00	824.20	824.20
South Carolina	0.00	0.00	0.00	0.00	0.00
Virginia	0.00	49.04	56,413.04	96,211.36	96,260.40
West Virginia	185.60	2,485.36	377,668.28	1,061,006.64	1,063,677.61
East South Central					
Alabama	400.20	556.62	16,079.78	76,751.53	77,708.35
Kentucky	214.60	925.87	166,981.96	650,263.67	651,404.14
Mississippi	1,513.80	1,184.10	0.00	0.00	2,697.90
Tennessee	0.00	6.47	5,365.76	25,851.54	25,858.01
West South Central					
Arkansas	614.80	2,318.40	2,067.35	16,948.72	19,881.92
Louisiana	24,516.60	70,265.48	0.00	0.00	94,782.08
Oklahoma	7,145.60	14,689.16	6,941.00	32,665.61	54,500.37
Texas	63,811.60	86,159.15	8,709.36	43,189.08	193,159.83

Table 7A-3 *(cont.)*

| Region and State | Petroleum | Gas | Coal | | Total Fuels |
			< 1 Percent Sulfur	All Coal	
Rocky Mountain					
Arizona	0.00	1.54	3,784.87	7,644.00	7,645.54
Colorado	1,676.20	2,064.22	166,853.16	331,880.54	335,620.97
Idaho	0.00	0.00	0.00	0.00	0.00
Montana	1,200.60	988.73	2,193,533.59	2,339,190.00	2,341,379.28
Nevada	0.00	0.00	0.00	0.00	0.00
New Mexico	3,625.00	13,103.56	82,517.92	101,431.98	118,160.54
Utah	1,455.80	1,131.45	50,078.64	102,841.20	105,428.45
Wyoming	5,237.40	4,297.39	699,271.63	1,099,790.37	1,109,325.16
Far West					
Alaska	58,545.20	34,957.66	261,939.02	266,213.85	359,716.70
California	20,630.60	5,698.48	0.00	0.00	26,329.08
Hawaii	0.00	0.00	0.00	0.00	0.00
Oregon	0.00	0.00	29.01	34.81	34.81
Washington	0.00	18.87	11,671.69	37,790.36	37,809.23

Appendix 7A

Table 7A-4
Reserves Relative to Production and Coal Production Ratios, 1974

Region and State	Petroleum	Gas	Coal	All Fuels	Coal Ratios Coal of <1 Percent Sulfur Relative to All Coal	All Coal Relative to All Reserves
New England						
Connecticut	0.00	0.00	0.00	0.00	0.0000	0.0000
Maine	0.00	0.00	0.00	0.00	0.0000	0.0000
Massachusetts	0.00	0.00	0.00	0.00	0.0000	0.0000
New Hampshire	0.00	0.00	0.00	0.00	0.0000	0.0000
Rhode Island	0.00	0.00	0.00	0.00	0.0000	0.0000
Vermont	0.00	0.00	0.00	0.00	0.0000	0.0000
Mid-Atlantic						
New Jersey	0.00	0.00	0.00	0.00	0.0000	0.0000
New York	12.28	33.17	0.00	22.99	0.0000	0.0000
Pennsylvania	14.38	18.06	385.28	367.25	0.2361	0.9977
East North Central						
Illinois	5.81	278.13	1,145.15	1,021.83	0.0164	0.9991
Indiana	4.88	364.19	447.72	425.88	0.0517	0.9992
Michigan	4.55	21.09	9,999.99[a]	26.97	0.0389	0.5734
Ohio	13.64	14.21	464.16	409.72	0.0064	0.9959
Wisconsin	0.00	0.00	0.00	0.00	0.0000	0.0000
West North Central						
Iowa	0.00	999.99[a]	4,821.86	4,829.57	0.0005	0.9984
Kansas	6.40	13.20	1,933.29	36.29	0.0000	0.6907
Minnesota	0.00	999.99[a]	0.00	9,999.99[a]	0.0000	0.0000
Missouri	0.00	536.22	2,052.20	2,045.06	0.0000	0.9999
Nebraska	4.08	21.51	0.00	5.26	0.0000	0.0000
North Dakota	8.78	13.87	2,144.31	901.80	0.3367	0.9936
South Dakota	0.00	999.99[a]	9,999.99[a]	2,240.75	0.2409	1.0000

Table 7A-4 *(cont.)*

Region and State	Petroleum	Gas	Coal	All Fuels	Coal Ratios	
					Coal of < 1 Percent Sulfur Relative to All Coal	All Coal Relative to All Reserves
South Atlantic						
Delaware	0.00	0.00	0.00	0.00	0.0000	0.0000
District of Columbia	0.00	0.00	0.00	0.00	0.0000	0.0000
Florida	8.34	8.10	0.00	8.30	0.0000	0.0000
Georgia	0.00	0.00	9,999.99[a]	9,999.99[a]	0.6000	1.0000
Maryland	0.00	315.77	448.52	448.20	0.1289	0.9983
North Carolina	0.00	0.00	9,999.99[a]	9,999.99[a]	0.0000	1.0000
South Carolina	0.00	0.00	0.00	0.00	0.0000	0.0000
Virginia	0.00	6.30	106.33	105.48	0.5863	0.9995
West Virginia	12.01	11.20	386.39	356.54	0.3560	0.9975
East South Central						
Alabama	5.18	18.21	150.41	125.72	0.2095	0.9877
Kentucky	4.72	11.74	186.16	180.08	0.2568	0.9982
Mississippi	5.14	13.70	0.00	7.08	0.0000	0.0000
Tennessee	0.00	346.87	130.84	127.98	0.2076	0.9997
West South Central						
Arkansas	6.41	17.05	1,463.08	81.67	0.1220	0.8525
Louisiana	5.73	8.26	0.00	7.42	0.0000	0.0000
Oklahoma	6.93	8.17	549.32	18.87	0.2125	0.5994
Texas	8.72	9.61	425.81	11.79	0.2017	0.2236
Rocky Mountain						
Arizona	0.00	6.25	54.28	52.60	0.4951	0.9998
Colorado	7.71	13.01	2,156.21	633.10	0.5028	0.9889
Idaho	0.00	0.00	0.00	0.00	0.0000	0.0000
Montana	5.99	16.43	7,684.40	4,143.91	0.9377	0.9991

Appendix 7A

Nevada	0.00	0.00	0.00	0.00	0.0000	0.0000
New Mexico	6.33	9.60	467.93	54.84	0.8135	0.8584
Utah	6.38	20.41	690.08	243.62	0.4870	0.9755
Wyoming	6.45	11.99	2,576.25	694.53	0.6358	0.9914
Far West						
Alaska	142.97	247.15	9,999.99[a]	634.49	0.9839	0.7401
California	11.01	14.22	0.00	11.58	0.0000	0.0000
Hawaii	0.00	0.00	0.00	0.00	0.0000	0.0000
Oregon	0.00	0.00	9,999.99[a]	9,999.99[a]	0.8333	1.0000
Washington	0.00	999.99[a]	499.36	499.61	0.3089	0.9995

[a]Some states with small 1974 production and some reserves had a very large ratio of reserves to production; for those cases, "artificial" ratios of 999.99 for gas and 9,999.99 for coal and total Btu were employed.

Appendix 7B:
1970 Earnings by SMSA for All Industries, Coal, Crude Petroleum and Natural Gas, Petroleum Refining, and Selected Manufacturing Industries

Table 7B-1
1970 Earnings by SMSA for All Industries

		1970 Earnings in Millions of 1967 Dollars						
		Mining			Manufacturing			
Region, State, SMSA	Total, All Industries	Coal	Crude Petroleum and Natural Gas	Petroleum Refining	Chemicals	Primary Metals	All Manufacturing	Population, Midyear, 1970 (000s)
New England								
Connecticut								
Bridgeport-Norwalk-Stamford	2,611.3	0.0	0.0[a]	3.9	49.2	56.2	1,084.2	794.6
Hartford-New Britain	3,159.2	0.0	0.0[a]	0.6	12.3	24.2	1,221.2	818.6
New Haven-Waterbury-Meriden	2,327.1	0.0	0.3[a]	1.3	44.0	115.9	819.7	746.6
Norwich-Groton-New London	639.2	0.0	0.0	0.0	D	D	238.2	231.2
Maine								
Lewiston-Auburn	193.9	0.0	0.0	0.0	0.4	1.7	77.9	91.4
Portland-South Portland	514.6	0.0	0.0	0.1	D	2.0	115.9	192.8
Massachusetts								
Boston	12,033.2	0.0	0.1	7.3	107.3	43.9	3,080.9	3,715.1
Fall River-New Bedford	1,029.9	0.0	0.0	0.2	5.1	29.7	459.1	445.1
Pittsfield	412.8	0.0	0.0	0.0	D	D	187.4	149.7
Springfield-Chicopee-Holyoke	1,517.2	0.0	0.0	0.3	D	30.0	510.4	584.0
Worcester-Fitchburg-Leominster	1,653.1	0.0	0.0	0.8	15.0	63.8	697.5	639.1
New Hampshire								
Manchester	639.2	0.0	0.0	0.1	3.9	6.7	255.8	225.3
Rhode Island								
Providence-Pawtucket-Warwick	2,090.8	0.0	0.0	1.1	D	60.0	726.6	770.8
Vermont								
Burlington	285.8	0.0	0.0	0.0	D	D	95.7	99.6

Appendix 7B

Mid-Atlantic						
New Jersey						
Atlantic City	423.4	0.0	0.0	D	61.7	175.7
Jersey City	2,134.8	0.0	13.6	86.4	875.6	611.5
Long Branch	968.0	0.0	0.3	8.1	147.9	461.1
New Brunswick	1,710.2	0.1	31.5	201.9	803.8	586.0
Newark	6,812.0	0.0[a]	24.8	477.6	2,177.3	1,863.5
Paterson-Clifton-Passaic	4,302.8	0.0	5.7	219.0	1,587.5	1,363.9
Trenton	1,021.5	0.1	2.7	25.1	321.7	305.1
Vineland	346.7	0.0	0.0	D	149.5	121.8
New York						
Albany-Schenectady-Troy	2,170.6	0.0	0.2	38.6	538.0	722.8
Binghamton	830.0	0.0	0.2	D	373.5	303.0
Buffalo	3,877.9	0.2	D	136.4	1,544.2	1,350.6
Elmira	275.5	0.0	0.0	D	115.4	101.6
New York City	42,837.1	15.5	154.1	634.8	8,882.9	11,587.6
Poughkeepsie	679.6	1.0[a]	18.2	0.1	318.1	222.5
Rochester	2,881.1	0.6[a]	0.5	13.1	1,412.6	883.6
Syracuse	1,734.9	0.0	0.5	51.2	536.2	637.3
Utica-Rome	917.4	0.0[a]	0.2	D	301.0	341.0
Pennsylvania						
Allentown	1,613.3	0.7	D	27.8	807.5	544.7
Altoona	311.6	0.1	0.2	0.4	98.0	135.6
Erie	714.9	0.0	2.6	1.4	349.2	264.2
Harrisburg	1,246.9	1.0[a]	0.6	D	284.6	411.4
Johnstown	542.7	54.9	0.2	D	186.1	263.3
Lancaster	881.1	0.0	0.4	D	400.3	320.3
Philadelphia	14,457.1	0.8[a]	224.7	348.4	4,617.6	4,829.0
Pittsburgh	6,877.9	84.8	D	1,128.0	2,564.2	2,405.9
Reading	854.7	0.0[a]	2.7	69.8	397.8	297.0
Scranton	523.1	3.3[a]	0.0	5.4	190.9	234.6
Wilkes-Barre	755.0	15.2	1.7	6.8	288.7	343.0
Williamsport	287.4	0.0	0.3	11.4	127.7	113.5
York	948.5	0.0[a]	2.7	D	418.2	330.2

Table 7B-1 (cont.)

1970 Earnings in Millions of 1967 Dollars

		Mining			Manufacturing			
Region, State, SMSA	Total, All Industries	Coal	Crude Petroleum and Natural Gas	Petroleum Refining	Chemicals	Primary Metals	All Manufacturing	Population, Midyear, 1970 (000s)

East North Central

Illinois

Bloomington	279.8	0.0[a]	0.0	5.6	0.0	0.0	49.3	104.6
Champaign-Urbana	449.2	0.0	0.2[a]	0.0	0.2	D	37.6	163.6
Chicago	24,893.6	1.5[a]	14.2[a]	118.2	504.9	646.7	8,126.0	6,993.4
Decatur	435.5	0.0	0.3	D	1.8	D	177.3	125.3
Peoria	1,088.9	4.0	0.1	0.0	4.1	62.5	469.1	342.7
Rockford	879.8	0.0	0.1	1.3	5.4	D	469.8	272.6
Springfield	503.0	0.0	0.4	D	D	D	88.2	161.7

Indiana

Anderson	384.3	0.0	0.0	0.0	D	D	227.0	138.8
Evansville	658.1	7.5	5.1	0.0	D	33.2	252.3	233.4
Fort Wayne	962.0	0.0	0.0	0.3	D	32.5	413.6	281.2
Gary-Hammond-E. Chicago	1,963.0	0.0	0.0	81.8	35.9	D	1,043.3	635.1
Indianapolis	3,458.7	0.0[a]	3.8[a]	6.2	132.8	D	1,135.7	1,112.9
Lafayette-W. Lafayette	321.9	0.0	0.0	1.1	D	10.5	95.4	109.7
Muncie	348.3	0.0	0.0	0.0	0.0	15.7	169.9	129.6
South Bend	744.4	0.0	0.0	3.1	D	9.6	280.3	280.8
Terra Haute	428.0	10.3	0.7	0.8	D	D	112.7	175.6

Michigan

Ann Arbor	837.4	0.0	0.1	0.0	7.8	D	348.5	234.8
Battle Creek	422.6	0.0	0.3[a]	D	0.8	D	181.3	142.4
Bay City	252.7	0.0	0.4[a]	5.8	D	4.6	100.4	117.7
Detroit	14,826.7	0.0	0.8[a]	18.2	194.7	548.0	6,446.2	4,212.2
Flint	1,493.9	0.0	0.4	0.0	D	D	802.2	498.1
Grand Rapids	1,514.3	0.0	0.4	0.0	24.9	D	604.3	540.8

Appendix 7B 319

Jackson	419.5	0.0	0.6	D	D	7.0	180.4	143.7
Kalamazoo	601.7	0.0	0.0	0.7	59.3	2.7	277.1	202.1
Lansing	1,078.8	0.0	0.1	0.0	5.5	D	385.7	379.5
Muskegon	418.2	0.0	0.4	1.8	9.2	60.3	218.6	157.9
Saginaw	626.3	0.0	0.0	0.1	6.1	D	317.0	220.4

Ohio

Akron	1,967.7	0.0	1.0[a]	0.8	D	D	908.6	681.5
Canton	1,045.1	0.5	3.0	4.4	4.7	182.6	512.3	373.5
Cincinnati	4,074.0	1.9[a]	0.4	24.5	195.2	31.2	1,502.9	1,389.1
Cleveland	6,884.2	6.0[a]	2.9	25.3	174.9	361.8	2,742.3	2,071.2
Columbus	2,778.0	0.7[a]	4.2	D	42.7	D	752.4	919.3
Dayton	2,847.7	0.0	1.1	D	14.6	D	1,186.2	853.1
Hamilton	593.5	0.0	0.1[a]	D	0.4	97.0	284.7	227.0
Lima	491.1	0.0	1.0[a]	12.2	D	D	207.0	172.0
Lorain-Elyria	674.3	0.0	0.2[a]	0.5	19.5	87.4	360.3	257.7
Mansfield	397.5	0.0	1.2	D	D	23.2	199.4	130.4
Springfield	402.8	0.0	0.0	0.4	D	3.1	160.1	157.6
Steubenville	483.3	12.4	0.3[a]	0.1	5.8	230.3	279.2	166.0
Toledo	2,019.1	0.1[a]	0.9[a]	28.7	20.6	54.7	787.5	694.9
Youngstown	1,570.3	0.8	1.5	0.1	2.3	432.5	788.9	537.8

Wisconsin

Appleton	759.1	0.0	0.0	0.0	1.0	26.2	327.5	277.8
Green Bay	396.8	0.0	0.0	0.0	0.9	1.4	141.6	158.8
Kenosha	309.1	0.0	0.0	0.0	0.2	14.2	138.3	118.3
La Crusse	215.1	0.0	0.0	0.0	0.4	D	60.6	80.7
Madison	824.4	0.0	0.0	0.0	D	D	145.6	291.3
Milwaukee	4,433.3	0.0	0.1	1.4	30.0	177.7	1,824.8	1,408.7
Racine	454.8	0.0	0.0	0.0	28.3	19.4	223.4	171.4

West North Central

Iowa

Cedar Rapids	493.7	0.0	0.0	0.1	0.4	0.0	219.6	163.5
Davenport-Rock Island-Moline	1,026.3	1.1[a]	0.1	0.0	D	D	390.4	363.3
Des Moines	967.3	0.0	0.0	0.2	5.6	0.3	218.0	286.6
Dubuque	257.8	0.0	0.0	0.0	1.4	0.0	119.5	90.8
Sioux City	299.4	0.0	0.0	0.0	D	D	74.8	116.4
Waterloo	366.0	0.0	0.0	0.0	0.1	D	158.7	133.2

Table 7B-1 *(cont.)*

		1970 Earnings in Millions of 1967 Dollars						
		Mining			Manufacturing			
Region, State, SMSA	Total, All Industries	Coal	Crude Petroleum and Natural Gas	Petroleum Refining	Chemicals	Primary Metals	All Manufacturing	Population, Midyear, 1970 (000s)
Kansas								
Topeka	434.5	0.0	0.0[a]	0.0	D	0.0	71.5	155.2
Wichita	1,080.9	0.0	23.3	20.6	9.4	5.2	330.2	389.2
Minnesota								
Duluth	641.6	0.0	0.2	2.7	D	D	84.2	266.5
Minneapolis	6,297.5	0.0	0.5[a]	23.0	42.6	37.6	1,842.4	1,821.7
Rochester	260.1	0.0	0.0	0.0	D	0.0	67.2	84.5
Missouri								
Columbia	171.4	0.0	0.0	0.0	1.2	0.0	15.1	81.2
Kansas City	4,034.0	0.1	0.6	11.2	83.5	56.9	1,067.8	1,259.3
St. Joseph	237.6	0.0	0.0	0.0	4.2	0.0	78.8	87.2
St. Louis	7,203.2	15.0[a]	1.8	D	279.7	D	2,498.1	2,370.2
Springfield	351.3	0.0	0.0[a]	0.0	3.0	0.8	96.1	153.4
Nebraska								
Lincoln	441.1	0.0	0.0	0.0	D	D	72.8	168.7
Omaha	1,622.5	0.0	0.2	D	D	D	306.2	545.3
North Dakota								
Fargo-Moorehead	282.9	0.0[a]	0.0[a]	0.1	D	0.0	19.8	120.5
South Dakota								
Sioux Falls	236.0	0.0	0.1	0.0	D	0.0	50.0	95.2
South Atlantic								
Delaware								
Wilmington	1,733.4	0.0	0.1[a]	D	D	D	739.7	501.3

Appendix 7B

District of Columbia						
Washington	10,087.2	0.0	0.0[a]	1.7	13.9	391.8	2,862.9
Florida						
Daytona Beach	293.7	0.0	0.0	0.0	D	40.7	170.9
Fort Lauderdale-Hollywood	1,259.3	0.0[a]	0.0	D	D	151.7	625.2
Fort Myers	186.1	0.0	0.1	0.0	D	10.6	106.1
Gainsville	230.8	0.0	0.0	0.0	D	21.7	105.6
Jacksonville	1,570.5	0.0	0.0	1.7	9.4	175.3	533.2
Lakeland-Winter Haven	568.1	0.0	0.0	0.0	26.9	100.7	229.1
Melbourne-Titusville-Cocoa	597.5	0.0	0.0	0.5	0.5	151.7	231.9
Miami	3,928.3	0.0	0.5[a]	2.5	16.5	474.0	1,278.2
Orlando	1,088.4	0.0	0.0	0.3	D	166.8	431.5
Pensacola	589.8	0.0	0.1[a]	0.1	D	115.9	245.1
Sarasota	245.4	0.0	0.0	0.0	D	24.9	121.4
Tallahassee	245.3	0.0	0.0	0.0	D	11.3	103.9
Tampa-St. Petersburg	2,205.8	0.0	0.2	1.4	8.2	375.7	1,020.9
West Palm Beach	881.8	0.0	0.0	D	2.5	D	176.8	351.6
Georgia						
Albany	205.8	0.0	0.0	0.0	2.8	42.1	89.9
Atlanta	4,664.7	0.0	0.3	1.5	45.2	976.1	1,393.9
Augusta	678.1	0.0	0.0	0.0	D	204.1	254.1
Columbus	607.6	0.0	0.0	0.0	1.0	D	239.2
Macon	545.7	0.0	0.0	0.4	1.7	87.3	206.9
Savannah	505.3	0.0	0.0	4.8	11.9	119.3	188.3
Maryland						
Baltimore	6,384.9	0.0[a]	0.4	7.8	104.3	1,611.4	2,078.4
North Carolina						
Asheville	346.4	0.0	0.0	D	D	113.0	145.3
Charlotte	1,377.9	0.0	0.0	0.1	36.3	299.8	410.1
Durham	477.7	0.0	0.0	0.0	D	125.7	190.7
Fayetteville	521.2	0.0	0.0	0.1	D	48.9	212.4
Gastonia	349.4	0.0	0.0[a]	0.0	3.9	196.8	148.7
Greensboro-W. Salem-H. Point	1,909.5	0.0	0.0	0.2	17.8	810.0	605.0
Raleigh	613.5	0.0	0.0	0.0	D	101.2	228.9
Wilmington	253.3	0.0	0.0	D	D	69.7	107.4

Table 7B-1 *(cont.)*

1970 Earnings in Millions of 1967 Dollars

Region, State, SMSA	Total, All Industries	Mining			Manufacturing			Population, Midyear, 1970 (000s)
		Coal	Crude Petroleum and Natural Gas	Petroleum Refining	Chemicals	Primary Metals	All Manufacturing	
South Carolina								
Charleston	679.1	0.0	0.0	2.1	4.8	D	109.9	304.5
Columbia	813.3	0.0	0.0	D	17.5	D	119.0	323.6
Greenville	779.2	0.0	0.0	0.0	28.4	D	317.5	300.1
Spartanburg	433.1	0.0	0.0	0.0	15.4	0.4	192.5	174.1
Virginia								
Lynchburg	349.3	0.1	0.0	0.0	27.3	D	160.9	123.6
Newport News	838.9	0.0	0.0	3.8	0.3	D	205.9	292.4
Norfolk	1,834.0	0.0	0.0	0.0	D	3.4	139.3	681.3
Petersburg	322.7	0.0	0.0	0.0	D	0.0	94.5	113.8
Richmond	1,659.0	0.0	0.0	0.0	87.6	D	394.5	533.9
Roanoke	525.8	0.0	0.0	0.0	1.3	5.9	125.3	181.6
West Virginia								
Charleston	653.7	33.2	5.8[a]	1.2	128.0	D	165.7	229.7
Huntington	635.5	1.3	2.6	11.1	19.5	106.6	232.4	254.2
Parkersburg	349.7	0.1[a]	1.7	0.7	65.8	D	151.0	144.3
Wheeling	438.3	49.5	1.4	1.3	D	D	119.9	183.1
East South Central								
Alabama								
Birmingham	1,865.8	30.0[a]	0.4	3.5	9.4	D	557.9	740.7
Florence	247.9	0.0	0.0	0.2	D	D	95.9	118.0
Gadsden	206.4	0.0	0.0	0.2	0.1	D	97.5	94.3
Huntsville	541.5	0.0	0.0	0.5	D	1.4	93.4	228.7
Mobile	779.3	0.0	0.8[a]	3.5	D	D	184.4	377.4
Montgomery	474.3	0.0	0.0	0.1	D	D	60.4	201.7
Tuscaloosa	224.1	1.0	0.0	2.4	D	D	69.4	116.3

Appendix 7B

Kentucky						
Lexington	528.7	0.0	0.1[a]	0.2	D	134.9	174.6
Louisville	2,530.7	0.1	0.3	1.6	93.5	962.5	827.9
Owensboro	182.4	1.2[a]	2.1	0.0	D	60.3	79.6
Mississippi						
Biloxi	309.5	0.0	0.9	0.5	1.6	24.4	134.5
Jackson	614.1	0.0	8.9	D	D	85.9	258.8
Tennessee						
Chattanooga	861.4	0.0	0.0	D	72.0	365.0	305.6
Knoxville	996.5	6.4	0.0	0.3	D	351.3	401.1
Memphis	2,003.7	0.4	0.0	6.5	47.2	444.7	771.6
Nashville	1,539.4	0.0	0.0	D	48.2	403.9	542.1
West South Central						
Arkansas						
Fort Smith	285.5	0.5[a]	2.4	0.0	0.7	91.9	160.8
Little Rock	845.6	0.0	0.1[a]	D	D	173.1	323.8
Pine Bluff	169.8	0.0	0.0	0.0	0.3	39.8	85.4
Louisiana						
Alexandria	217.3	0.0	1.0	0.4	1.8	27.5	118.1
Baton Rouge	744.0	0.0	3.2	44.3	99.8	191.0	285.2
Lafayette	240.0	0.0	47.6	0.0	0.0	13.1	111.8
Lake Charles	336.3	0.0	13.2[a]	52.7	33.8	101.0	145.4
Monroe	230.2	0.0	1.6[a]	0.0	9.3	49.3	115.4
New Orleans	2,766.5	0.0	151.9	14.5	13.7	414.3	1,046.7
Shreveport	688.4	0.0	29.5[a]	5.6	3.2	105.1	294.0
Oklahoma						
Lawton	284.9	0.0	0.1	0.2	0.0	9.1	108.7
Oklahoma City	1,815.9	0.0	64.8	7.6	D	279.1	644.1
Tulsa	1,362.2	0.0	122.9	17.8	3.2	334.8	478.4

Table 7B-1 *(cont.)*

		1970 Earnings in Millions of 1967 Dollars						
		Mining			Manufacturing			
Region, State, SMSA	Total, All Industries	Coal	Crude Petroleum and Natural Gas	Petroleum Refining	Chemicals	Primary Metals	All Manufacturing	Population, Midyear, 1970 (000s)
Texas								
Abilene	275.3	0.0	9.7[a]	D	0.2	0.0	26.3	114.5
Amarillo	359.0	0.0	9.2	2.6	D	D	43.1	145.1
Austin	703.0	0.0	0.8	D	D	0.1	87.3	297.0
Beaumont	904.0	0.0	11.8[a]	188.0	98.8	D	381.9	317.6
Brownsville	230.2	0.0	0.6[a]	0.1	2.4	0.0	22.3	141.1
Bryan	110.4	0.0	0.2	0.0	D	0.0	9.7	58.3
Corpus Christi	635.1	0.0	37.5	22.3	D	D	100.7	286.3
Dallas	5,006.3	0.0	82.5[a]	D	45.8	17.3	1,259.3	1,563.9
El Paso	801.7	0.0	0.0[a]	5.8	0.9	D	126.1	361.1
Fort Worth	2,040.7	0.0	22.5	D	12.0	17.3	737.0	766.0
Galveston	400.7	0.0	2.2[a]	33.0	D	2.0	127.6	170.7
Houston	5,882.6	0.0	318.3	191.3	278.6	79.6	1,366.3	1,995.2
Killeen	438.0	0.0	0.1[a]	0.0	D	D	24.8	160.6
Laredo	122.8	0.0	0.8	0.1	0.0	D	5.1	73.2
Lubbock	431.1	0.0	0.7[a]	0.3	D	1.0	44.4	180.2
McAllen	236.0	0.0	6.9[a]	0.3	D	0.1	16.0	182.5
Midland	205.1	0.0	71.6	D	0.8	0.0	9.7	65.8
Odessa	232.6	0.0	42.1[a]	1.3	16.8	0.4	31.6	92.3
San Angelo	165.4	0.0	2.1	0.0	0.0	D	19.4	71.4
San Antonio	2,000.8	0.0	11.2	3.3	D	D	203.3	868.4
Sherman-Denison	199.0	0.0	2.5	0.0	0.1	0.7	64.4	83.7
Texarkana	242.2	0.0	0.7[a]	0.2	D	0.0	63.2	101.6
Tyler	237.5	0.0	9.7	3.2	D	D	75.5	97.6
Waco	344.8	0.0	0.1	0.1	D	0.0	81.7	148.3
Wichita Falls	340.0	0.0	28.3[a]	D	D	1.4	D	128.3
Rocky Mountain								
Arizona								
Phoenix	2,681.9	0.0	0.3	0.7	5.0	33.9	603.6	979.2
Tucson	838.2	0.0	0.6	0.0	D	1.1	71.3	355.5

Appendix 7B

Colorado							
Colorado Springs	599.7	0.0	0.0	0.0	D	D	43.9	237.9
Denver	3,797.5	0.9	56.9	7.6	15.6	9.3	714.7	1,237.4
Pueblo	273.9	0.0	0.0	0.0	0.2	D	77.6	119.2
Idaho							
Boise City	290.8	0.0	0.0	0.0	0.1	D	37.8	112.9
Montana							
Billings	211.6	0.5[a]	2.7	13.1	D	0.0	29.8	87.7
Great Falls	209.6	0.0	0.1	0.0	0.0	D	21.3	82.1
Nevada							
Las Vegas	880.8	0.0	0.2	0.0	D	10.1	38.7	275.7
Reno	445.2	0.0	0.3	0.0	D	D	22.6	122.1
New Mexico							
Albuquerque	786.0	0.0	1.2[a]	0.0	0.4	D	61.6	316.4
Utah							
Ogden	302.4	0.0[a]	0.0[a]	0.0	D	D	34.7	127.2
Provo	233.9	0.0	0.2[a]	0.0	D	D	73.8	139.9
Salt Lake City	1,452.1	0.1	4.0[a]	10.2	6.1	25.8	229.9	563.0
Wyoming							
Cheyenne	151.0	0.0	1.0	2.1	D	0.0	7.9	56.6
Far West							
Alaska							
Anchorage	730.5	0.1[a]	3.0[a]	1.1	2.7	0.0	36.6	178.1
California							
Anaheim	3,422.6	0.0	15.7	11.0	25.5	13.8	1,164.0	1,431.1
Bakersfield	881.3	0.0	53.7	9.9	2.6	D	69.4	329.7
Fresno	1,043.7	0.0	6.9	1.5	5.9	2.0	127.5	413.6
Los Angeles-Long Beach	25,317.6	0.0	96.9	209.6	266.3	248.1	7,364.7	7,029.4
Modesto	471.0	0.0	0.1	0.0	D	0.2	111.0	195.2
Oxnard-Simi V.-Ventura	771.4	0.0	16.7	1.8	3.6	D	116.6	378.8
Riverside	2,571.1	0.0	0.1[a]	0.6	D	D	451.9	1,147.7

Table 7B-1 *(cont.)*

Region, State, SMSA	Total, All Industries	Mining		1970 Earnings in Millions of 1967 Dollars	Manufacturing			Population, Midyear, 1970 (000s)
		Coal	Crude Petroleum and Natural Gas	Petroleum Refining	Chemicals	Primary Metals	All Manufacturing	
Sacramento	2,214.7	0.0	1.0[a]	1.8	7.4	D	210.3	804.0
Salinas	839.0	0.0	1.5	0.0	4.1	D	54.7	251.4
San Diego	3,775.6	0.0	0.0[a]	D	D	D	611.9	1,357.6
San Francisco	11,556.1	0.0	7.1	175.0	127.8	141.1	1,946.6	3,112.0
San Jose	3,230.8	0.0	0.1	1.0	21.0	7.2	1,265.3	1,071.8
Santa Barbara	648.2	0.0	9.1	1.3	D	D	90.8	264.5
Santa Cruz	253.7	0.0	0.1	0.1	0.1	D	45.9	124.0
Santa Rosa	420.5	0.0	0.1[a]	0.0	0.3	D	52.8	205.1
Stockton	778.2	0.0	0.3[a]	0.2	6.1	1.2	137.6	291.2
Vallejo	678.3	0.0	1.2[a]	5.5	D	D	60.5	251.0
Hawaii								
Honolulu	2,187.6	0.0	0.0	2.6	3.8	1.8	129.5	621.8
Oregon								
Eugene	468.2	0.0	0.0	0.7	D	D	150.0	214.4
Portland	2,949.1	0.0	0.0	4.5	16.2	62.9	706.0	1,013.8
Salem	405.1	0.0	0.0	0.2	D	0.0	74.7	187.6
Washington								
Richland-Kennewick	265.8	0.0	0.0	0.1	D	D	50.4	93.5
Seattle-Everett	4,453.7	0.2[a]	0.2	D	D	D	1,297.6	1,423.9
Spokane	746.7	0.0	0.0	0.2	1.1	D	107.1	287.9
Tacoma	1,132.9	0.0	0.0	2.1	10.1	17.3	173.3	411.6
Yakima	322.9	0.0	0.0	0.0	0.4	0.2	46.6	145.2

Source: Bureau of Economic Analysis, Department of Commerce, *Population and Economic Activity in the United States and Standard Metropolitan Statistical Areas*, EPA, HUD, Washington, 1972.
D: data deleted in source document to avoid disclosure of information pertaining to an individual establishment. Earnings are on a where-earned basis.
[a]Inference based on available data.

**Part V
Regional Differences in
Metropolitan and Rural
Areas**

8 Regional Variations in Metropolitan Growth and Development

Charles L. Leven

The notion that regional shifts in population are heavily, if not mainly, influenced by the magnetism of the Sunbelt has gained considerable currency in the 1970s. According to a recent study, for example, "The South's dramatic population growth during the 1970's reflects the overall national trend of the rise of the Sunbelt and a relative decline of the older, industrial regions of the Northeast and Midwest."[1] The implication of this and other observations on recent growth is that the shift to the Sunbelt represents an important, new phenomenon.

The original intent of this chapter was to demonstrate that the attractive image of the Sunbelt was more mirage than substance; not that the South and West had not shown higher rates of population growth than the rest of the United States in the 1970s, but that the higher rates were due to compositional differences in the types of settlement areas in the various regions, with the higher observed rates in the so-called Sunbelt simply an artifact of aggregation. My supposition was that the South and West were underrepresented in really big metropolitan areas and very remote rural population which were growing slowly everywhere, and overrepresented in small town and medium metropolitan area populations, which were doing well everywhere. As data were compiled, however, they failed to support that view. However they were disaggregated, by big SMSA or small, by central city or suburbs, by metropolitan versus nonmetropolitan areas, the South showed more and the West still more rapid growth in the 1970s than did similar areas in the rest of the United States. And these results held up more or less for Census subregions as well. Many other observers have reached substantially the same conclusions. For example, "The South is experiencing net migration gains from all other regions in numbers that are more than double the net gain to the West—the only other region now experiencing a net migration gain."[2]

However, when the analysis for this chapter was extended back to 1940, two additional and somewhat surprising conclusions emerged. First, while the shift to the South and West does seem real, it is not new; it extends at least to the period of World War II. Second, and even more surprising, to the extent that population shifts have been influenced by the pull of the so-called Sunbelt, that pull has become a less, not more, dominant factor during the 1970s.

This raises a most intriguing point: Why it is that we see as "new" an explanation that seems to have been with us for at least thirty-five years and, if anything, has become a weaker influence than it was. In part this may be due to a growing tendency to explain population shifts, of whatever kind, in terms of a shift in the supposed underlying preferences for various life-styles, ignoring more simple economic determinism. Just why this should be so is probably best left to other economists to explain; but if it is so, it would seem to have important implications for the issues raised by this symposium surrounding potential regional confrontation. Accordingly, in the next section the shift in perspective on regional population shift is outlined. Within that perspective we then look at the specific post-1940 experience in the following section, particularly in terms of its metropolitan and nonmetropolitan dimensions. In the concluding section, some implications for potential regional confrontation are discussed.

Changing Perceptions of Regional shifts

The first kind of major shift in U.S. population was a general westward movement, beginning with the opening of the Cumberland Gap and accelerating to the territories beyond the Missouri after the Civil War. For the most part this was a purely regional shift, not a by-product of urbanization, although it was accompanied by increasing urbanization, especially in the post-Civil War period. In any event there was little doubt as to just why it was that Horace Greeley was urging young men to "Go West!" It was not to see the "Wonders of the West" or to seek out the superior life-style of river and prairie; it was to seek superior economic opportunity, in technical terms a response to interregional disequilibrium in land-labor ratios.

By the same token, the flocking of youth from countryside to city during the latter part of the nineteenth and early part of the twentieth century was seen not as the outcome of their search for a better way of life, but as a response to their redundancy in the rural labor force.[3] True, by about the time of World War I there was concern with, "How ya' gonna' keep 'em down on the farm, after they've seen Paree!" but at least until then moving to the city was seen more as an unfortunate social consequence of personal economic necessity.

And the reverse migration of the 1930s was not seen as reflecting a triumph of rural virtue over urban vice; it was simply that rural subsistence became the only economically meaningful alternative to industrial unemployment for large numbers of people.

It was probably somewhere in the early post-World War II period that we began to develop serious astigmatism in viewing regional and rural-urban shifts in population. An early example of confusing simple economic optimization with the pursuit of deeper social values was the view of suburbanization as the pursuit by middle-class citizens of what they apparently saw as a necessarily superior life-style.[4] It may have really been so that people preferred "the little white house in suburbia," but as I have pointed out elsewhere, it was also true that "it was the cheapest housing deal in town" for the middle class.[5]

As we moved further into the postwar period, our perceptions of the forces behind population shift took on still more romantic overtones. The early and middle 1960s saw the emergence of what might be called a "lemming theory" of regional preferences, a transfer of population from interior to periphery. But in the 1970s, as Alonso has shown, the aggregate of counties on or within 50 miles of the sea has actually *lost* population.[6]

By the late 1960s population press accounts would have had us believe that glimmers of population growth in nonmetropolitan areas were a manifestation of an emergent preference for a more natural life-style. But again, Alonso has pointed out all too clearly that "flower children" seeking their version of Walden Pond have always been with us, but alas, or fortunately as the case may be, never in very large numbers.[7]

Most recently, we see yet another "pursuit of life-style" argument to explain population shift, in this instance a pursuit of the sun. Like all the post-1940 explanations of regional and rural-urban shift, it contains an element of economic rationale, but as in other recent explanations, the Sunbelt mystique is one couched in social and environmental values as much as or more than in economic motive.[8]

In the next section we present some data which show that the apparent relative attractiveness of the South and/or West, if real, has been with us at least since 1940, with any real preference for the sunny climates *less* apparent since 1970.

Regional and Metropolitan Population Shift since 1940

As indicated above, table 8-1 shows clearly that the South and West have had much more rapid population growth than the Northeast and North Central regions since 1940. And the relative differentials have remained remarkably constant. From 1940 to 1950 the South grew at 1.2 and the West at 1.7 times the national rate; in 1970-1975 the South grew at 1.3 and the West at 1.8 times the U.S. average.[9] Even at a subregional level there has been little change in the general pattern of differential changes. In the 1940s the West South Central region was the fastest-growing region in the South, but since 1950 it has been the South Atlantic region. More recently, since 1970 the decline in growth in the East North Central region has been much greater than the decline for West North Central, and the Mountain region

Table 8-1
U.S. Population Growth by Region
(annual percentage rate of change)

	1940-1950	1950-1960	1960-1970	1970-1975
United States[a]	2.14	2.26	1.36	1.30
Northeast	1.09	1.56	1.01	0.64
New England	1.26	1.64	1.18	0.78
Middle Atlantic	0.90	1.47	0.82	0.57
North Central	1.56	1.76	1.06	0.55
East North Central	1.84	1.90	1.16	0.48
West North Central	1.03	1.50	0.88	0.71
South	2.61	2.59	1.38	1.69
South Atlantic	2.67	3.02	1.86	1.97
East South Central	1.70	1.46	0.86	1.42
West South Central	3.05	2.71	1.10	1.51
West	3.75	3.50	2.37	2.40
Mountain	3.25	3.71	2.09	3.18
Pacific	4.28	3.27	2.60	1.81
F statistic for region	24.8	11.7	9.5	28.4
F statistic for subregion	11.9	6.0	5.2	13.4

[a]Continental United States, excluding Hawaii and Alaska.

has replaced the Pacific region as the fastest-growing area in the West. Throughout, however, an analysis of regional changes shows that both region and subregion have been significant differentiators of growth rates in all periods. (Note all F statistics are greater than 4.0.)

In table 8-2 we can see that metropolitan scale also is associated with differences in growth rates, but not as consistently as is "region." Note that there is no significant difference in growth rates in SMSAs of different size (F statistics less than 4.0) after 1960. Also note the consistent strengthening of non-SMSA to SMSA growth trends since 1940. In 1940-1950, for example, SMSAs grew at more than 3 times the rate for non-SMSA sections of the country. In 1950-1960 the SMSA rate was only 2.5 times and in 1960-1970 well under 2 times the rate for non-SMSAs. In the 1970-1975 period, nonmetropolitan areas actually grew more rapidly than did all SMSAs combined.

Table 8-3 reveals some interesting interactions between the impact of metropolitan scale and region on rates of population increase. An analysis of variance that keeps size class of metropolitan area constant always shows the variance by region to be significant for each of the four periods for each of the three SMSA classes. On the other hand, the significance of SMSA size class, keeping region constant, shows size class sometimes significant

Table 8-2
U.S. Population Growth by Class of Metropolitan Area
(annual percentage rate of change)

	1940-1950	1950-1960	1960-1970	1970-1975
U.S.	2.14	2.26	1.36	1.30
Non-SMSA	0.68	0.95	0.78	1.46
SMSA, all (central-city part)	2.22 (2.15)	2.31 (1.96)	1.38 (1.07)	1.26 (0.58)
Class I	2.67 (2.74)	2.72 (2.67)	1.52 (1.38)	1.34 (1.06)
Class II	2.18 (1.42)	2.28 (1.46)	1.30 (0.96)	1.36 (0.74)
Class III	0.84 (0.90)	1.33 (−0.45)	1.19 (−0.18)	0.93 (−0.91)
F statistic for SMSA class	14.5	8.2	1.1	2.2
F statistic for central-city size	11.7	10.8	5.6	14.5

Note: Classes I and II are for SMSAs of less than 1 million with central cities of less than or more than 100,000, respectively. Class III is SMSAs of over 1 million. Change for 1940-1950 and 1950-1960 is based on 1960 boundaries. Change for 1960-1970 is based on 1970 boundaries. Change for 1970-1975 is based on 1975 boundaries.

but frequently not, although size has emerged more consistently as significant in recent periods, especially in the older Northeast and North Central regions. This seems to reflect the stagnation of the older metropolises which are more concentrated in the older industrial regions of the nation. In any event, region does appear to be a more consistent and more persistent influence than metropolitan population size.

Perhaps even more interesting is the convergence between non-SMSA and SMSA growth rates by region. With the exception of a rise in the SMSA growth rate in the South for the 1970-1975 period as compared with 1960-1970, SMA growth rates have declined consistently since 1940 in all four major Census regions. Non-SMSA growth rates, on the other hand, generally have been increasing; moreover, non-SMSA growth was faster in the Northeast than in any of the other three major regions from 1950-1970 and even for 1970-1975 was slightly faster in the Northeast than in the South.

In the North Central region non-SMSA growth has increased very rapidly, albeit from a very small base, and is now about the same as the rate for SMSAs which has slowed notably. In the South and West non-SMSA growth has accelerated very rapidly since 1970 and is approaching or exceeding SMSA growth rates. Thus, in spite of the substantial variation in regional growth rates in all major regions, SMSAs have shown growth rates that have been declining relative to the non-SMSA rates ever since 1940, with only a few exceptions. Also, for SMSAs the rank order of growth by region has remained exactly the same since 1940: Northeast slowest, then North Central, then South, then West in all four periods. The ranking of

Table 8-3
Metropolitan Growth Rates, by Region
(annual percentage rate of change)

	1940-1950	1950-1960	1960-1970	1970-1975
Northeast	1.09	1.56	1.01	0.64
Non-SMSA	1.34	1.76	1.67	1.45
SMSA	1.05	1.52	0.92	0.51
Class I	1.13 (0.70)	1.77 (0.28)	1.14 (0.29)	0.92 (0.86)
Class II	0.96 (0.25)	1.35 (−0.42)	0.67 (−0.50)	0.58 (−0.04)
Class III	1.00 (0.39)	1.33 (−0.78)	1.06 (−0.63)	0.30 (−1.68)
North Central	1.56	1.76	1.06	0.55
Non-SMSA	0.06	0.44	0.21	0.54
SMSA	1.82	1.99	1.19	0.55
Class I	1.92 (1.61)	1.94 (1.72)	1.27 (1.09)	1.02 (1.05)
Class II	1.72 (1.14)	2.03 (1.07)	1.13 (0.76)	0.58 (−0.10)
Class III	0.75 (0.71)	1.19 (−0.69)	0.79 (−0.99)	0.24 (−1.91)
South	2.61	2.59	1.38	1.69
Non-SMSA	0.35	0.50	0.62	1.42
SMSA	3.02	2.98	1.50	1.73
Class I	2.96 (3.70)	3.07 (3.46)	1.54 (1.30)	1.28 (0.70)
Class II	3.17 (2.32)	2.71 (2.41)	1.28 (1.42)	1.92 (1.29)
Class III	0.18 (1.01)	1.15 (−0.30)	1.45 (0.93)	1.52 (−0.44)
West	3.75	3.50	2.37	2.40
Non-SMSA	1.43	1.62	1.03	2.65
SMSA	4.55	4.14	2.74	2.34
Class I	4.85 (4.14)	4.51 (5.22)	2.76 (3.58)	2.61 (2.65)
Class II	4.04 (2.77)	3.77 (3.32)	2.88 (2.07)	2.59 (2.02)
Class III	3.99 (2.50)	3.12 (0.97)	1.83 (0.65)	1.78 (0.51)

Note: Figures in parenthesis are for central-city portions.

non-SMSA regions is not quite so consistent: North Central, South, Northeast, West for 1940-1950; North Central, South, West and Northeast for 1950-1960 and 1960-1970; North Central, South, Northeast, West, again, for 1970-1975. It is difficult to see the emergence of any recent dominance of the Sunbelt in these figures. For SMSAs it is as much a move West as South, and really just a move to regions with newer metropolitan regions, in general. For non-SMSA regions no consistent pattern emerges at all, and for 1970-1975 the Northeast is growing faster than the South. The general nature of the 1970-1975 growth pattern seems to be continuing through 1976.[10] The growth rate for the United States declined from about 0.96 percent a year in 1970-1976 to 0.76 percent in 1975-1976. But for the Northeast and North Central growth rates dropped much more: from 0.16 to 0.10 percent and from 0.37 to 0.18 percent, respectively. The total growth rate for the South fell from 1.67 to 1.20 percent between 1970-1975 and 1975-1976; the West stayed constant at about 1.75 percent per year.

At a subregional level, as shown in table 8-4, it is even more difficult to see the emergence of any particular recent regional dominance. For example, the three regions where non-SMSA growth has outstripped SMSA growth since 1970 are the Middle Atlantic, the East North Central, and the West. The regions where non-SMSA growth rates actually fell of since 1970 are New England and East North Central; the regions with more rapid SMSA growth since 1970 are East South Central, West South Central, and Mountain. This SMSA pattern probably could be made to look something like a shift to the Sunbelt, but it seems a rather odd Sunbelt with the South Atlantic and Pacific regions missing. If we were ever a nation of lemmings, it seems that since 1970 we have become more a nation of mountain goats. This is reflected even more strongly in recent projections for the 1975-2025 period now being prepared for Resources for the Future, Inc., which show the fastest growth projections for regions conforming roughly with the Rockies and Appalachia.[11]

Growth rates for central-city parts of SMSAs also are shown in tables 8-2 and 8-3. The comparison between SMSA and central-city rates in table 8-2, as expected, shows the emergence of suburban growth in the 1950s, especially in class III SMSAs, although in all three classes the central city

Table 8-4
Non-SMSA Growth by Subregion
(annual percentage rate of change)

	1940-1950	1950-1960	1960-1970	1970-1975
United States	2.14	2.26	1.36	1.30
Non-SMSAs	0.68	0.95	0.78	1.46
New England	1.45	1.75	1.77	0.88
Middle Atlantic	1.13	1.78	1.51	2.41
East North Central	0.54	0.97	0.71	0.67
West North Central	−0.28	0.06	−0.15	0.44
South Atlantic	0.98	1.25	0.88	1.80
East South Central	−0.06	−0.26	0.29	1.13
West South Central	−0.51	−0.23	0.41	0.93
Mountain	0.73	1.58	0.94	2.83
Pacific	3.06	1.70	1.22	2.24
SMSAs	2.22	2.31	1.38	1.26
New England	1.22	1.62	1.07	0.75
Middle Atlantic	0.88	1.43	0.75	0.41
East North Central	1.98	2.00	1.20	0.46
West North Central	1.47	1.98	1.18	0.78
South Atlantic	3.03	3.39	2.04	2.00
East South Central	2.12	1.87	0.98	1.47
West South Central	3.50	3.08	1.17	1.56
Mountain	4.61	4.86	2.66	3.35
Pacific	4.51	3.56	2.78	1.76

was growing slower than the SMSA. In the 1960s total SMSA growth slowed markedly, but not much more in the central cities than in the suburbs, and losses in class III areas actually were lower than in the 1950s. More serious central-city declines in all classes have appeared since 1970. With some slight exceptions in the Northeast, in all periods in all regions, the larger the SMSA, the slower the growth or greater the decline of its central city.

Systematic trend variations among the four regions are apparent, but in a longer-run historical dynamic sense they may be more apparent than real. For example, it is possible to view central cities in the North Central region as showing growth patterns more or less like those in the Northeast about a decade earlier, at least up until 1970. Central cities in the South grow rather like those in the North Central region, also with about a decade lag until 1970. Since 1970 the patterns in the South and North Central have converged on that in the Northeast, with central cities only in the West showing continued strong growth rates, but even there not in class III SMSAs. Increasingly, the "problem of the central city" is seen as a generalized national, rather than significantly regional, problem.[12]

This is not to say that there is no discernible regional dimension (say, Southern or Western) to the pattern of metropolitan growth. For example, of the twenty fastest-growing SMSAs from 1970 to 1975, eleven are in the South (Ft. Lauderdale, Orlando, Tampa, Houston, Miami, Atlanta, Jacksonville, San Antonio, Oklahoma City, Nashville, and Dallas), six are on the Pacific Coast (Anaheim, San Diego, San Jose, Sacramento, Portland, and Riverside), only three are in the Mountain states (Phoenix, Denver, and Salt Lake City), and none are in the Northeast or North Central regions. On the other hand, fast growth is far from universal in the South and West. Population in many Southern SMSAs is growing only at about national average rates for all SMSAs from 1970 to 1975 (for example, New Orleans, Memphis, Norfolk, Greensboro-High Point-Winston-Salem, Charlotte-Gastonia, Tulsa, and Richmond), and there are SMSAs in both the South and the West that have had very slow population growth (for example, Los Angeles and San Francisco, and Louisville and Birmingham).

The preceding discussion has been meant to illustrate the view that a regional bias favoring growth in the South and/or West has been with us at least since 1940, although, if anyhthing, has become less clear-cut, especially since 1970. This is indicated perhaps more clearly, if less precisely, in tables 8-5 through 8-8.

Population growth experience by state for 1940-1950 is shown in table 8-5. The very fast growth is in and only in the classic Sunbelt—California, Florida, and the desert Southwest. No state in the West or South (except for Arkansas) shows the slow growth rates associated with heavy outmigration. The basis for really rapid growth seems entirely metropolitan; no state has rapid growth in its non-SMSA area. All the fast-growth states have fast-growth SMSAs, with fast SMSA growth also in Maryland and Oregon; slow

Table 8-5
Fast- and Slow-Growth States,[a] 1940-1950
(ratio of growth rate to U.S. average)

		Total Population		Nonmetropolitan Population		SMSA Population
F	Nevada	2.4	None		Nevada	3.6
a	Arizona	2.1			New Mexico	3.5
s	California	2.1			Arizona	2.9
t	New Mexico	2.1			Florida	2.2
	Florida	2.1			Maryland	2.2
					California	2.2
					Oregon	2.1
S	Iowa	0.5	Massachusetts	0.5	Maine	0.4
l	South Dakota	0.5	Utah	0.5	Arkansas	0.4
o	Minnesota	0.4	Indiana	0.4	West Virginia	0.3
w	Maine	0.4	New Hampshire	0.4	Massachusetts	0.3
	Missouri	0.4	New York	0.4	Pennsylvania	0.2
	Nebraska	0.4	Ohio	0.4		
	Massachusetts	0.3	Vermont	0.3		
	West Virginia	0.3	Maine	0.3		
	Arkansas	0.3	South Carolina	0.3		
	Pennsylvania	0.2	Wisconsin	0.2		
	North Dakota	0.1	Michigan	0.2		
			Tennessee	0.2		
			West Virginia	0.2		
			Louisiana	0.1		
			Illinois	0.1		
			Montana	0.1		
			Pennsylvania	*		
			Minnesota	*		
			Nevada	*		
			Georgia	*		
			Iowa	*		
			South Dakota	*		
			Colorado	*		
			Texas	*		
			Alabama	*		
			Mississippi	*		
			Kentucky	*		
			Kansas	*		
			North Dakota	*		
			Arkansas	*		
			Nebraska	*		
			Missouri	*		
			Oklahoma	*		

[a]Fast growth is defined as twice and slow growth as half or less than the U.S. rate of 2.14 percent per year.
*Less than 0.05 or absolute decline.

SMSA growth is very limited. Thirty-two states show very slow non-SMSA growth, with fourteen of these indicating actual population loss.

Except for a slowdown in New Mexico, the fast-growth states in 1950-1960 are the same as in 1940-1950 (see table 8-6). Again fast growth seems entirely metropolitan and in the same states as before, except no

Table 8-6
Fast- and Slow-Growth States,[a] **1950-1960**
(ratio of growth rate to U.S. average)

		Total Population		Nonmetropolitan Population		SMSA Population
F				None		
a	Florida	3.0			Nevada	3.3
s	Nevada	2.5			Florida	3.2
t	Arizona	2.4			Arizona	2.9
	California	2.0			New Mexico	2.6
					California	2.0
S	South Dakota	0.5	Louisiana	0.5	Massachusetts	0.4
l	Pennsylvania	0.5	New York	0.5	Arkansas	0.3
o	Massachusetts	0.4	Washington	0.4	West Virginia	0.1
w	Missouri	0.4	North Carolina	0.4	Maine	0.1
	North Dakota	0.3	Montana	0.4		
	Maine	0.2	Maine	0.3		
	Arkansas	0.1	Wisconsin	0.3		
	West Virginia	*	Vermont	0.2		
			South Carolina	0.2		
			Minnesota	0.2		
			Colorado	0.1		
			South Dakota	0.1		
			Texas	0.1		
			Pennsylvania	0.1		
			Georgia	*		
			North Dakota	*		
			Iowa	*		
			Tennessee	*		
			Missouri	*		
			Kansas	*		
			Illinois	*		
			Mississippi	*		
			Alabama	*		
			Nebraska	*		
			Kentucky	*		
			Oklahoma	*		
			Arkansas	*		
			West Virginia	*		

[a]As in table 8-5, relative to U.S. rate of 2.26 percent per year.
*Less than 0.05 or absolute decline.

longer in Maryland or Oregon. The number of very slow-growth states has dropped from eleven to eight, but all eight with slow growth in 1950-1960 were slowly growing in 1940-1950 as well, and Arkansas and West Virginia continue to be as close to the South or West as any of the slow-growth states. Except for Pennsylvania, the slow-SMSA-growth states also are the same. Again, there are no fast non-SMSA growth states, but the number of slow non-SMSA growth states has dropped to twenty-eight, with the number with absolute declines reduced to twelve.

In the 1960-1970 period the pattern becomes a little less South/West-SMSA dominated (see table 8-7). The District of Columbia and Maryland join the fast SMSA growth list, and the deep South (Alabama) and the West

Table 8-7
Fast- and Slow-Growth States,[a] 1960-1970
(ratio of growth rate to U.S. average)

	Total Population		Nonmetropolitan Population		SMSA Population	
F	Nevada	3.3	New Jersey	2.5	Nevada	4.1
a	Florida	2.4	Florida	2.3	Arizona	2.4
s	District of				Florida	2.4
t	Columbia	2.3			District of	
	California	2.2			Columbia	2.3
	Arizona	2.1			California	2.2
					Maryland	2.1
S	Alabama	0.5	Utah	0.4	Alabama	0.5
l	Pennsylvania	0.3	Ohio	0.4	Pennsylvania	0.3
o	North Dakota	0.2	North Carolina	0.4	Maine	0.2
w	Maine	0.2	Arkansas	0.4	West Virginia	*
	South Dakota	0.2	Idaho	0.3	New Hampshire	*
	Wyoming	*	Alabama	0.3		
	West Virginia	*	Vermont	0.3		
	New Hampshire	*	Illinois	0.2		
			Minnesota	0.2		
			South Carolina	0.2		
			Maine	0.2		
			New Mexico	0.1		
			Oklahoma	0.1		
			Pennsylvania	0.1		
			Texas	0.1		
			Kentucky	0.1		
			Missouri	0.1		
			Montana	*		
			Wyoming	*		
			Mississippi	*		
			Iowa	*		
			Kansas	*		
			Nebraska	*		
			North Dakota	*		
			South Dakota	*		
			West Virginia	*		

[a]As in table 8-5 relative to U.S. rate of 1.36 percent per year.
*Less than 0.05 or absolute decline.

(Wyoming) show up on the slow-growth list for the first time. Fast growth shows up in non-SMSA areas of states for the first time, but as much outside the Sunbelt (New Jersey) as in it (Florida). On the other hand, many other states in the South and West continue to show up on the slow non-SMSA growth list, although the list has now edged down to twenty-six states with only seven showing absolute declines, despite the substantial drop in the national growth rate between 1950-1960 and 1960-1970.

The changes for the 1970-1975 period are shown in table 8-8. They show a marked contrast to change in the previous three decades, and the change is really toward a growth pattern which is less clearly a Sunbelt phenomenon.

Table 8-8
Fast- and Slow-Growth States,[a] **1970-1975**
(ratio of growth rate to U.S. average)

		Total Population		Nonmetropolitan Population		SMSA Population	
F	Arizona	3.4	New Jersey	4.2	Arizona	3.5
a	Florida	3.3	Florida	3.6	Florida	3.3
s	Nevada	2.9	Arizona	3.4	Idaho	3.1
t	Idaho	2.5	Massachusetts	3.4	Nevada	2.9
	Colorado	2.5	Nevada	3.0	Colorado	2.7
	Utah	2.4	Utah	2.4	Mississippi	2.6
	Wyoming	2.3	Wyoming	2.3	Utah	2.4
	Mississippi	2.2			New Mexico	2.4
S	Wisconsin	0.4	Indiana	0.4	New Jersey	0.5
l	Connecticut	0.4	North Dakota	0.3	Georgia	0.4
o	Iowa	0.4	South Dakota	0.3	Iowa	0.4
w	Pennsylvania	0.4	Kansas	0.3	Wisconsin	0.4
	Indiana	0.4	Louisiana	0.2	Pennsylvania	0.4
	Illinois	0.3	Nebraska	0.1	Indiana	0.4
	Ohio	0.3	Iowa	0.1	Connecticut	0.3
	West Virginia	0.1	Illinois	*	Massachusetts	0.3
	Kansas	*	Rhode Island	*	Illinois	0.3
	Rhode Island	*			Ohio	0.2
					New York	0.1
					West Virginia	*
					Rhode Island	*
					Kansas	*

[a]As in table 8-5, relative to U.S. rate of 1.30 percent per year.
*Less than 0.05 or absolute decline.

The classical Sunbelt states of Arizona, Florida, and Nevada are still on the fast-growth list, but so are four Mountain states and Mississippi; California has dropped off the list. Almost the same eight states (Utah is exchanged for Wyoming) show fast SMSA as show fast total growth.

The number of slow SMSA growth states has jumped substantially, from five to fourteen and at least one Deep South state (Georgia) is on the list. In sharp contrast, there are now only nine slow non-SMSA growth states (only one with absolute decline); and there are now seven states with very fast non-SMSA growth, including New Jersey and Massachusetts. True, much of this non-SMSA growth may be more ex-urban than rural in character, but it did no occur before, and it is not explainable as following the sun. The mountains seem as much a magnet, in general, as does the seashore.

Metropolitan Growth and Development

Obscurity of Preferences for Region

What all this evidence seems to add up to is that it is much harder to generalize about the apparent reasons for people favoring one region over another than many would like to believe. Part of the problem is that it is so easy to find differences in growth trends that correspond closely with macrogeographic characteristics. In fact, it is too easy. As illustrated by this chapter, if we do not stop looking when we find the "explanation" we want, we keep finding new ones. This leaves us in one of two uncomfortable positions.

First, we can continue to believe that shifts of population between non-SMSA and SMSA regions in various parts of the country are primarily dictated by people seeking bright lights, picket fences, the sea, the sun, or the mountains. But if we believe this, then we have to deal with people in the United States as perfidious as well as peripatetic. Who knows what they will be seeking a decade or less from now—isolation, valleys, trees, or what?

The other possibility, of course, is that while environmental preferences are not unimportant, they probably are dominated by regional differences in economic opportunity, with people learning rather quickly not only to adapt their preferences to the character of their destination, but also to shift their articulated preferences as well. We seem rather good at inventing socially acceptable reasons for explaining why what is most profitable is also most wholesome. For example, a whole generation of Americans seeking cheaper housing said they were moving to the suburbs for their children. My friends in Belgrade tell me the suburbs are much better because of the air!

The obvious answer to this apparent dilemma is that we must look at regional and metropolitan differences in economic opportunity as well as livability. Chapter 3 clearly is a step in that direction. But it is only a step, since there probably is supply-demand interaction in the sense that the pattern of economic opportunities is influenced by the pattern of population settlement; and, as suggested above, preceptions of regional livability may themselves be altered by materialistic ambitions or necessity. The puzzle, of course, is that the "new South" may be a product of the South's ability to exploit a new (or simply income-elastic) demand for life in a mild climate. Alternatively, innovations in air conditioning and the passage of Civil Rights legislation may have removed prior barriers to economic development having nothing to do with the natural environment. This is a puzzle which is still not untangled, and untangling it does have important consequences for the nature and extent of likely regional conflict.

Accordingly, something like a simultaneous-equation approach to regional differences in demand for and supply of population is needed. This has been understood for some time, as in Muth's important article on interregional migration and multipliers some years ago.[13] The problem in getting practically useful models of this type, however, is the absence of usable estimates of regional differences in unit factor costs. Wage rate data, even fully adjusted for differences in working hours and cost of living, would not be good enough; estimates which took factor productivity differences into account would be required. This would be necessary to distinguish among three possible explanations of higher real wages in, say, California: (1) California is a superior environment; it attracts labor at a discount; having such an advantage it uses a lot of high-quality labor in its production function and so attracts the highest quality (highest paid) labor at a substantial discount from what it would receive elsewhere. (2) California actually is an inferior living environment which can attract workers only by paying wage premiums. (3) Because of imperfections in the labor market and lack of information, workers in California are receiving quasi-rents which will tend to disappear, but we cannot predict in how long a time.

In any event, on the assumption that the picture of regional and metropolitan change outlined in this chapter would stand up under more rigorous analysis, it does have important consequence for the main question raised, namely whether regional confrontation is becoming a more serious issue in our society, perhaps requiring federal intervention. In particular, it seems that while regional differences in metropolitan growth have been notable at least since 1940 and likely will continue, they are less clear and less correlated with any durable regional distinctions. Sectionalism and the problems caused by it have played a very important role in our history. No doubt there are many areas in which sectional conflict may heighten. But at least in one respect, chapter has demonstrated that it likely that regional differences as a potential for confrontation are becoming blunted.

Notes

1. Southern Growth Policies Board, "Southern Growth Trends 1970-1976" (Research Triangle Park, North Carolina, June 1977), p. 17.

2. Brian J.L. Berry and Donald C. Dahmann, "Population Redistribution in the United States in the 1970s" (Washington: National Academy of Sciences, 1977), p. 35.

3. James Coleman, "Human Resource Strategies and the Economy and Ecology of Metropolis," in C. Leven, ed., *Mature Metropolis* (Lexington, Mass.: D.C. Heath, forthcoming).

4. Editors of Fortune, *Exploding Metropolis* (New York: Fortune Magazine, 1958).

5. C.L. Leven, J.T. Little, H.O. Nourse, and R.B. Read, *Neighborhood Change* (New York: Praeger, 1976).

6. William Alonso, "The Current Halt in the Metropolitan Phenomenon," in Leven, et al., ibid.

7. Ibid.

8. "The Shift to the Sunbelt: What It Means for Cities," *National Urban Coalition Network*, Summer 1976, p. 1.

9. Unless otherwise noted data sources are: 1960 Census of population for 1940-1950 and 1950-1960 changes, 1970 Census of population for 1960-1970 change, *Current Population Reports*, Series P-26, nos. 75-1 to 75-50 for 1970-1975 change.

10. Changes for 1975-1976 computed from 1976 population estimates in *Current Population Reports*, Series P-20, no. 307, April 1977.

11. Benjamin H. Stevens and Glynnis A. Trainer, "Distribution of Population and Economic Activity among the BEA Regions of the United States in the Year 2025" (draft) (Philadelphia: Regional Science Research Institute, October 1976)

12. "The Current Fiscal Condition of Cities, A Survey of 67 of the 75 Largest Cities," a Joint Committee Print of the Joint Economic Committee, 95th Cong., 1st Sess. (July 28, 1977), classifies cities in a variety of ways—by unemployment rate, by growth rate, and so on—but, interestingly, not by region.

13. Richard F. Muth, "Economic Growth; Chicken or Egg?" *Southern Economic Journal*, January 1971.

9 Rural Conditions and Regional Differences

Kenneth L. Deavers

Introduction

Recent concern about national regional development policy has focused attention on differences among regions in rates of growth of population, income, and employment. These differences are seen by many as signaling a long-term regional imbalance and requiring federal intervention in order to correct it. While growth rates or trends in development are important, they cannot alone form the basis for a national regional policy. At least three other elements of development need to be considered.

1. *Level of Development.* How well off people (and areas) are depends more on the level of development that has been achieved than on the rate at which it is changing. Per capita or median family income is used in this chapter as a measure of both absolute and relative levels. Although income is an economic measure, it is closely associated with (and, therefore, a good proxy for) other noneconomic characteristics of development such as education, public and private infrastructure, and so on.

2. *Structure of Development.* The mix of activities (economic base) on which development is based is another important element. Often future prospects for development depend more on this factor than on current levels of development or recent changes. Measures of employment opportunity and the structure of employment are used here as indicators of development structure.

3. *Spatial Composition of Development.* Within and between regions important differences exist in the development experience and prospects of rural and urban areas. Even more important, perhaps, than the existence of these differences is the extent to which public policy discussions have tended to ignore rural areas. Thus the major focus of this chapter is on differences among rural areas that are critical in designing a national regional development policy.

This chapter is divided into six major sections. The first section defines rural as it is used throughout the chapter, indicating the difficulties with the definition as well as the data problems facing analysts of rural conditions. Because so little of the recent public debate has recognized the importance of rural areas, and rural-urban interactions, the next two sections provide a brief history of the urban growth and rural population decline and economic readjustment that characterized the United States during most of the post-World War II period. They also highlight the trend toward population redistribution in favor of rural areas since 1970.

The fourth through sixth sections describe important differences in rural conditions among regions. Differences in demographic trends, level of development, and structure are highlighted. Finally, the last section draws some conclusions about the implications of these differences for national regional development policy.

Rural Conditions and Regional Differences

Definition of Rural

For measurement purposes, most research on "rural" areas uses either the Census rural-urban delineation or the OMB definition of metropolitan-nonmetropolitan areas. Rural territory, as defined by the Census, consists of open country and towns of less than 2,500 population, except those that are within the "urbanized area" surrounding metropolitan central cities. Nonmetropolitan territory consists of all counties whose largest urban nucleus is less than 50,000 people that are also nonmetropolitan in "character" and do not have a sizable proportion of workers commuting to a metropolitan center.[1] These definitions yield 1970 estimates of the number of "rural" people ranging from 54 to 64 million, or between 25 and 30 percent of the U.S. population. There are important differences in composition and data availability. Table 9-1 summarizes these compositional differences.

The Census definition of rural areas can be used only in conjunction with a full federal Census of Population, and therefore data based on it have been available only at ten-year intervals.[2] In general, rural territory includes almost all agricultural enterprise, which is certainly consistent with the popular conception of rural. However, it excludes many smaller to medium-sized towns in which significant nonfarm employment and income growth have occurred, which most observers would count as a rural component of development.

A wide array of nonmetropolitan data from sample surveys is available annually at the national level. And there is a growing list of county-level data available annually which can be aggregated to nonmetropolitan measures for state, regions, and the United States. Generalizing metropolitan areas to county boundaries has a number of important effects. It excludes from (rural) nonmetropolitan territory extensive open country located on the outer fringe of metropolitan areas, as well as many of the major agricultural counties in the nation. However, nonmetropolitan areas

Table 9-1
1970 Population of the United States by Residence: Metropolitan, Nonmetropolitan, Rural Nonfarm, and Rural Farm[a]
(million)

	Metropolitan	Nonmetropolitan	Total
Urban places of at least 50,000	73.3	—	73.3
All other urban population	49.7	26.3	76.0
Rural nonfarm (places of less than 2,500, open country, except farm)	14.9	30.7	45.6
Rural farm[b]	1.5	6.8	8.3[c]
Total	139.4	63.8	203.2

Source: 1970 Census of Population.

[a]Residence definitions used in the 1970 Census. Rural figures are corrected and do not correspond to originally published Census figures.

[b]A farm is a rural place of at least ten acres that sold at least $50 worth of agricultural products in the reporting year of any rural place that sold at least $250 worth of agricultural products in the reporting year.

[c]Note that this estimate of the farm population is significantly less than the 1970 count of 9.7 million as given in *Current Population Reports, Farm Population, Series P-27*, no. 44, June 1973, p. 1.

have been occurring and which have been the recipient of considerable federal development assistance.
include many towns in which nonfarm employment and income growth

As this discussion suggests, neither definition leads to a fully satisfactory identification of rural areas. However, measurement is basic to both understanding the condition of rural people and communities and assessing whether some action is required to improve these conditions. To provide such measures, this chapter adopts primarily the nonmetropolitan definition of rural areas. This is done for two reasons: (1) Data are available more recently than for the Census definition. (2) The metropolitan-nonmetropolitan concept is often a more useful way of identifying crucial relationships in the economic geography of the United States. It is important to recognize, however, that in many cases there are differences among nonmetropolitan areas as significant as those between metropolitan and nonmetropolitan areas. For example, the prospects for development of a nonmetropolitan area are likely to be different depending on the degree of metropolitanization of its region; the health, education, skill mix, and racial composition of its labor force; its current (and historic) economic structure; and so on. Thus the definition of rural needs to be flexible, to fit the unique needs of particular rural areas and the context in which rural questions are being considered.

Dominant Demographic and Economic Trends Affecting Rural Areas

Migration of people from rural areas to cities did not begin in the United States during World War II, but the scale of movement then and for the next thirty years made earlier movements seem minor. The trend was not limited to this country. It characterized every advanced nation of the world and most of the developing ones as well. People move for different (not well-understood) reasons, but the overwhelming economic force at work during this period was the displacement of workers from traditional rural employment—especially farming—as capital was substituted for labor. Combined with this was the attraction of the cities for people seeking jobs in expanding manufacturing and service industries, higher incomes, and better and more interesting living conditions.

In the United States, the farm population dropped from 30 million in 1940 to 9 million in 1970. The consequences of this human exodus from farming were devastating for many rural communities. Not only did the demand for farm labor decrease, but so did the demand for many of the goods and services the rural communities provided. These rural (service centers) communities have had to adapt to changes in their environments, and adaptation has not always been easy or successful.

All through the 1940s and 1950s, farms lost a net of one million people annually through outmigration. Some of these people remained in rural and small-town areas in other work, and the rural nonfarm population grew. But most rural communities faced with declining farm job opportunities, and with nonfarm job growth unable to fill the gap, lost people. Millions from farms and these communities went to the major cities, contributing to the metropolitanization of the nation. There was little negative public reaction to this influx for many years. Labor was needed and the cities did not have an ample local supply, because of low urban birthrates in the 1920s and 1930s and the relative absence of immigration.

In the same period, the zone of urban influence grew substantially. Suburban development occurred in a sprawling manner around all the major cities, and expanded highway programs permitted many indigenous rural people to commute to cities for work. Ironically, the massive rural-urban migration was beginning to play itself out before it became a major topic of congressional and public debate. The sources of migration from rural farming, mining, and forest areas were not unlimited. Once a relative correction between labor supply and labor needs occurred, the flow diminished. For example, the last year that the farm population lost 1 million people through net outmigration was 1962. There had been nine such years in the previous fifteen. And by 1969 the drop in coal mining employment had ended completely.

Rural Conditions and Regional Differences

By 1970 the urbanization process had left only a little more than one-fourth of the total U.S. population in rural areas. Since 1920 when urban people first became more than half of the total, all net growth of population had gone into the cities and urban towns. While the rural total had remained numerically nearly constant, it too had undergone a vast redistribution and recomposition. Huge, predominately agricultural areas had become largely depopulated because of farm enlargement. This was true in much of the Great Plains, the Corn Belt, and the old Cotton Belt. It was also the case in the coal mining regions. But simultaneously other rural and small-town communities had grown. There was a filling in of territory between the major metropolitan areas of the Northeast and lower Great Lakes, by people who residentially were rural—and deliberately so—but whose employment was urban in nature. There was sustained growth in rural and small-town industrialization in the southern Piedmont. And, a host of individual nonmetropolitan counties were growing from military work, college expansion, or emergence as recreational and retirement centers.

Important changes in the economic structure of the rural United States accompanied urbanization. Beginning in the early 1960s, manufacturing emerged as a more important growth sector for rural areas than for urban areas. During this decade manufacturing in rural areas grew at a rate nearly five times that for metropolitan areas. In fact, for the 1960s the absolute change in manufacturing employment in nonmetropolitan areas exceeded that for metropolitan areas. This has continued into the 1970s. Also, in the early 1970s rural employment diversified with growth of government, services, and trade. (See table 9-2.)

Trend toward Redistribution since 1970

Although it was apparent in the late 1960s that the net flow of rural-urban migration was ebbing, no one seems to have foreseen the rapidity and scope of the turnaround in the relative growth rates of urban and rural areas that occurred. As Figure 9-1 shows, the turnaround is not simply explained by metropolitan sprawl. Even nonmetropolitan counties not adjacent to SMSAs are experiencing substantial population growth.

From 1970 to 1975, net migration has been *into* nonmetropolitan areas. As a consequence, nonmetropolitan population increased by 6.6 percent in the first half of the decade, compared with a metropolitan growth of 4.1 percent. (During roughly the same period, nonmetropolitan areas absorbed 40 percent of the total increase in nonfarm employment, expanding their share of total employment to over 25 percent.) This is the first time such a trend has been observed in the modern history of the nation. Even in the

Table 9-2
Job Changes in Nonmetropolitan and Metropolitan Areas, 1970-1977[a]

Designation	Nonmetropolitan[b]		Metropolitan	
	(000)	*Percent*	*(000)*	*Percent*
Mining	148	36	43	20
Construction	262	32	-47	-2
Manufacturing	323	6	-1,085	-7
TCU[c]	126	14	-14	e
Trade	1,007	30	2,039	18
FIRE[d]	183	34	547	18
Services	916	39	2,562	28
Government	858	23	1,770	20
Total	3,823	22	5,815	11

Source: C.C. Haren, "Where the Jobs Are," *Farm Index*, Economic Research Service, U.S. Department of Agriculture, August 1977, pp. 20-21.

[a]Adapted from state employment security agency estimates for March in respective years, seasonally adjusted.

[b]Includes some fifty smaller standard metropolitan statistical areas, but excludes approximately 330 rural and other fringe counties.

[c]Transportation, communications, and utilities groups.

[d]Finance, insurance, and real estate groups.

[e]Less than 0.5 percent decrease.

1930s during the Depression, metropolitan growth continued to be somewhat faster than nonmetropolitan growth. None of the superior recent nonmetropolitan growth is explained by rates of natural increase (excess of births over deaths). Instead, it stems from a net inmigration of nearly 2 million people over the five-year period. To understand how extraordinary this is, these same areas experienced a 3.0 million net outmigration in the 1960-1970 decade as a whole.

Regional Trends since 1970

Along with the overall nonmetropolitan population increase of 1970-1975, net inmigration has been occurring in these areas of four regions. In fact, in every region except the South, the nonmetropolitan population grew more rapidly than the metropolitan population. (See figure 9-2 and table 9-3.) The range of nonmetropolitan growth, from 1970 to 1975, was from 3.4 percent in the North Central states to 13.4 percent in the West. The latter growth rate is so high, nearly 2.5 percent compounded per year, that it is likely to create many problems for communities in providing needed facilities and services.

The circumstances associated with nonmetropolitan growth vary

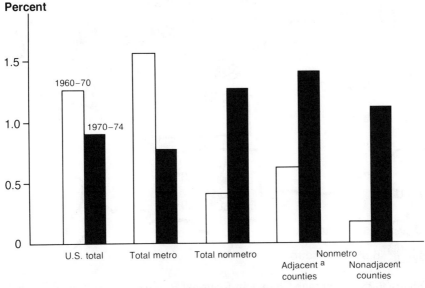

a. Counties adjacent to standard metropolitan statistical areas as defined in 1974.

Figure 9-1. Population Changes

somewhat regionally. In the Northeast, it is primarily a movement into the picturesque and most habitable areas of northern New England, the Catskills and Poconos, and the Atlantic Coast. It is not typically associated with antecedent economic development and often involves people of above average socioeconomic status seeking a small town or rural locale. Some of it involves retirement and interstitial filling in between metropolitan areas.

In the North Central states, the areas of high nonmetropolitan growth are typically north country woods and lakes counties plus the Missouri Ozarks. These are areas of relatively cheap land prices in areas of relatively low income. Recreation, retirement, and manufacturing growth are associated with much of the population increase. Most counties in the Great Plains and Western Corn Belt continue to lose population, unable to generate adequate alternatives to farm employment.

The South has a number of rapid-growth nonmetropolitan areas. The larger of these include the northern Piedmont, the southern Appalachian Coal Fields, the Blue Ridge area, the Florida Peninsula, the lower Tennessee Valley, the Arkansas Ozarks, the Oachita Mountains, and the central Texas hill country. Various elements are present: revived coal mining, manufacturing, recreation, and retirement. Contrary to the general

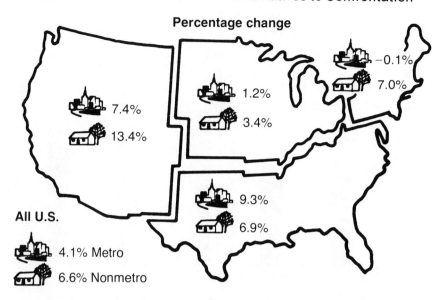

Figure 9-2. Percentage Change in Regional Population Growth, 1970-1975

nonmetropolitan growth, counties with high proportions of black residents continue to experience net outmigration. This includes areas such as those in the Delta and the Alabama Black Belt.

Most nonmetropolitan counties in the West are growing rapidly. Coal, oil, and other mining developments produce some of the growth. In other areas it stems from retirement and recreation or from the high natural growth of Indian and Alaskan native populations. New England-like outmovements of highly motivated people into rural and small-town settings are clearly evident, too, for example, in the Sierra Nevada foothills or in southwestern Oregon.

Finally, it might be worthwhile commenting on the Sunbelt-Snowbelt growth. While it is generally true that the Southern half of the country is experiencing much more rapid population growth than the Northern half, the notion is meaningless when extended across the West. (This was done by *Fortune* magazine in its June 1977 issue. In mapping the Sunbelt, *Fortune* divided California along North-South lines, despite the fact that the Northern half has had more rapid growth than the Sunbelt half.) The important point is that the entire West is growing, not just the sunny and mild sections. The major contrasting regional population growth pattern is between growth of the West, the Southeastern quarter of the United States and the north country of the Upper Great Lakes and northern New England on one hand, contrasted with the declining industrial-metropolitan belts of the Northeast and the agricultural areas of the prairies and Great Plains on the other.

Rural Income Levels and Regional Underdevelopment

In 1970 approximately 44 percent of the nation's poverty population resided in nonmetropolitan areas. However, as shown in table 9-4 and figure 9-3, there were important regional variations. In particular, the portion of the poor in nonmetropolitan areas of the South was much higher than for other regions. In fact, the higher incidence of poverty in the nonmetropolitan South means that nearly 60 percent of all the nonmetropolitan poor in the United States reside in the South. While the percentage of the poverty population residing in nonmetropolitan areas decreased by about 20 percent between 1970 and 1974, the relationship between the South and North and West did not change.

The composition of the nonmetropolitan poverty population differs significantly from that of the metropolitan poverty population. In 1970 a much higher proportion of nonmetropolitan poor families were male-headed. (See table 9-5.) Because of the stronger attachment to the labor force of male-headed families, there is a high incidence of "working poor" in nonmetropolitan areas. These poor are excluded from many of the current categorical welfare programs, and would be among the most benefited by proposals for welfare reform such as Family Assistance Plan, Allowance for Basic Living Expenses, and so on.

Finally, in the South, nearly 40 percent of the nonmetropolitan poor are black. But, outside the South the nonmetropolitan poverty population is almost exclusively white. (See table 9-6.) The continued existence of a large concentration of low-income rural areas in the South Census region, despite generally improving rural conditions even within the region, is a particularly serious development problem. While poverty causes serious human problems wherever it occurs, nowhere are these problems more severe than in rural areas where poverty is so endemic as to be reflected in areawide data on per capita or family incomes. Because of the inadequacy of local resources to support needed facilities and services, communities in these areas chronically underinvest in human resources—inadequate educational opportunity, worker training, and health care are continuing problems. For similar reasons, these areas are also short on basic community facilities and amenities that are typically found in other rural areas—poor housing, lack of public water and sewer systems, inadequate fire protection, and other such conditions are prevalent.

The geographic concentration of areas experiencing the effects of chronic underinvestment in human and physical resources means that local opportunities for individuals to escape from poverty are limited. Therefore, these areas have historically experienced very high rates of outmigration. Often even the outmigrants have been unable to separate themselves from the problems they left, suffering in their new areas from the results of poor

Table 9-3
Population Change and Net Migration by Metropolitan Residence, 1970-1975 and 1960-1970

		Regions				Divisions						
Item	United States	North-east	North Central	South	West	New England	Middle Atlantic	East North Central	West North Central	South Atlantic	East South Central	West South Central	Mountain	Pacific
Population														
Total														
1975	213,053	49,455	57,665	68,102	37,831	12,188	37,267	40,978	16,687	33,703	13,544	20,855	9,646	28,185
1970	203,304	49,061	56,593	62,812	34,838	11,847	37,213	40,266	16,328	30,679	12,808	19,325	8,290	26,549
Percentage change 1970-1975	4.8	0.8	1.9	8.4	8.6	2.9	0.1	1.8	2.2	9.9	5.7	7.9	16.4	6.2
Metropolitan[a]														
1975	155,037	42,412	39,593	43,023	30,009	10,046	32,365	31,151	8,442	22,153	6,812	14,058	5,618	24,391
1970	148,881	42,481	39,110	39,350	27,939	9,853	32,628	30,836	8,275	20,050	6,459	12,842	4,772	23,167
Percentage change 1970-1975	4.1	-0.1	1.2	9.3	7.4	2.0	-0.8	1.0	2.0	10.5	5.5	9.5	17.7	5.3
Nonmetropolitan														
1975	58,016	7,043	18,072	25,079	7,822	2,142	4,902	9,827	8,245	11,550	6,732	6,797	4,028	3,794
1970	54,424	6,580	17,483	23,462	6,899	1,995	4,585	9,430	8,053	10,630	6,349	6,483	3,518	3,381
Percentage change 1970-1975	6.6	7.0	3.8	6.9	13.4	7.4	6.9	4.2	2.4	8.7	6.0	4.8	14.5	12.2
Nonmetropolitan adjacent counties[b]														
1975	30,074	5,180	9,116	12,578	3,199	1,113	4,067	6,522	2,595	5,766	2,842	3,971	1,231	1,968
1970	28,033	4,822	8,805	11,642	2,763	1,036	3,786	6,303	2,502	5,227	2,637	3,778	1,026	1,736
Percentage change 1970-1975	7.3	7.4	3.5	8.0	15.8	7.3	7.4	3.5	3.7	10.3	7.8	5.1	20.0	13.4
Nonmetropolitan nonadjacent counties														
1975	27,942	1,863	8,955	12,501	4,623	1,028	835	3,305	5,650	5,784	3,891	2,826	2,797	1,826
1970	26,391	1,758	8,677	11,819	4,136	959	799	3,127	5,550	5,403	3,712	2,704	2,491	1,645
Percentage change 1970-1975	5.9	6.0	3.2	5.8	11.8	7.3	4.4	5.7	1.8	7.1	4.8	4.5	12.3	11.0

Rural Conditions and Regional Differences 355

Net Migration

Total														
1970-1975	2,466	-695	-883	2,623	1,421	60	-755	-773	-110	1,854	204	565	833	588
1960-1970	3,001	319	-757	590	2,850	310	9	-153	-604	1,332	-699	-44	305	2,544
Metropolitan[a]														
1970-1975	625	-999	-1,046	1,835	835	-33	-966	-890	-156	1,337	64	434	531	305
1960-1970	5,997	307	127	2,494	3,069	287	20	28	99	2,070	-86	509	598	2,472
Nonmetropolitan														
1970-1975	1,841	304	163	788	586	93	211	117	46	517	140	131	302	283
1960-1970	-2,996	12	-884	1,904	-220	23	-11	-181	-703	-738	-613	-553	-292	73

Source: U.S. Bureau of the Census *Current Population Reports*.
Note: Population and net migration figures are rounded to the nearest thousand without adjustment to group totals.
[a]Metropolitan status as of 1974.
[b]Nonmetropolitan counties adjacent to Standard Metropolitan Statistical Areas.

Table 9-4
Percentage of Poverty Population Residing in Nonmetropolitan Areas

	1970
United States	44
Northeast	21
North Central	44
West	29
North and West	33
South	58

Source: Census of Population, 1970.

educational systems. Outmigration has further complicated the problem of designing programs to stimulate development in such areas, since the age structure and other characteristics of the local population may make undertaking a new enterprise appear to be a very risky venture.

Another aspect of the problem of rural underdevelopment is the extent to which the rural poor, and blacks in particular, are failing to share in the general trend toward decentralization of industry to rural areas.[3] A number of studies indicate that the extent to which jobs are provided for the local poor is a function of the nature and the kind of industry and its skill requirements. Although state and local development agencies often concen-

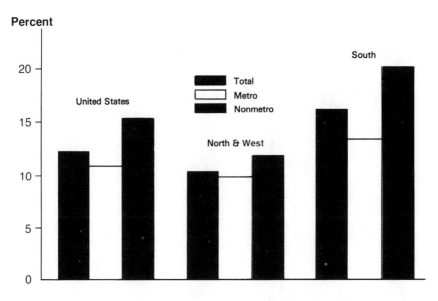

Figure 9-3. Persons in Poverty, 1975

Table 9-5
Percentage of Poor Families which Are Male-headed, 1970

	Metropolitan	Nonmetropolitan
Northeast	56	70
North Central	57	81
West	60	77
North and West	57	77
South	63	76

Source: Census of Population, 1970.

trate their efforts on the attraction of high-wage industry, such enterprises often have the fewest employment effects for local people. Other studies seem to indicate that industry avoids rural places with heavy concentrations of blacks, and that blacks overshare in employment declines and undershare in growth.

At least one implication of this discussion seems clear. For many of the rural poor, the most important federal program likely to affect their future well-being is not within the tools of agencies with developmental responsibilities. That program is welfare reform—the assurance to each citizen that wherever he/she chooses to live there will be an opportunity to have an income level sufficient to maintain a socially satisfactory minimum level of living.

Employment Opportunities and Regional Evidence of Structural Change

In a dynamic economy which adopts innovative and new economic techniques, has a highly mobile labor force, and is generally characterized by change, there are some areas that prove considerably less able than others to adapt to such change without serious employment disturbances. Structural changes may show up as chronic high unemployment rates, slow growth in

Table 9-6
Proportion of the Poverty Population which Is Black, 1970

	Metro	Nonmetro
Northeast	26	4
North Central	32	3
West	14	2
North and West	24	3
South	44	39

Source: Census of Population, 1970.

employment coupled with declines in relative income levels, declines in labor force participation or the relative availability of employment opportunities, and so on. Only one of these measures, unemployment rates, is generally used in legislation or in public discussion as an indicator of the need for development assistance.

Among the more important kinds of structural change which may give rise to adjustment problems are the following.

1. *Technological change.* In an advanced innovative nation which makes substantial investments in research and development, technological change will be a continuing process. Some kinds of technological change will be translated into the marketplace in a gradual fashion, and may be absorbed by affected regions without serious adjustment problems. Other kinds of technological change may be so significant and require such rapid or extensive adjustment as to be a truly traumatic shock to the economy of an area.

2. *Resource depletion.* Areas whose primary economic advantage consists of an exhaustible natural resource face two kinds of problems. Obviously, when the resource is gone, so is the primary justification for continued economic activity in the region. Furthermore, extractive industries typically require that the product be exported to another region before conversion, processing, or use. Thus, little secondary activity develops in such regions to serve as a base for development once the resource is gone.

3. *Changes in the international competitive position of U.S. industries.* These are changes that reduce the ability of specific U.S. industries to compete successfully in international trade, independent of any changes in trade or tariff policy. The impacts are primarily on exporting industries or on import substitute-producing industries. Particular areas, which are heavily dependent on such industries for employment and income generation, may find it difficult to adapt to such changes.

4. *Shifts of federal government expenditures, policies, or regulations.* Because of the scale of its expenditure programs and the pervasiveness of its policies and regulations, the federal government has significant impacts on relative economic advantage. Often the execution of various programs is independent of any consideration of area impact, which means that these federal programs may intensify disparities that result from market forces. The two federal expenditure programs which seem to have the most significant impacts are defense spending and the federal highway construction program. A relatively new area for federal policy formulation is the result of the passage of the National Environmental Policy Act of 1969. Decisions required by that act concerning air quality and water quality appear likely to have significant effects on the growth prospects of particular areas, and consequently on their comparative economic advantage. An even newer area is federal efforts to develop strategies for meeting future energy needs.

Several regulatory agencies, through their powers to establish prices, also may have very significant effects. Examples include such agencies as the Interstate Commerce Commission and Civil Aeronautics Board.

5. *Changes in the composition of final demand.* Rising family incomes, changes in family size, and changes in consumers' tastes all give rise to consumption of different bundles of goods and services over time. Such changes will impact on the growth or decline of particular industries, and this in turn may result in significant adjustment problems for particular areas.

Because of the structural diversity among areas, the varied types of structural change which may be occurring simultaneously, and the different capability which areas have to adjust to such changes without prolonged distress, there is not the same geographic concentration of rural areas with structural adjustment problems as can be observed for underdeveloped areas. In addition, the areas that are identified as experiencing distress will be somewhat different, depending on the measure of employment opportunities chosen to identify them.

Two examples make this point rather dramatically. Based on average annual unemployment rates for 1976, the 350 rural areas with rates above 7.9 percent are widely scattered across the Northeast, South, and West. With the exception of Michigan, the North Central region is remarkably free of such serious unemployment problems. Furthermore, except for areas seriously impacted by seasonal factors or with extractive industry as the primary employment base, few concentrations of unemployment problems are apparent. However, if one uses another measure of employment opportunity, the ratio of percentage change in employment to percentage change in population (1970-1974), the picture is quite different. Nearly half of the 350 rural counties with relatively inadequate employment opportunity (as measured by a low ratio) are located within the North Central region. The only other region with a significant concentration of rural development problems by this measure is the South. Interestingly, these same two regions also contain an overwhelming majority of the 350 rural areas that are doing well in capturing employment opportunities by the same measure. In many cases, success and failure are juxtaposed in groups of adjacent counties.

Conclusions

In light of the preceding discussion, a number of conclusions can be drawn about a successful national regional development policy. First, no such policy is possible that does not recognize the important interaction between rural and urban areas and the recent strong trend toward less dense settlement patterns that has made rural areas among the most dynamic growth areas in the nation.

Second, a regional development policy needs to be able to respond to the unique problem of rural areas that differentiate them from urban areas. Three critical factors combine to differentiate rural settlements and activities for public policy purposes.

    1. Rural areas are diverse. There are some which hold little promise of providing an acceptable level of living for their residents. Still others including those undergoing various structural adjustments appear to require significant public policy intervention to bring about satisfactory levels of living.

    2. Rural communities are smaller and their populations are more widely dispersed. Because of this there are institutional obstacles to development that require special recognition if they are to be overcome effectively and economically.

    3. Technologies (both physical and social-organizational) developed for densely settled urban populations are often inappropriate for rural areas. Adaptation and new technologies are essential to economically meet the needs of less densely settled rural areas and smaller communities.

Third, probably the most pressing national rural development problem is the continued existence of a large concentration of low-income rural areas in the South Census region. A crucial aspect of the Southern rural underdevelopment is the extent to which minorities, blacks, and Hispanics, in particular, are disproportionately represented among the rural poor. In addition, many areas in which minorities are concentrated have failed to share proportionately in the general trend toward decentralization of industry within that region.

To increase access to economic opportunity for people in *poor rural areas*, a heavy emphasis on human resource development, especially remedial education and job skill training, is likely to be an appropriate major component of public policy. Such human resource investments will need to be closely linked to efforts to encourage private job creation, however, since in many cases the jobs are unlikely to locate in the most seriously distressed communities. For many people in these communities it may, in fact, be a more sensible and efficient economic development strategy to encourage them to move to jobs, rather than the reverse. Given the extraordinary and widespread success of many rural areas in capturing new industrial and service industry growth, this need *not* mean that these people give up their rural life-style.

Finally, the urbanization and economic development process of the last thirty to forty years has seen the growth and decline of many communities and areas, in both absolute and relative terms. The adjustments required as a result have been enormously large, but for the most part public policy has allowed individuals acting on the basis of market signals to make these adjustments as they saw fit. Serious questions are being raised about the appropriateness of the market signals people continue to receive and about whether individual marketplace-determined solutions should be allowed to shape our sociodemographic and economic futures. The first question is one that needs to be addressed in formulating a national regional development policy, while the latter appears likely to be an unresolved issue of public debate for many years.

Notes

1. For exact definitions of these concepts see U.S. Bureau of the Census, "1970 Census Users' Guide: Part I" (Washington: Government Printing Office, 1970).
2. Legislation now provides for a Census of population every five years which will cut this time span in half.
3. See Ray Marshall, *Rural Workers in Rural Labor Markets*, (Salt Lake City, Utah: Olympic Publishing Co., 1974), pp. 73-76.

Part VI
Regional Differences in Public Services

10 Public Services and Economic Development

Robert L. Lineberry

Introduction

Public services are what we buy with the 34 percent of our gross national product which the U.S. governments spend. Political scientists have told us much about the way the public budgets are established and the fiscale impact and incidence of that budget. Yet we know more about the budgets themselves than about the public services they purchase. This chapter is a preliminary effort to link public services with the significant transformation of population and economy now evident in the United States. In it I address three broad issues. First, after deploring the neglect of attention to questions of public services in general, I examine several emergent conceptions of supply and delivery of public services. Second, I examine variations in service needs and delivery from neighborhoods to regions. Third, I offer several hypotheses about the links between public services and patterns of growth and decay. I conclude by offering a dual agenda, one for knowledge and one for action, which would make services a critical variable in explaining economic and demographic change.

Neglect and Emergence: Perspectives on Public Services

The distinguished regional scientist Michael Tietz observes that modern man is affected by public-service decisions concerning public health, transportation, communication, water and sewerage, police and fire protection and recreation.[1] One could add other public services to this very abbreviated list. But one could also stress that the supply and distribution, and perhaps the cost, of public services affect locational decisions of the businesses and industries which provide modern man and woman with jobs. Yet, as far as I know, there is no systematic body of research or theory on public services in any way comparable to the research and theory on public expenditures. We know a good deal about what government costs and who pays for it, but little about what people get for their money. We know a good deal about who bears the burdens of government,[2] but relatively little about who gets the benefits.[3] Even less do we know about the impacts of public-service decisions on family and firm locational choices. We can identify, however, five emergent perspectives on public services. We describe these as (1) services as socialism for the rich, (2) services as socialism for the poor, (3) services as socioeconomic infrastructure, (4) services as life-quality determinants, and (5) services as a public-private mix.

Services as Socialism for the Rich

The rich get richer, it is said, and the poor get poorer. It is not an uncommon hypothesis that public services do more for the rich than for the poor. Miller and Roby describe public services as "hidden multipliers of income" and argue that the rich receive more multiplier effects than the poor.[4] Public services, this perspective suggests, follow the Biblical axiom that "to everyone who has will more be given, but from him who has not, even what he has will be taken away" (Matt. 25:29, RSV). There is considerable evidence that the advantaged classes are further advantaged in at least a few service areas. Higher education is one area where net tax/expenditure effects are essentially redistributive upward.[5] Suburbanites are said to enjoy much higher levels of schools and services than central-city residents. A New Trier Township High School is not often duplicated in core city areas. Airports are not likely to benefit directly the half of the population who have never flown.

Yet the accumulating evidence casts doubt on any simple perspective of public services as merely socialism for the rich. We should be extremely wary of research which assumes direct links between government expenditures and services. But the evidence from the Tax Foundation and from Reynolds and Smolensky finds public expenditures decidedly progressive, that is, redistributive upward.[6] A growing body of research on the intramunicipal distribution of services also takes issue with the long-standing hypothesis that richer and whiter neighborhoods universally receive superior services.[7] But questions of the final redistributive impact of public expenditures and services are far from settled and cut to the core of Marxist critiques and non-Marxist defenses of U.S. political economy.

Services as Socialism for the Poor

Marxist critics of the U.S. economy are wont to describe it as "socialism for the rich and capitalism for the poor." Yet perhaps this aphorism fails to examine the degree to which public services constitute a form of socialism for the poor. The poor are plainly more dependent on the public sector than the rich. Lacking spacious yards and the time and money for travel, they rely on public recreation; lacking easy access to medical care in the health-for-profit sector, they utilize public health services; lacking access to private schools, they use the public schools exclusively; lacking often the wherewithal for adequate housing, they rely on public housing or food stamps. For these critical social and economic values—recreation, health, schooling, transportation, housing, and food—and for others (even legal services, job training, and scholarships) the poor may receive public services in a range to rival the Scandinavian democracies.

There is evidence that we are seriously misstating the poverty problem in this country by focusing on incomes alone. Poverty economist Robert Haveman stresses that in-kind "payments" to the poor have made poverty a very different problem in the 1970s than it was in 1964:[8] in-kind services, particularly food stamps, Medicaid, and housing subsidies, are major contributors to this altered condition of the poverty class. It may well be, of course, that we are merely trading one problem for another by creating massive *service dependencies* by the poor. In the 1960s, it was fashionable to be concerned about welfare dependency, an issue which troubled liberals and conservatives alike. Yet in the 1970s, it is possible that we have exchanged welfare dependency for service dependency. Services, to some extent at least, constitute socialism for the poor.

Services as Socioeconomic Infrastructure

Public services provide the infrastructure for economic and demographic growth. In the extreme case, public services are an absolute delimiter of growth. In the absence of water, sewers, airports, roads, utilities, and police and fire protection, neither families nor firms will find an area attractive. Service levels are not insignificant considerations in locational decisions although how significant they are, I discuss later). The president of the San Antonio Economic Development Association gloomily reported the story of his visit with a prospective corporate mover in Chicago. The Chicago executive demanded to know if San Antonio (the nation's tenth largest city) had an airport. Assured that it did, the executive stressed that "now by an airport, I mean one with regularly scheduled commercial flights."[9] Public-service provisions usually precede any realistic economic growth. If they do not precede it, their promise must be very believable.

For this reason, debate about public services has become intertwined with the "progrowth-no-growth" conflicts in cities and states.[10] Marin County, California, for example, found its current drought exacerbated because it declined to build more reservoir capacity, a decision motivated in part by the desire to minimize growth. The antigrowth advocates of Marin County, Santa Barbara, Boulder, and Oregon remain exceptions, however. More commonly cities and states rush to provide incentives for growth.

Public Services and the Quality of Life

The quality of life in the United States has been much at issue lately. There is scattered but consistent evidence that overall life satisfaction has declined in recent years particularly and, interestingly, among upper-income groups.[11] The declines are not precipitous, but they are steady. There are many reasons advanced to explain this decline in life satisfaction, including economic factors, the aging process itself, and changes in family structure. When such a large share of our national income is devoted to the public sector, though, public policy and public services may also be implicated. People in the United States have watched police expenditures increase while crime increases, school expenditures rise while Scholastic Aptitude Test scores decline, and municipal costs escalate while streets get dirtier.

Both popular[12] and scholarly[13] studies have examined variations in quality of life from city to city, state to state, and region to region. Some of, but not all, these surveys single out the Southern region and its cities and states as the "worst" region. Others single out cities in the Northeastern region as particularly disadvantaged. Arthur Louis' overall ranking of urban life quality in the fifty largest cities (see table 10-1) finds places like Newark, St. Louis, Chicago, and Detroit near the bottom. If this ranking has any credence, then it is sobering to note that the Spearman correlation between percentage black and quality ranking is a substantial -0.68. The worse the city, in other words, the greater its concentration of nonwhites. Either blacks have very different ideas about the components of life quality than Louis does, or they are concentrated in cities with the most severe social and economic problems. In those very cities the cost of public services tends to be high and their quality at least debatable. In Newark, for example, taxes now are $2,000 on a $20,000 home. Mayor Gibson's last budget message announced further service and personnel cutbacks.

Public Services as Public-Private Mix

The disciplines of political science and economics are built on the assumption of clear distinctions between the public and private sectors. It may be more useful, though, to see "public services" as coproduced by both the public and private spheres. This perspective suggests important linkages between political scientists and economists using public choice or political economy models of policy.

Very probably, no "public" service is performed wholly by the public sector. Almost any service (public health, public education, public safety, public recreation) has private sector counterparts. If we think of a public "function" being performed rather than merely a public service, then the

Table 10-1
The "Worst American City": Overall Mean Scores on Twenty-four Indicators[a]

1. Seattle (average rank: 14.0)	26. Oakland (25.9)
2. Tulsa (14.8)	27. Washington (26.5)
3. San Diego (14.9)	28. Houston (27.4)
4. San Jose (15.6)	29. Buffalo (28.2)
5. Honolulu (16.4)	30. Louisville (28.2)
6. Portland (17.8)	31. Pittsburgh (28.4)
7. Denver (18.2)	32. New York (28.5)
8. Minneapolis (18.8)	33. Memphis (29.3)
9. Oklahoma City (19.1)	34. Boston (29.6)
10. Omaha (19.3)	35. Miami (29.9)
11. San Francisco (19.46)	36. Atlanta (30.0)
12. Nashville (19.54)	37. El Paso (30.7)
13. St. Paul (19.6)	38. New Orleans (30.7)
14. Columbus (19.75)	39. Philadelphia (31.0)
15. Toledo (19.79)	40. Tampa (31.1)
16. Indianapolis (20.6)	41. San Antonio (31.7)
17. Long Beach (21.2)	42. Norfolk (31.9)
18. Milwaukee (21.9)	43. Cleveland (32.0)
19. Kansas City (22.6)	44. Jacksonville (32.2)
20. Dallas (23.25)	45. Birmingham (32.5)
21. Phoenix (23.33)	46. Baltimore (32.7)
22. Los Angeles (24.6)	47. Detroit (33.0)
23. Fort Worth (24.7)	48. Chicago (33.7)
24. Cincinnati (24.9)	49. St. Louis (35.3)
25. Rochester (25.3)	50. Newark (41.6)

Source: Arthur Louis, "The Worst American City," *Harper's Magazine*, January 1975, p. 71. Copyright © 1974 by *Harper's Magazine*. All rights reserved. Reprinted from the January 1975 issue by special permission.
[a] Cities ranked from "best" to "worst" on combined index of twenty-four indicators.

functions of health, safety, recreation, and education are really performed by a mix of public and private institutions. To put the matter another way, the activities of a given service agency seldom, if ever, solely determine the degree to which the environmental state which is the agency's objective (a secure community, an educated citizenry, and so on) prevails. With respect to police departments and their manifest goal of a secure community, Elinor Ostrom notes that while the activities of a police department

> contribute to the security of a community, *it is almost never the sole contributor to this state of affairs.* The degree of security enjoyed by a community is created by individuals interacting with one another within a set of institutional arrangements.... Included among these are employment markets, product markets, housing markets, welfare programs, educational systems, community organizations, court systems, penal systems, and the police.[14]

Recognition of the coproduction function has stimulated debate about the relative efficiency of public sector in comparison to private sector service provision. Savas, for example, condemns public service "monopolies" and argues that many "public" services could be produced more efficiently and cheaply through competitive contracting. Particularly in sanitation, and fire protection, but perhaps also in schooling, policing, or other services, the private sector might produce more efficient public services.[15] In any case, ignoring the coproduction aspect of public-service provision unrealistically focuses attention only on the public side of the production function.

Variations in Service Needs and Levels

Public-service "needs" are variables, and so are the "levels" of public services. The concept of need is a tricky one, existing partly in the eye of the beholder. But it cannot be dismissed out of hand as an important criterion for evaluating policy. If attention to the quality of life is to mean anything at all, it must imply that there are objectively identifiable "needs" which policy can help attain. Brooklyn needs no farm supports, and Iowa counties need little mass transit aid. Newark needs more police protection than Nashville. Florida needs more old-age assistance than North Carolina. Dismissing the notion of service needs will only permit political power and group strength to determine allocations of governmental largesse.

Spatially, we can think of needs as varying from neighborhood to neighborhood, city to city, state to state, and region to region. Neighborhood variation is commonly seen in cities, large and small. Contrasts between the ghetto and the Gold Coast are real and nearly universal. Recently, the courts have toyed with the idea of enforcing equal services from neighborhood to neighborhood, in both schools[16] and general municipal services.[17] City-to-city variation is most evident in contrasting service levels in poor cities versus their rich suburbs, or poor versus rich suburbs.

State-to-state variation in service needs is marked. It also varies sharply from one service area to another. Table 10-2 identifies the range and ratios of selected need indices in the areas of income, schooling, crime, and health. This is an exceedingly crude tabulation, as there is no common metric for measuring needs across indicators. Even so, the difference between the best educated state and the worst educated state is on the order of 1.5:1. The range between the most crime-ridden state and the least is on the order of 63:1. Needs like education and health are less variable than needs in the crime and poverty areas. If we are committed to a goal of evening out service needs among the states, there is much more evening out to be done in some services areas than others.

Table 10-2
Variations in Public Service Needs by State

Income and Welfare

1. Percentage of persons below poverty line in 1970:
   High: Mississippi, 47.0%
                                    Ratio: 7:1
   Low: Connecticut, 7.2%

2. Average monthly public-assistance payments (ADC), 1973:
   High: Massachusetts, $324
                                    Ratio: 6:1
   Low: Mississippi, $52

Schooling

Percentage of high school graduates:
   High: Utah, 67.3%
                                    Ratio: 1.5:1
   Low: South Carolina, 37.8%

Crime

Robberies known to police per 100,000 population:
   High: New York, 439.6
                                    Ratio: 63:1
   Low: North Dakota, 7.3

Health

1. Infant death rates per 1,000 live births, 1970 (white):
   High: New Mexico, 30.9
                                    Ratio: 1.7:1
   Low: Delaware, 17.8

2. Deaths from heart disease, per 100,000 population, 1970:
   High: West Virginia, 458.4
                                    Ratio: 2:1
   Low: New Mexico, 200.8

Regional variation has come more to public attention with disputes about the "regional" shift. The South and North seem almost to vie with each other over who is the more disadvantaged and deserving of federal subsidies. Southern states are eager to use income-in-aid formulas, but not percentage urban; Northern states are eager to incorporate tax effort into formulas, while their Southern neighbors oppose it; Southern states can claim subsidies for their rural poor, while Northern states can claim it for their urban poor. The index one chooses to measure service needs virtually determines which region will fare as most needy.

Although there is variation in all these components—neighborhood, city state, and region—it is clear that the more discrete the unit, the sharper the range of variation. The variation from neighborhood to neighborhood is considerably larger than the variation from region to region. Yet in our haste to develop a belated regional policy, we may ignore a much sharper set of need variations, that from one neighborhood to another in the same community. It would be unfortunate if a regional policy crowded out policies to equalize service needs at the neighborhood, city, and state levels.

Services and Economic Growth: Some Hypotheses

Two Kinds of Growth

There is probably cause to suspect that public services are related to growth and decay. It is, at least, a hypothesis I examine in this section. But what kind of growth and decay? Growth is of two kinds. The first may be called economic growth, or *economic development*. In taking an initial stock of people, their wealth is magnified. Without benefit of migration or influx, this selfsame stock of population increases its capital, its job opportunities, and its wealth. A second form of growth is *demographic growth*. Populations are mobile, and demographic growth results from the influx of persons and production to new regions. There is no necessary connection between these two forms of growth. A population may remain fixed and grow in economic wealth. Or it may remain fixed and, whether through natural reversals or artificial ones, may suffer economic decay. Similarly, an influx of population may bring new economic growth, or it may drain existing economic arrangements. It could be argued that the recent movement to the South brought an influx of better-trained and higher-paid Northerners who raised the economic levels of the South. Northerners, conversely, frequently argue that the South's export of people to them has been mostly high-cost, low-productivity citizens. In any event, these are alternative notions of growth, neither of which is dependent on the other.

Here I examine three different hypotheses about the links between public services and both types of growth. I consider three hypotheses. The first holds that "the federal government did it," that the federal policies which have favored the South provided economic infrastructure for growth there. The second is the "Tiebout hypothesis" generalized. It holds that families and firms move specifically in response to the mix of taxes and services they might enjoy. The third is in effect a "no-effect hypothesis," holding that the link between services and growth is minimal.

The Federal Government Did It

The role of fedeal government as culprit in regional shifts is frequently discussed. The power of Southern congressmen and senators to "deliver" rivals that of a Chicago mayor. The location of NASA in Houston and Florida, the dominance of Southern members of congress in defense appropriations, and the military bases of the South are testimonials to Southern ability to deliver defense and defense-related dollars back home. Not only in defense, but also in social services, federal dollars have gone disproportionately to Sunbelt states and cities. The most careful analysis of the distribution of the federal largesse is the *National Journal*'s 1976 report on "Where the Funds Flow."[18] Table 10-3 summarizes their estimates of the "net dollar flow," the ratio of federal taxes paid to federal dollars received. In terms of dollars per capita, the worst off are those in the Great Lakes region; the best off are the Mountain states, with the South Central, Pacific, and South Atlantic regions not far behind.

The hypothesis that federal policy played a significant role in regional migration has some plausibility. Not only the power of senior members of congress but also a conscious policy of equalizing income inequalities through grants-in-aid formulas can be implicated. The most significant single public "service," of course, is public jobs, and federal largesse provided significant Sunbelt employment increases. Federal policy also provided roads critical for Western and Southern redevelopment, energy policy (like the TVA) to encourage resource exploitation, and defense plants for Southern and Western jobs.

Yet, however fashionable it is to blame everything on the federal government, there are reasons not to overstate the federal role. Contrary to widespread opinion, federal policies which redistribute from North to South are not of recent vintage. Davis and Legler provide evidence on per capita federal expenditures by region from 1815 to 1895.[19] Table 10-4 highlights some contrasts between expenditures in New England and the South during the last century. Throughout most of U.S. history, North-to-South redistribution has been evident. If the rise of the Sunbelt is of recent origin, the South's expenditure advantage is not. This does not dismiss entirely the possibility that federal policies contributed to a regional shift, but it does require that we seek other explanations.

The Tiebout Hypothesis Generalized

One of the very few theoretical statements explicitly linking public services to growth and decline is the classic paper by Charles M. Tiebout.[20] This elegant formulation, rooted in Samuelson's "pure theory" of public expenditures, holds that consumer-citizens select locations on the basis of the location's service-taxation mix. In the Tiebout model, there are several working assumptions. It is assumed that consumer-citizens are fully mobile and fully informed about differences in services and taxing patterns; that there are several governments, each offering alternative packages of services and taxes; that citizens have variable preference schedules for services or

Table 10-3
Net Dollar Flow by State, Fiscal 1975

Region/State	Value	Region/State	Value
Northeast	$0.86	*West*	$1.20
New England	0.96	Mountain	1.30
Maine	1.12	Montana	1.28
New Hamsphire	1.00	Idaho	1.25
Vermont	1.17	Wyoming	1.21
Massachusetts	0.95	Colorado	1.20
Rhode Island	0.92	Utah	1.35
Connecticut	0.92	Nevada	0.96
Mid-Atlantic	0.83	Arizona	1.41
New York	0.89	New Mexico	1.93
New Jersey	0.66	Pacific	1.17
Pennsylvania	0.87	California	1.11
		Oregon	0.94
Midwest	0.76	Washington	1.40
Great Lakes	0.70	Alaska	2.60
Ohio	0.70	Hawaii	1.58
Indiana	0.73		
Illinois	0.72	*District of Columbia*	7.67
Michigan	0.65		
Wisconsin	0.73	Total, U.S.	1.00
Great Plains	0.94		
Minnsota	0.83		
Iowa	0.69		
Missouri	1.10		
Kansas	0.98		
Nebraska	0.84		
South Dakota	1.29		
North Dakota	1.35		
South	1.14		
South Atlantic	1.12		
Delaware	0.66		
Maryland	1.20		
Virginia	1.34		
West Virginia	1.21		
North Carolina	0.98		
South Carolina	1.19		
Georgia	1.16		
Florida	1.00		
South Central	1.17		
Kentucky	1.21		
Tennessee	1.13		
Alabama	1.34		
Mississippi	1.76		
Louisiana	1.16		
Arkansas	1.37		
Oklahoma	1.22		
Texas	1.03		

Source: "Federal Spending: The North's Loss is the Sunbelt's Gain," *National Journal*, June 26, 1976, p. 881. Reprinted with permission.

taxes; and that governments attract or repel individuals according to their tax-service mixes. The formulation was originally proposed to explain local government expenditure policy. It has subsequently figured prominently in several defenses of the fragmentation of metropolitan government by those who contend that a multiplicity of governments maximizes citizen choice. The logic of the model, however, need not be confined to the urban level. Indeed, it is applicable to citizen choices in any system of multiple governments where information and mobility are options and where service-tax packages vary. Nor, for that matter, is there any need to restrict its applicability to the individual consumer-citizen. Firms, like families, may seek an optimal service-taxation mix in making location choices.

If there is evidence to support this generalized version of the Tiebout theory, then the service-taxation mix would be intimately linked with growth and decline. But the evidence that taxes and services are crucial determinants of mobility is limited.

There is clear evidence, for example, that neither families nor firms weigh tax considerations very highly in choosing a location. Barlow, Brazer, and Morgan studied economic behavior of the affluent. They found that about half of their sample of well-to-do people in the United States believed that taxes were higher in their own than in other states and cities, but that only "one sixth had thought of moving in order to save taxes and only one-sixth of that sixth said they probably would move."[21] Similar evidence about the effects of taxation policies on firms has been reviewed by John Due, who concluded that location shifts were only weakly related to fiscal policies of state and local governments.[22] A recent survey of corporate headquarters movement reported that moving firms ranked taxes as far less significant a factor than other costs.[23] To be sure, decisions on mobility are often based on factors at the margins. But no evidence suggests that taxes and services are more critical than factors like wage costs, transportation access, and market and supplier access.

The movement of families has been less carefully investigated than the migratory behavior of firms. People are probably less calculating about taxes and services than corporations. They are likely to confuse the "conditions" of a city or neighborhood with the quality or quantity of public services targeted to improve those conditions. People may move to escape crime, vaguely sensing that police protection is inadequate; they may move because streets are unsightly, vaguely blaming municipal services. A trail of reasons is left behind on the road to suburbs and an even longer trail on the road to another region. It is doubtful that careful tax-service calculations are made. Rather, individual families are more likely to move based on a gestalt of past and future neighborhoods, cities, and regions. Aspects of life

Table 10-4
Per Capita Federal Expenditures by Region, 1815-1895

Year	New England	Middle Atlantic	South Atlantic
1815	$3.28	$3.51	$ 5.03
1835	1.31	0.77	2.20
1855	1.79	1.67	6.88
1875	7.55	5.62	11.48
1895	6.54	5.18	12.21

Source: Lance Davis and John Legler, "Government in the American Economy, 1815-1902: A Quantitative Study," *Journal of Economic History* 26 (1966):514-52.

quality and service quality may be too intermingled to separate their independent effects.

The No-Effect Hypothesis

One could also make a case for the proposition that public services bear very little connection to economic growth and decay. Services in the United States may simply exhibit too little variation to be associated with growth or decay. No state or city has, according to this hypothesis, so little in the way of services that growth is discouraged. The order of magnitude of service differentiation is simply too small to explain variation in economic growth. As the famous dictum of sociologist W.F. Ogburn has it, "one cannot explain a variable with a constant." Wage rates, market access, employment opportunities, and other production factors vary more, and explain more variance. Given our scanty knowledge about the role of public services in economic growth, it is impossible to dismiss the no-effect hypothesis out of hand.

Two Agendas

A Research Agenda

We know, to repeat, very little about the public service sector and less about its impact on growth and decay. To describe our ignorance as mere lacunae is to understate considerably what we do not know. I suggest, therefore, a three-pronged research agenda about which would move us a few steps toward knowledge about the public-service sector.

Measuring Variation in Service Outputs. Both economists and political scientists have done more to explain the size of public budgets than in the services they procure. Yet it is extremely risky to assume a close correspondence between the level of public spending and the level of public services. Sharkansky, among others, has demonstrated that dollars spent are only weakly correlated with the scope and quality of services purchased.[24] We cannot continue to assume that "more money buys more service." To the contrary, there is accumulating research on the productivity of public services to show that productivity variations are subtle, complex, and often unrelated to dollars. If we could better measure what is being purchased, we could do more to connect service dollars to service outputs. There is a real need to know more about what police services do to crime rates, what school services do to learning, and what social services do for social welfare.

Linking Service Variations with Mobility. Once we can identify variations in public services in quality and quantity, we can take a second step and link these to mobility variation. The economist's model of mobility, especially the exemplary work of Tiebout, hypothesizes strong ties between service-taxation mixes and the movement of people and production. We can hypothesize that people migrate from areas with poor police protection to those with good police protection, but we do not know for sure. We can hypothesize that people move from areas of poor schools to areas of good schools, but we do not know for sure. We can hypothesize that firms from areas with poor public services move to areas with good public services, but we do not know for sure. Until we can better specify what dimensions of service delivery account for good or poor services, we cannot very well link service variations to locational choices.

Linking Service Infrastructure to Economic Growth. In the extreme case, public services are a necessary and sufficient condition for economic growth. Public services represent an economic infrastructure, without which economic growth would be negligible. Public services are a neglected form of economic capital. Schultz has pointed out that one public service—education—can function as human capital.[25] If there were solid evidence on the production functions of other public services, we could relate them to variations in economic growth. In recent years, perhaps the most advanced area of productivity research has been law enforcement. Research in some of the "hard" capital-intensive services, whose connection with growth seems more obvious, lag. If we knew more about how dollars translate into outputs, we could also know more about how outputs affect growth.

A Policy Agenda

Services have become so important in our semiwelfare state that the policy agenda could be very long indeed. Let me focus only on two issues on which conscious, rather than inadvertent, policy might be determined. One of these is an age-old question still seeking an answer: Who should perform what services? The other is a question we answer half-consciously, but rarely address directly: Do we want to equalize?

Who Should Perform What Services? The 79,000 governments in the United States form a complicated network of service providers. Few public services do not involve two to three government contributors, often federal and local or state and federal. Some services may involve interlocking provision by numerous governments. This is clearly not the place to enter the hoary debate about the fragmentation of urban governments in the United States. But if public services are to be elevated to a matter of national attention, it is proper to ask which governments should be responsible for which services. Nor is it beyond reason to question whether some public services could be performed more economically and efficiently by private sector providers.

Responsibility for public services has drifted recently toward state and federal governments. States have increased their share of state and local expenditures.[26] The federal government, once well behind the state-local sector in domestic spending, has increased its share of spending for domestic services. Yet, instead of simplifying it, this bewildering array of government transactions has made service delivery more complex, so complex that one book on a job-creation program in Oakland was subtitled "How Great Expectations in Washington Are Dashed in Oakland, or, Why It's Amazing that Federal Programs Work at All. . . ."[27] The multiplicity of service providers has rendered implementation difficult and impact problematic. A national policy for public service would first settle the question of who is to do what.

These nationalizing trends have run counter to other trends to "localize" and "privatize" public services. Even if community control is an idea whose time has passed, it left in its wake profound feelings of dissatisfaction with bureaucratically centralized service provision. Liberal and black demands for community control and voucher systems have paradoxically met conservative demands for voucher systems and private contracting of public services. Stephen Savas of the Columbia Business School remarks that "it is easy to lead to the unwarranted implication that public goods paid for by the public through payments to the public tax collector must be provided *to* the public *by* a public agency through *public* employees."[28] Savas presents evidence that sanitation, fire protection, and other traditional services might be more effectively and cheaply produced, under public contract, by private sector institutions. In considering a national policy for public services, it is possible that contracting out some services is an idea whose time may have come.

Do We Want to Equalize? The quantity and quality of public services may vary from neighborhood to neighborhood, city to city, state to state, and region to region. Few unqualified generalizations about service variation are possible, but variation there is. Local citizens have challenged service inequalities on constitutional grounds. In *Hawkins* v. *Shaw*, the Fifth Circuit held that intramunicipal services inequalities were unconstitutional; in *Hobson* v. *Hansen*, a federal court held that intradistrict educational inequalities were unconstitutional; but the Supreme Court rejected Dimitrio Rodriguez' challenge to statewide school finance inequalities. As yet, there is no definitive judicial resolution of the question of service inequalities from place to place.

On the congressional front, we have followed a half-conscious policy of "creeping equalizaiton" from state to state. Almost all federal programs have contained seeds of redistribution. Poorer states and cities typically get an extra slice of the fiscal pie. Yet there are lingering issues: If we equalize from state to state, why not equalize from school district to school district or neighborhood to neighborhood? Just how much equalization do we want? Ambivalence about equality appears in the service area, as in other policy domains. We have substituted, one might argue, service dependency for genuine income opportunities for the poor. Perhaps we do the same thing at the government level.

Notes

1. Michael B. Tietz, "Toward a Theory of Public Facility Location," *Papers of the Regional Science Association* 21 (1968):35.

2. See, for example, Joseph Pechman and Benjamin Okner, *Who Bears the Tax Burden?* (Washington: Brookings Institution, 1974).

3. The early and path-breaking work on the distribution of government benefits was the Tax Foundation's *Allocating Tax Burdens and Government Benefits by Income Class* (New York, 1967). More recently Morgan Reynolds and Eugene Smolensky have explored more systematic data on the burden/benefit ratio. See their "The Post Fisc Distribution: 1961 and 1970 Compared," *National Tax Journal*, December 1974, pp. 515-30.

4. S.M. Miller and Pamela Roby, *The Future of Inequality* (New York: Basic Books, 1970).

5. W. Lee Hansen and Burton Weisbrod, *Benefits, Costs and Finances of Public Higher Education* (Chicago: Markham, 1969).

6. Tax Foundation, *Allocating Tax Burdens*; Reynolds and Smolensky, "The Post Fisc Distribution."

7. Frank S. Levy, Arnold Meltsner, and Aaron Wildavsky, *Urban Outcomes* (Berkeley: University of California Press, 1974); Robert L. Lineberry, *Equality and Urban Policy* (Beverly Hills, Calif.: Sage, 1977).

8. Robert Haveman, "Poverty, Income Distribution, and Social Policy: The Last Decade and the Next," Discussion paper prepared for the Center for the Study of Democratic Institutions, June 27, 1977, pp. 9, 13.

The study cited by Haveman is T. Smeeding, *Measuring the Economic Welfare of Low-Income Households* (New York: Academic Press, forthcoming). The economist Edgar Browing, in "How Much More Equality Can We Afford?" *The Public Interest*, Spring 1976, p. 92, contends that with transfer payments and in-kind services factored in, the average poor family had an income of 130 percent of the "poverty line."

9. *San Antonio Light*, July 24, 1977, p. 12A.

10. On the growth issue, see William Alonzo, "Urban Zero Population Growth," *Daedalus* 102 (Fall 1973):191-206.

11. Angus Campbell, Philip E. Converse, and Willard L. Rodgers, *The Quality of American Life* (New York: Russell Sage, 1976), pp. 26-30 and figure 2-1.

12. See, for example, H.L. Mencken, "The Worst American State," *American Mercury* 24 (1931):1-16, 177-88, 355-71; Arthur Louis, "The Worst American City," *Harper's*, January 1975, pp. 67-71.

13. Mark Schneider, "The Quality of Life in Large American Cities," *Social Indicators Research* 1 (1975):495-509; David M. Smith, *The Geography of Social Well-Being in the United States* (New York: McGraw-Hill, 1973); Ben-Chieh Liu, *Quality of Life Indicators in U.S. Metropolitan Areas 1970* (Washington: Environmental Protection Agency, 1975).

14. Elinor Ostrom, "On the Meaning and Measurement of Output and Efficiency in the Production of Urban Police Services," *Journal of Criminal Justice* 1 (June 1973):97.

15. E.S. Savas, "Municipal Monopolies versus Competition in Delivering Urban Services," in Willis D. Hawley and David Rogers, eds., *Improving the Quality of Urban Management* (Beverly Hills, Calif.: Sage, 1972), chapter 15. See also Savas' "Municipal Monopoly," *Harper's Magazine*, December 1971, pp. 55-60.

16. Hobson v. Hansen, 269 F. Supp. 401 (D.D.C. 1967).

17. Hawkins v. Shaw, 437 F.2d 1286 (5th Cir. 1971).

18. "Where the Funds Flow," *National Journal*, June 26, 1976, pp. 878-91.

19. Lance Davis and John Legler, "Government in the American Economy, 1815-1902: A Quantitative Study," *Journal of Economic History* 26 (1966):514-52.

20. C. Tiebout, "A Pure Theory of Local Expenditures," *Journal of Political Economy* 64 (October 1956):416-24.

21. Robin Barlow, Harvey Brazer, and James Morgan, *The Economic Behavior of the Affluent* (Washington: Brookings Institution, 1966), p. 169.

22. John F. Due, "Studies of State and Local Tax Influences on the Location of Industry," *National Tax Journal* 14 (1961):163-73.

23. Leland S. Burns and Wing Ning Pang, "Big Business in the Big City: Corporate Headquarters in the Central Business District," *Urban Affairs Quarterly* 12 (June 1977):533-44.

24. Ira Sharkansky, "Governmental Expenditures and Public Services in the American States," *American Political Science Review* 66 (December 1967):1066-77.

25. Theodore W. Schultz, "Capital Formation by Education," *Journal of Political Economy* 68 (December 1960):571-83.

26. G. Ross Stephens, "State Centralizaiton and the Erosion of Local Autonomy," *Journal of Politics* 36 (February 1974):44-76.

27. Jeffrey L. Pressman and Aaron Wildavsky, *Implementation* (Berkeley: University of California Press, 1974).

28. Stephen Savas, "Municipal Monopolies versus Competition in Delivering Public Services," in Hawley and Rogers, *Improving the Quality of Urban Management*, p. 483.

List of Contributors

Roy Bahl, Professor, Maxwell School of Citizenship and Public Affairs, Syracuse University

David L. Birch, Professor, Program on Neighborhood and Regional Change, Massachusetts Institute of Technology

Kenneth L. Deavers, Director, Economic Development Division, Economic Research Service, U.S. Department of Agriculture

Irving Hoch, Research Associate, Resources for the Future, Washington, D.C.

Charles L. Leven, Professor and Chairman, Department of Economics, Washington University

Robert L. Lineberry, Professor, Political Science Department, Northwestern University

William H. Miernyk, Benedum Professor of Economics and Director, Regional Research Institute, West Virginia University

Peter A. Morrison, Member, Senior Staff, Social Sciences, The Rand Corporation, Santa Monica, California

Thomas Muller, Research Associate, The Urban Institute, Washington, D.C.

George Peterson, Senior Research Associate, The Urban Institute, Washington, D.C.

Jeffrey C. Williamson, Professor, Department of Economics, University of Wisconsin

About the Editor

Victor L. Arnold is a member of the faculty of the Lyndon B. Johnson School of Public Affairs, The University of Texas at Austin and teaches in the areas of strategic planning and quantitative analysis. He received the B.S. and M.S. degrees from Colorado State University and the Ph.D. degree from the University of Wisconsin at Madison. Dr. Arnold has taught on the faculty of The University of Minnesota and has also served in various professional capacities in the federal government and the state governments of Minnesota and Texas.